Circuits and Networks: Analysis, Design and Synthesis

Circuits and Networks: Analysis, Design and Synthesis

Venkata Krishna Reddy

RANDOM

RANDOM PUBLICATIONS

NEW DELHI (INDIA)

Circuits and Networks: Analysis, Design and Synthesis

ISBN 978-93-5111-571-7

Published in 2014 in India by

RANDOM PUBLICATIONS

4376-A/4B, Gali Murari Lal, Ansari Road
New Delhi-110 002
Phone : +9111-43580356, 011-23289044, 011-43142548
e-mail: sales@randompublications.com,
info@randompublications.com, randomexports@gmail.com

Reprinted 2022

Type Setting by : Friends Media, Delhi-110089
Digitally Pritnted at : Replika Press Pvt. Ltd.

Preface

A network, in the context of electronics, is a collection of interconnected components. Network analysis is the process of finding the voltages across, and the currents through, every component in the network. There are many different techniques for calculating these values. However, for the most part, the applied technique assumes that the components of the network are all linear. The methods described in this book are only applicable to linear network analysis, except where explicitly stated.

Today we live in a predominantly electrical world. Electrical technology is a driving force in the changes that are occurring in every engineering discipline. For example, surveying is now done using lasers and electronic range finders. Circuit analysis is the foundation for electrical technology. An indepth knowledge of circuit analysis provides an understanding of such things as cause and effect, feedback and control and, stability and oscillations. Moreover, the critical importance is the fact that the concepts of electrical circuit can also be applied to economic and social systems. Thus, the applications and ramifications of circuit analysis are immense.

A useful procedure in network analysis is to simplify the network by reducing the number of components. This can be done by replacing the actual components with other notional components that have the same effect. A particular technique might directly reduce the number of components, for instance by combining impedances in series. On the other hand it might merely change the form into one in which the components can be reduced in a later operation. A resistive circuit is a circuit containing only resistors, ideal current sources, and ideal voltage sources. If the sources are constant (DC) sources, the result is a DC circuit. Analysis of a circuit consists of solving for the voltages and currents present in the circuit. The solution principles outlined here also apply to phasor analysis of AC circuits.

This book gives comprehensive treatment to the course of Circuits and Networks also known as Network and Systems or Circuit and Systems in various universities. The book provides the applications of various tools and techniques for analysis and synthesis of circuits.

The knowledge and understanding of cost accounting is an integral part of learning the overall nature of accounting and is one of the varied aspects of accounting available for careers in this field. I would like to thank my team for standing beside me throughout my career and writing this book. My special thanks go to "Random Publications" who have published the book.

– Venkata Krishna Reddy

Contents

1

Design of Biasing Circuits

A bipolar junction transistor, (BJT) is very versatile. It can be used in many ways, as an amplifier, a switch or an oscillator and many other uses too. Before an input signal is applied its operating conditions need to be set. This is achieved with a suitable bias circuit, some of which I will describe. A bias circuit allows the operating conditions of a transistor to be defined, so that it will operate over a pre-determined range. This is normally achieved by applying a small fixed dc voltage to the input terminals of a transistor.Bias design can take a mathematical approach or can be simplified using transistor characteristic curves.

The characteristic curves predict the performance of a BJT. There are three curves, an input characteristic curve, a transfer characteristic curve and an output characteristic curve. Of these curves, the most useful for amplifier design is the output characteristics curve. The output characteristic curves for a BJT are a graph displaying the output voltages and currents for different input currents. The linear (straight) part of the curve needs is utilized for an amplifier or oscillator. For use as a switch,a transistor is biased at the extremities of the graph, these conditions are known as "cut-off" and "saturation".

CURVES

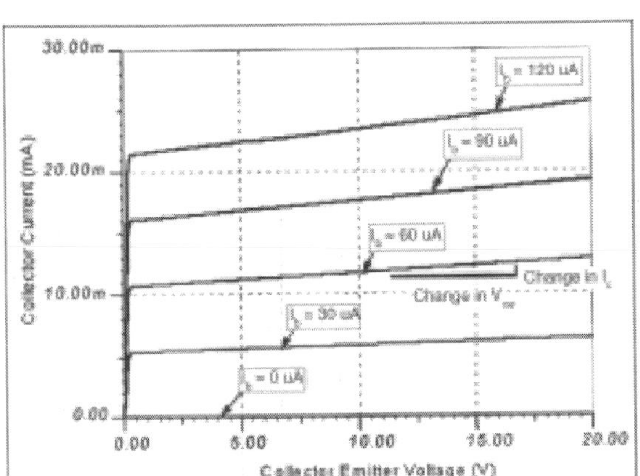

For each transistor configuration, common emitter, common base and emitter follower the output curves are slightly different. A typical output characteristic for a BJT in common emitter mode are shown above:

After the initial bend, the curves approximate a straight line. The slope or gradient of each line represents the output impedance, for a particular input base current. So what has all this got to do with biasing ? Take, for example the middle curve. The collector emitter voltage is displayed up to 20 volts. Let's assume that we have a single stage amplifier, working in common emitter mode, and the supply voltage is 10 volts.

The output terminal is the collector, the input is the base, where do you set the bias conditions? The answer is anywhere on the flat part of the graph. However, imagine the bias is set so that the collector voltage is 2 volts. What happens if the output signal is 4 volts peak to peak ? Depending on whether the transistor used is a PNP or NPN, then one half cycle will be amplified cleanly, the other cycle will approach the limits of the power supply and will "clip". This is shown below:

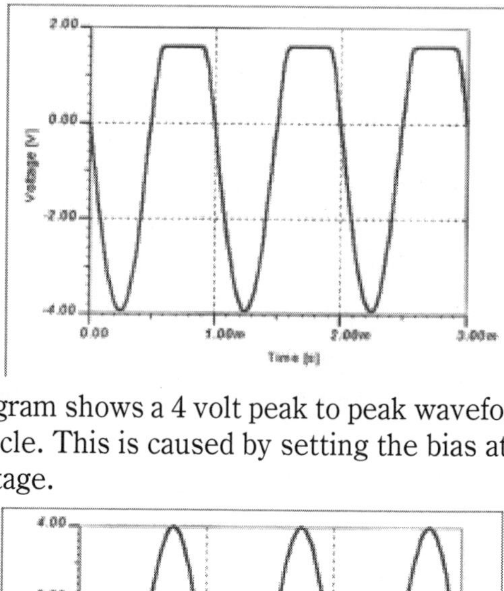

The above diagram shows a 4 volt peak to peak waveform with clipping on the positive half cycle. This is caused by setting the bias at a value other than half the supply voltage.

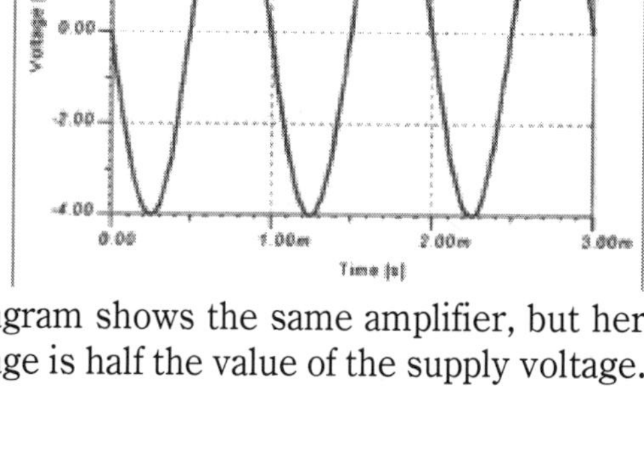

The lower diagram shows the same amplifier, but here the bias is set so that collector voltage is half the value of the supply voltage. Hence, it is a good idea to set the bias for a single stage amplifier to half the supply voltage, as this allows maximum output voltage swing in both directions of an output waveform.

CHARACTERISTIC

Before describing the bias circuits, it is worthwhile looking at a typical input characteristic curve for a small signal BJT. The following is the input characteristic for a transistor in common emitter mode, it is a plot of input base emitter voltage verses base current. It is shown with both x and y axis slightly zoomed.

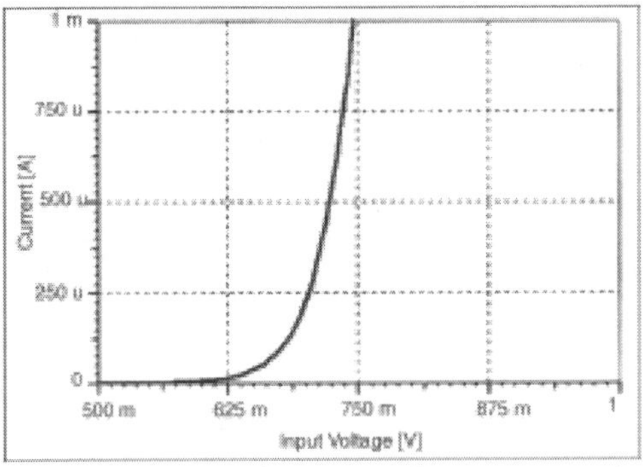

The base emitter voltage, Vbe is quoted in most text books as either 0.6 V or 0.7 V Both values are an approximation, and as can be seen from the above graph the value of Vbe varies between this range. For small signal work with base currents of 50uA or below a value for Vbe of 0.6 volts is a reasonable quote. For higher base currents, a Vbe of 0.7 V is a better approximation. In fact, in a large power transistor, the Vbe value can be even higher. The value of Vbe also varies widely with temperature change.

TRANSISTOR TERMINALS

The essential usefulness of a transistor comes from its ability to use a small signal applied between one pair of its terminals to control a much larger signal at another pair of terminals. This property is called gain. A transistor can control its output in proportion to the input signal; that is, it can act as an amplifier. Alternatively, the transistor can be used to turn current on or off in a circuit as an electrically controlled switch, where the amount of current is determined by other circuit elements.

There are two types of transistors, which have slight differences in how they are used in a circuit. A *bipolar transistor* has terminals labeled base, collector, and emitter. A small current at the base terminal (that is, flowing from the base to the emitter) can control or switch a much larger current between the collector and emitter terminals. For a *field-effect transistor*, the terminals are labeledgate, source, and drain, and a voltage at the gate can control a current between source and drain.

The image to the right represents a typical bipolar transistor in a circuit. Charge will flow between emitter and collector terminals depending on the current in the base.

Since internally the base and emitter connections behave like a semiconductor diode, a voltage drop develops between base and emitter while the base current exists. The amount of this voltage depends on the material the transistor is made from, and is referred to as V_{BE}.

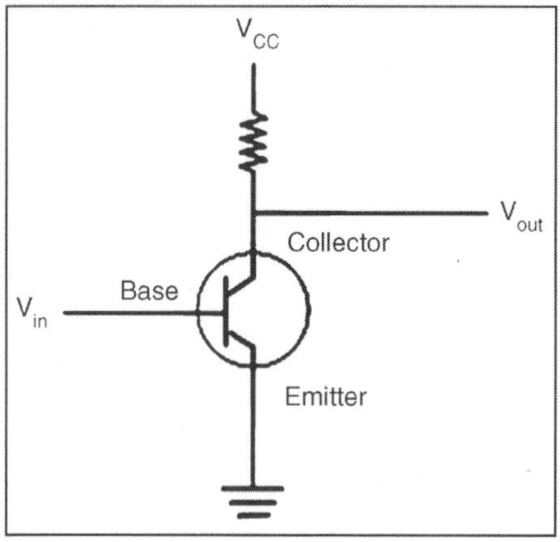

TRANSISTOR ACTION

Bipolar transistors are made from 3 sections of semiconductor material (alternating P-type and N-type), with 2 resulting P-N junctions. Schematically, a bipolartransistor can be thought of in this fashion:

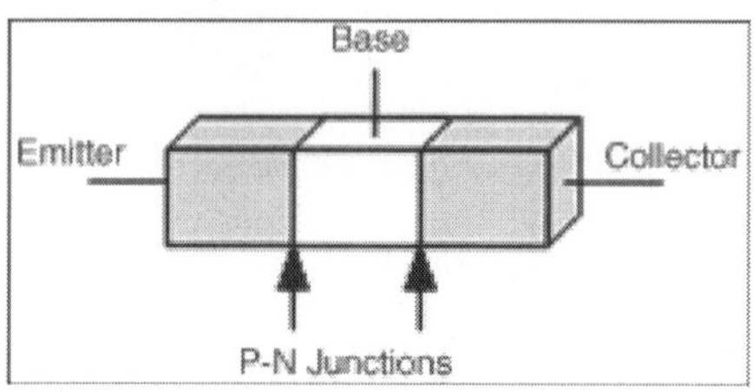

One P-N junction is between the emitter and the base; the other P-N junction is between the collector and the base. Note that the emitter and collector are usually doped somewhat differently, so they are rarely electrically interchangeable. While the terms "collector" and "emitter" go back to vacuum tube days, the base derives its name from the first point-contact transistors—here the center connection also formed the mechanical base for the structure. In modern practice, the base region is made as thin as possible to achieve reasonable levels of current gain; it is often only about one millionth of a meter thick.

Bipolar transistors are classified as either NPN or PNP according to the arrangement of their N-type and P-type materials. Their basic construction and chemical treatment is implied by their names. So an NPN transistors is formed by introducing a thin region of P-type material between two regions of N-type material.

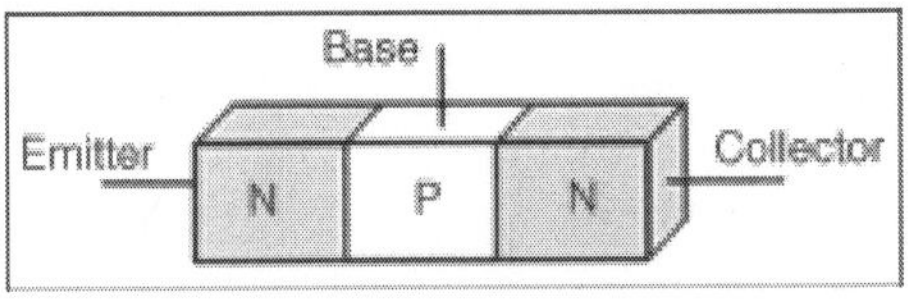

On the other hand, a PNP transistor is formed by introducing a thin region of N-type material between two regions of P-type material.

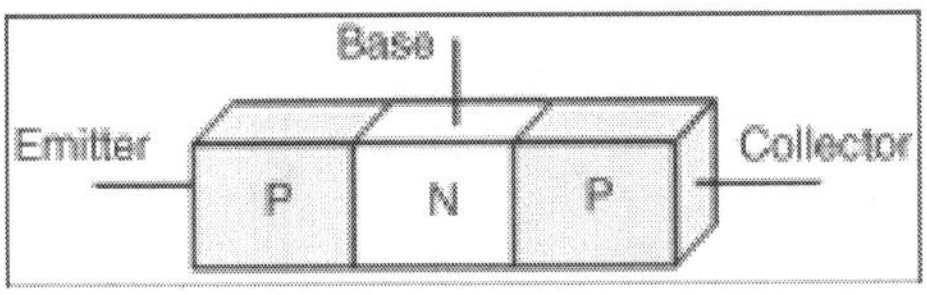

Since the majority and minority current carriers are different for N-type and P-type materials, it stands to reason that the internal operation of the NPN and PNPtransistors will also be different. These two basic types of transistors along with their circuit symbols are shown here:

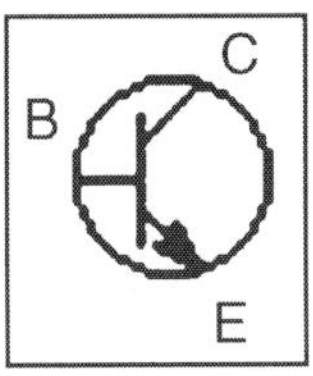

Fig. NPN

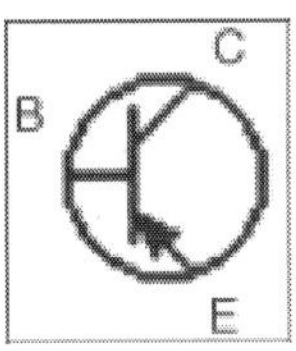

Fig. PNP

Note that the two symbols are subtly different. The vertical line represents the base (B), the angular line with the arrow on it represents the emitter (E), and the other angular line represents the collector (C). The direction of the arrow on the emitter distinguishes (graphically) the NPN from the PNP transistor. If the arrow points in, (Points iN) the transistor is a PNP. On the other hand if the arrow points out, the transistor is an NPN (Not Pointing iN).

Bear in mind that the arrow always points in the direction of hole flow (current), or from the P-type to N-type sections, no matter whether the P-type section is theemitter or base.

On the other hand, electron flow is always "against" the arrow, just like in the junction diode. As a result, a PNP transistor is "triggered" when its base is pulled low; an NPN transistor is "triggered" when its base is brought high.

Note that the bipolar transistor is a current-amplifying device, unlike the vacuum tube and the field-effect transistor (FET), both of which depend upon voltage changes to operate. It is the amount of current flowing in the base circuit that controls the amount of current flowing in the collector circuit.

BJT AMPLIFIER CONFIGURATIONS

There are plenty of texts around on basic electronics, so this is a very brief look at the three basic ways in which a bipolar junction transistor (BJT) can be used. In each case, one terminal is common to both the input and output signal. All the circuits shown here are without bias circuits and power supplies for clarity.

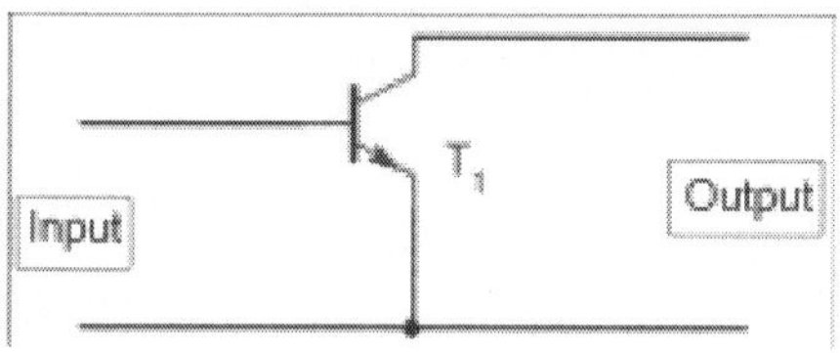

Fig. Common Emitter Configuration

Here the emitter terminal is common to both the input and output signal. The arrangement is the same for a PNP transistor. Used in this way the transistor has the advantages of a medium input impedance, medium output impedance, high voltage gain and high current gain.

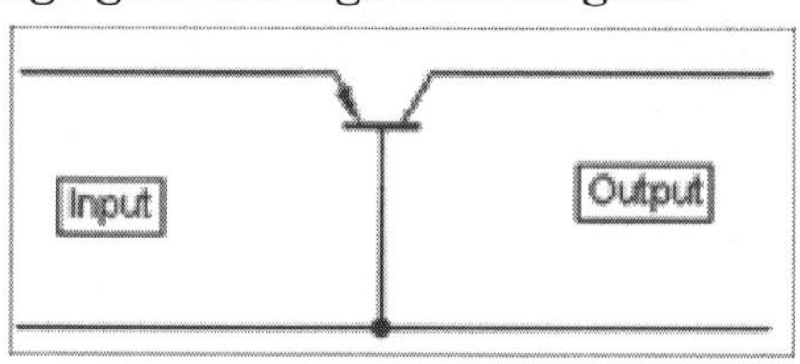

Fig. Common Base Configuration

Here the base is the common terminal. Used frequently for RF applications, this stage has the following properties. Low input impedance, high output impedance, unity (or less) current gain and high voltage gain.

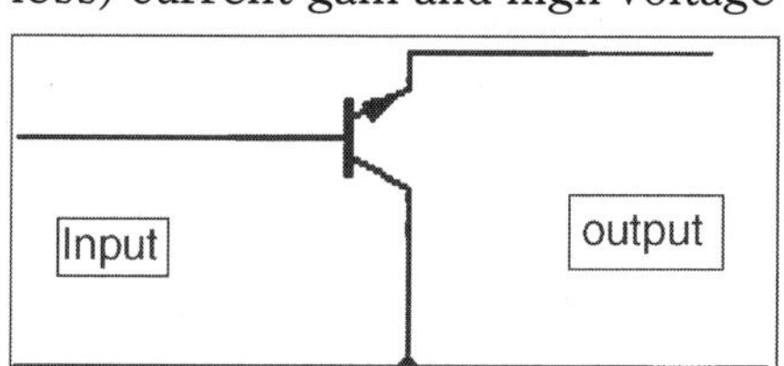

Fig. Common Collector Configuration

This last configuration is also more commonly known as the emitter follower. This is because the input signal applied at the base is "followed" quite closely at the emitter with a voltage gain close to unity.

The properties are a high input impedance, a very low output impedance, a unity (or less) voltage gain and a high current gain. This circuit is also used extensively as a "buffer" converting impedances or for feeding or driving long cables or low impedance loads.

Amplifier Type	Common Base	Common Emitter	Common Emitter (Emitter Resistor)	Common Collector (Emitter Follower)
Input/ Output Phase Relationship	0°	180°	180°	0°
Voltage Gain	High $\frac{\alpha R_C}{R_s + r_e}$	Medium $\frac{\beta(R_C \| r_o)}{R_s + r_\pi}$	Medium $\frac{\beta R_C}{R_s + (\beta+1)(r_e + R_E)}$	Low $\frac{(\beta+1)(R_L \| r_o)}{R_s + (\beta+1)[r_e + (R_L \| r_o)]}$
Current Gain	Low α	Medium $\beta \frac{r_o}{R_C + r_o}$	Medium β	High $(\beta+1)\frac{r_o}{r_o + R_L}$
Power Gain	Low	High	High	Medium
Input Resistance	Low r_e	Medium $r_\pi = (\beta+1) r_e$	Medium $(\beta+1)(r_e + R_E)$	High $(\beta+1)[r_e + (r_o \| R_L)]$
Output Resistance	High R_C	Medium $R_C \| r_o$	Medium R_C	Low $r_o \| \left[r_e + \frac{R_s}{(\beta+1)}\right]$

COMMON EMITTER AMPLIFIER CIRCUIT

In the Bipolar Transistor tutorial, we saw that the most common circuit configuration for an NPN transistor is that of the Common Emitter Amplifier and that a family of curves known commonly as the Output Characteristics Curves, relates the Collector current (I_c), to the output or Collector voltage (V_{ce}), for different values of Base current (Ib).

All types of transistor amplifiers operate using AC signal inputs which alternate between a positive value and a negative value so some way of "presetting" the amplifier circuit to operate between these two maximum or peak values is required. This is achieved using a process known as Biasing. Biasing is very important in amplifier design as it establishes the correct operating point of the transistor amplifier ready to receive signals, thereby reducing any distortion to the output signal.

We also saw that a static or DC load line can be drawn onto these output characteristics curves to show all the possible operating points of the transistor from fully "ON" to fully "OFF", and to which the quiescent operating point or Q–point of the amplifier can be found. The aim of any small signal amplifier is to amplify all of the input signal with the minimum amount of distortion possible to the output signal, in other words, the output signal must be an exact reproduction of the input signal but only bigger (amplified).

To obtain low distortion when used as an amplifier the operating quiescent point needs to be correctly selected. This is in fact the DC operating point of the amplifier and its position may be established at any point along the load line by a suitable biasing arrangement. The best possible position for this Q–

point is as close to the centre position of the load line as reasonably possible, thereby producing a Class A type amplifier operation, ie. Vce = 1/2Vcc. Consider the Common Emitter Amplifier circuit shown below.

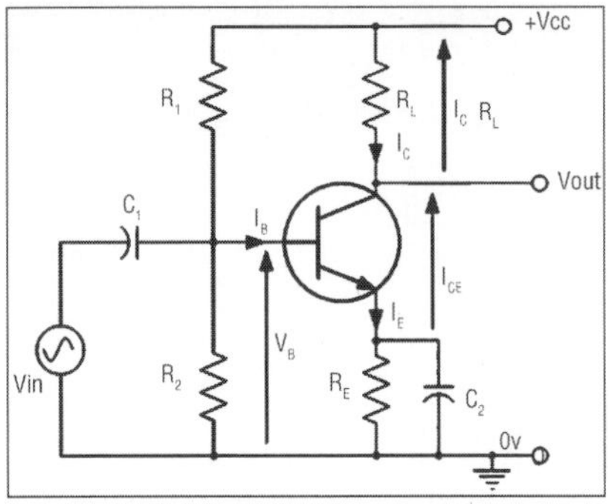

Fig: The Common Emitter Amplifier Circuit

The single stage common emitter amplifier circuit shown above uses what is commonly called "Voltage Divider Biasing". This type of biasing arrangement uses two resistors as a potential divider network and is commonly used in the design of bipolar transistor amplifier circuits. This method of biasing the transistor,

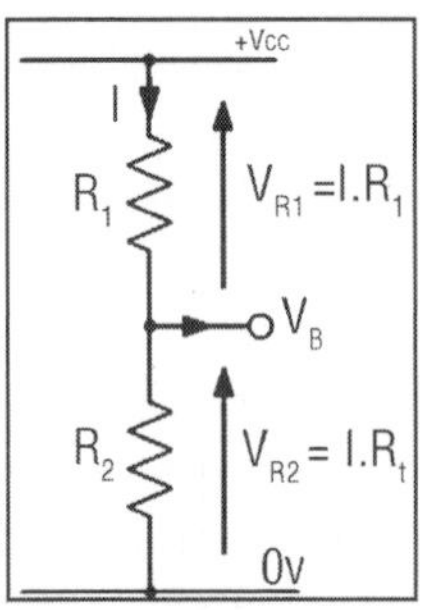

greatly reduces the effects of varying Beta, (b) by holding the Base bias at a constant steady voltage level allowing for best stability. The quiescent Base voltage (V_b) is determined by the potential divider network formed by the two resistors, R_1, R_2 and the power supply voltage Vcc as shown with the current flowing through both resistors. Then the total resistance RT will be equal to $R_1 + R_2$ giving the current as i = Vcc/RT. The voltage level generated at the junction of resistors R_1 and R_2 holds the Base voltage (V_b) constant at a value below the supply voltage. Then the potential divider network used in the common emitter amplifier circuit divides the input signal in proportion to the resistance.

This bias reference voltage can be easily calculated using the simple voltage divider formula below:

$$V_B = \frac{VccR_2}{R_1 + R_2}.$$

The same supply voltage, (Vcc) also determines the maximum Collector current, I_C when the transistor is switched fully "ON" (saturation), $V_{ce} = 0$. The Base current Ib for the transistor is found from the Collector current, Ic and the DC current gain Beta, b of the transistor.

$$\beta = \frac{\Delta I_C}{\Delta I_B}.$$

Beta is sometimes referred to as h_{FE} which is the transistors forward current gain in the common emitter configuration. Beta has no units as it is a fixed ratio of the two currents, Ic and Ib so a small change in the Base current will cause a large change in the Collector current.

One final point about Beta. Transistors of the same type and part number will have large variations in their Beta value for example, the BC107 NPN Bipolar transistor has a DC current gain Beta value of between 110 and 450 (data sheet value) this is because Beta is a characteristic of their construction and not their operation.

As the Base/Emitter junction is forward–biased, the Emitter voltage, Ve will be one junction voltage drop different to the Base voltage. If the voltage across the Emitter resistor is known then the Emitter current,Ie can be easily calculated using Ohm's Law. The Collector current, Ic can be approximated, since it is almost the same value as the Emitter current.

Example

A common emitter amplifier circuit has a load resistance, R_L of 1.2kWs and a supply voltage of 12v. Calculate the maximum Collector current (Ic) flowing through the load resistor when the transistor is switched fully "ON" (saturation), assume Vce = 0.

Also find the value of the Emitter resistor, R_E with a voltage drop of 1v across it. Calculate the values of all the other circuit resistors assuming an NPN silicon transistor.

$$I_{C(MAX)} = \frac{Vcc - V_{RE}}{R_L} \frac{12-1}{1200} = 9.2mA$$

$$V_{CE} \quad = 0(Saturation)$$

This then establishes point "A" on the Collector current vertical axis of the characteristics curves and occurs when Vce = 0. When the transistor is switched fully "OFF", their is no voltage drop across either resistor R_E or R_L as no current is flowing through them. Then the voltage drop across the transistor, Vceis equal to the supply voltage, Vcc.

This then establishes point "B" on the horizontal axis of the characteristics curves. Generally, the quiescent Q–point of the amplifier is with zero input signal applied to the Base, so the Collector sits half–way along the load line between zero volts and the supply voltage, (V_{cc}/2). Therefore, the Collector current at the Q–point of the amplifier will be given as:

$$I_{C(Q)} = \frac{\frac{12-1}{2}}{1200}$$
$$= \frac{5.5}{1200}$$
$$= 4.58mA$$

This static DC load line produces a straight line equation whose slope is given as:–1/(R_L + R_E) and that it crosses the vertical Ic axis at a point equal to Vcc/(R_L + R_E). The actual position of the Q–point on the DC load line is determined by the mean value of Ib.

As the Collector current, Ic of the transistor is also equal to the DC gain of the transistor (Beta), times the Base current (b × I_b), if we assume a Beta (b) value for the transistor of say 100, (one hundred is a reasonable average value for low power signal transistors) the Base current Ib flowing into the transistor will be given as:

$$\beta = \frac{I_C}{I_B}$$
$$\therefore I_B = \frac{I_C}{\beta} = \frac{4.58mA}{100} = 45.8\mu A$$

Instead of using a separate Base bias supply, it is usual to provide the Base Bias Voltage from the main supply rail (V_{cc}) through a dropping resistor, R_1. Resistors, R_1 and R_2 can now be chosen to give a suitable quiescent Base current of 45.8mA or 46mA rounded off. The current flowing through the potential divider circuit has to be large compared to the actual Base current, Ib, so that the voltage divider network is not loaded by the Base current flow. A general rule of thumb is a value of at least 10 times Ib flowing through the resistor R_2. Transistor Base/Emitter voltage, Vbe is fixed at 0.7V (silicon transistor) then this gives the value of R_2 as:

$$R_2 = \frac{V_{(RE)} + V_{(BE)}}{10 \times I_B}$$
$$= \frac{1+0.7}{458 \times 10^{-6}}$$
$$= 3.71k\Omega$$

If the current flowing through resistor R_2 is 10 times the value of the Base current, then the current flowing through resistor R_1 in the divider network must be 11 times the value of the Base current. The voltage across resistor R_1

is equal to Vcc–1.7v (V_{RE} + 0.7 for silicon transistor) which is equal to 10.3V, therefore R_1 can be calculated as:

$$R_1 = \frac{V_{CC} - \left(V_{(RE)} + V_{(BE)}\right)}{11 \times I_B}$$

$$= \frac{12 - 1.7}{504 \times 10^{-6}}$$

$$= 20.45k\Omega$$

The value of the Emitter resistor, R_E can be easily calculated using Ohm's Law. The current flowing through R_E is a combination of the Base current, Ib and the Collector current Ic and is given as:

$$I_E = I_C + I_B = 4.58mA + 45.8\mu A = 4.63mA\,.$$

Resistor, R_E is connected between the Emitter and ground and we said previously that it has a voltage of 1 volt across it. Then the value of R_E is given as:

$$R_E = \frac{V_{RE}}{I_E} = \frac{Iv}{4.63mA} = 216\Omega\,.$$

So, for our example above, the preferred values of the resistors chosen to give a tolerance of 5per cent (E24) are:

$$R_1 = 20k\Omega, R_2 = 3.6k\Omega, R_L = 1.2k\Omega, R_E = 220\Omega$$

Then, our original Common Emitter Amplifier circuit above can be rewritten to include the values of the components that we have just calculated above.

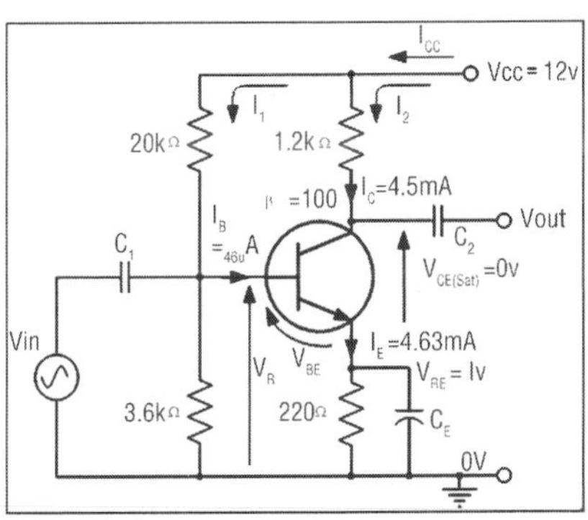

Fig. Completed Common Emitter Circuit

TYPES OF BJT'S

Transistors are three terminal active devices made from different semiconductor materials that can act as either an insulator or a conductor by the application of a small signal voltage. The transistor's ability to change between these two states enables it to have two basic functions: "switching" (digital electronics) or "amplification" (analogue electronics). Then bipolar transistors have the ability to operate within three different regions:

- *Active Region*: The transistor operates as an amplifier and Ic = β.Ib
- *Saturation*: He transistor is "fully-ON" operating as a switch and Ic = I (saturation)
- *Cut-off*: The transistor is "fully-OFF" operating as a switch and Ic = 0

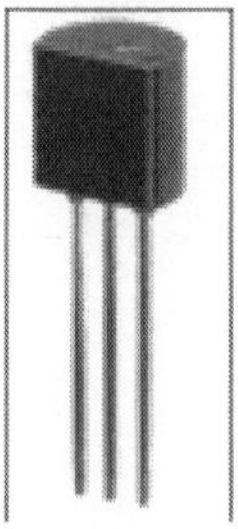

Fig. Typical Bipolar Transistor.

The word Transistor is an acronym, and is a combination of the words Transfer Varistor used to describe their mode of operation way back in their early days of development. There are two basic types of bipolar transistor construction, PNP and NPN, which basically describes the physical arrangement of the P-type and N-type semiconductor materials from which they are made.

The Bipolar Transistor basic construction consists of two PN-junctions producing three connecting terminals with each terminal being given a name to identify it from the other two. These three terminals are known and labelled as the Emitter (E), the Base (B) and the Collector (C) respectively. Bipolar Transistors are current regulating devices that control the amount of current flowing through them in proportion to the amount of biasing voltage applied to their base terminal acting like a current-controlled switch.

The principle of operation of the two transistor types PNP and NPN, is exactly the same the only difference being in their biasing and the polarity of the power supply for each type.

BIPOLAR TRANSISTOR CONSTRUCTION

The construction and circuit symbols for both the PNP and NPN bipolar transistor are given above with the arrow in the circuit symbol always showing the direction of "conventional current flow" between the base terminal and its emitter terminal. The direction of the arrow always points from the positive P-type region to the negative N-type region for both transistor types, exactly the same as for the standard diode symbol.

A BJT consists of three differently doped semiconductor regions, the emitter region, the base region and the collector region. These regions are, respectively, p type, n type and p type in a PNP, and n type, p type and n type in a NPN transistor.

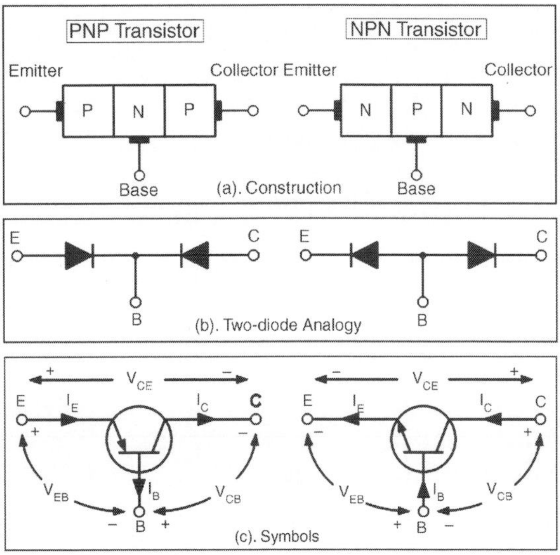

(a). Construction

(b). Two-diode Analogy

(c). Symbols

Each semiconductor region is connected to a terminal, appropriately labeled: emitter (E), base (B) and collector (C). The base is physically located between the emitter and the collector and is made from lightly doped, high resistivity material. The collector surrounds the emitter region, making it almost impossible for the electrons injected into the base region to escape being collected, thus making the resulting value of a very close to unity, and so, giving the transistor a large β. A cross section view of a BJT indicates that the collector—base junction has a much larger area than the emitter—base junction.

The bipolar junction transistor, unlike other transistors, is usually not a symmetrical device. This means that interchanging the collector and the emitter makes the transistor leave the forward active mode and start to operate in reverse mode. Because the transistor's internal structure is usually optimized for forward-mode operation, interchanging the collector and the emitter makes the values of a and β in reverse operation much smaller than those in forward operation; often the α of the reverse mode is lower than 0.5.

The lack of symmetry is primarily due to the doping ratios of the emitter and the collector. The emitter is heavily doped, while the collector is lightly doped, allowing a large reverse bias voltage to be applied before the collector–base junction breaks down. The collector–base junction is reverse biased in normal operation.

The reason the emitter is heavily doped is to increase the emitter injection efficiency: the ratio of carriers injected by the emitter to those injected by the base. For high current gain, most of the carriers injected into the emitter–base junction must come from the emitter.

TRANSISTOR BIASING

The simplest bias circuit is shown below. It consists only of a fixed bias resistor and load resistor. The BJT is operating in common emitter mode. The dc current gain or beta, h_{FE} is the ratio of dc collector current divided by dc base current. The BJT is a BC107A. The values of Rb and Rc can be determined by either mathematical approach or by using the output characteristic curves for the BC107A.

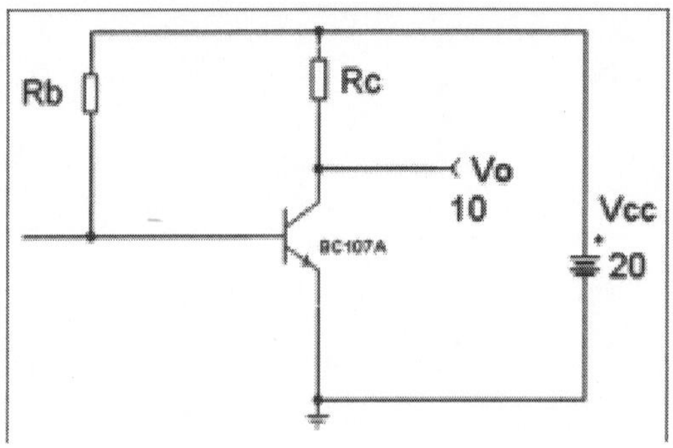

DESIGN

The collector voltage Vc for the simple bias design is 10 volt. The dc current gain, h_{FE} for the BC107A is obtained from the manufacturers data sheets and varies between devices. A typical beta is around 290. Taking a base current of 20uA and reading values direct from the output curves, the collector current, for a collector emitter voltage of 10 volts is around 3.9mA. As h_{FE}= Ic/Ib then a BC107A must have a beta of at least 3.9mA/20uA = 195 to work with this circuit. Also, the base emitter voltage, Vbe is typically 0.6v. Knowing the above data and using ohm's law, values for Rb and Rc can be determined:

Rb = Vcc–Vbe/Ib = (20–0.6)/20u = 970k use (1M)

Rc = Vc/Ic = 10/3.9m = 2.56K use (2.7K)

MATHEMATICAL APPROACH

Without using the output characteristic curve, values for Rb and Rc can still be calculated. A value for h_{FE} must be estimated first and a desired collector current. As h_{FE} varies in each transistor the value chosen should be the lowest value from the manufacturers data sheets. *The equations to use are*:

$$Rc = \frac{Vc}{Ic}$$

$$Ib = \frac{Ic}{h_{FE}}$$

Rb = Vcc–0.6/Ib.

Using the example above with Vcc=20 and h_{FE} =195 yields the same values.

NEED FOR BIASING A TRANSISTOR

A BJT (Bipolar Junction Transistor) require a voltage normally in the range of 0.7V for the internal junctions to become conductive. It is a fixed parameter of Silicon (Si) due to the amount of 1.1eV required to get electrons from the valence energy band into a conductive band. To be able jump the energy gap which is a forbidden band for electrons or to raise the Fermi energy level in the atom. The energy, whether it is electrically applied, thermally or optically, is required to be able change the state of a semiconductor from an insulator to an conductor. You can read more on "semiconductor theory" for better understanding.

Then with a non-linear relationship the conductivity will increase as one increase the forward bias current through the base to emitter junction. Biasing is used for classical transistor amplifier applications.

Biasing is required to have the transistor half way saturated for Class-A amplification or barely switched on for Class-B power amplifiers. If a Class-B amplifier is not biased, then the lower 0.7V of the audio or sine wave will not be amplified causing crossover distortion. When you bias it correctly, the distortions will be gone, since the entire half wave will then fit into the on state of the transistor.

If a Class-A amplifier is not correctly biased, premature clipping on the positive or negative part of the wave will occur.Biasing may be used for other applications as well, such as photo transistors, internal construction of IC's such as op-amps.

This is not a problem for FM operation, where the drive level is constant, and always enough to 'turn on' the PA transistor. FM-only PAs usually have an RF choke to provide a DC return path directly from the base to ground.

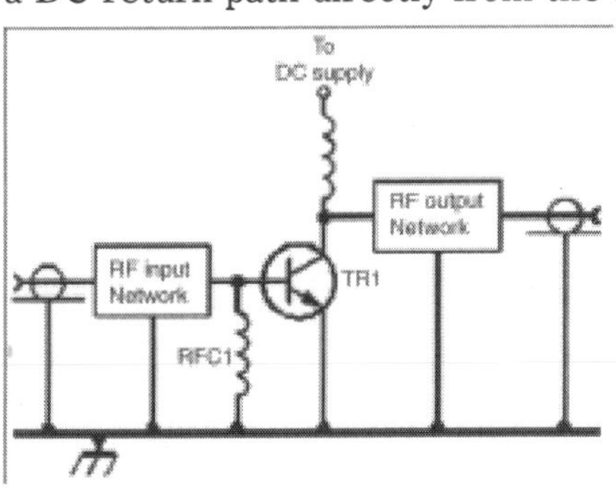

This circuit has no bias—only RFC1 from base to ground. This is sometimes called 'Class C' operation. But for SSB modulation, the delayed turn-on at low drive levels is disastrous: it means that all the low-level parts of the modulation are severely distorted and the amplifier only delivers RF power in bursts on speech peaks. This is why you must never use an FM-only power amplifier on SSB - the splatter is horrendous!

For linear operation, it is *essential* to use a fixed DC bias to make sure that the transistor is already 'turned on' before any RF drive arrives (so-called 'Class AB'). Typically, if the transistor is drawing about 100mA in 'standing current' with no RF drive, distortion will be quite low, provided that the bias supply can also cope with the higher demands of peak modulation. (There's nothing magical about 100mA, by the way; it's just a good figure to aim for in many cases.)

The standing collector current in the RF power transistor depends on the current flowing from the bias supply into the base, and on the beta of the transistor (ratio of collector current to base current). For a typical RF power transistor with a beta of 50, a standing collector current of 100mA needs a base current of 2mA. However, beta is not a very well-controlled value, so the next transistor might need only 2mA; or maybe 7mA. That's one reason why the bias supply must be adjustable. Another reason is that a very small change in base voltage can make a very big difference to the collector current.

WHAT HAPPENS IF A TRANSISTOR IS NOT BIASED

Concept

It is very important that the bias supply maintains a fairly constant output voltage, even at maximum current demand; if it fails to do that, the RF drive will bias the transistor back towards Class C, causing heavy splatter on speech peaks.

Another difficulty is that the base-emitter voltage of a transistor decreases as the transistor heats up. If the bias supply voltage remains constant, the transistor will draw more base current - and hence more collector current - when it's hot than when it's cool. More collector current means more power dissipation, which means more heat, which means more base current, which means more collector current... which can very quickly mean "goodbye transistor" due to thermal runaway. The solutions to this problem are a large, effective heatsink and a base bias supply whose output voltage *decreases* with temperature so that the PA collector current remains fairly constant.

BAD CIRCUITS

Here are two variants of a very common circuit - which generally works very badly.The voltage drop across the forward-biased diode D_1 is approximately the same as the voltage drop across the transistor's base-emitter junction, and by adjusting RV1 to send the correct current through D_1, the collector current in TR1 can be set to the desired value.

If D_1 is in thermal contact with TR1, the voltage across D_1 will drop as TR1 and D_1 heat up together, maintaining a constant and temperature-stabilized current through TR1.

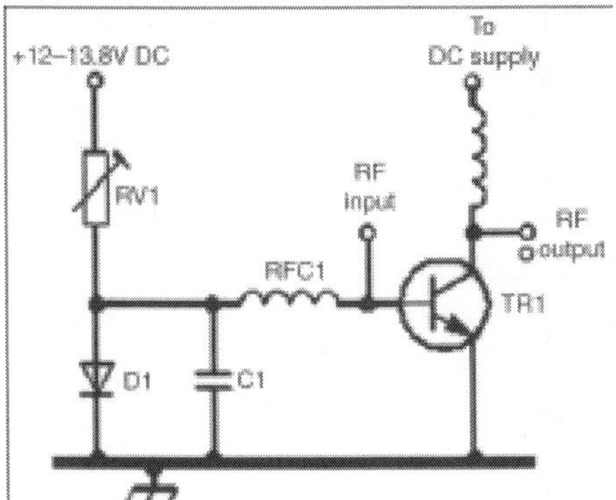

(a)

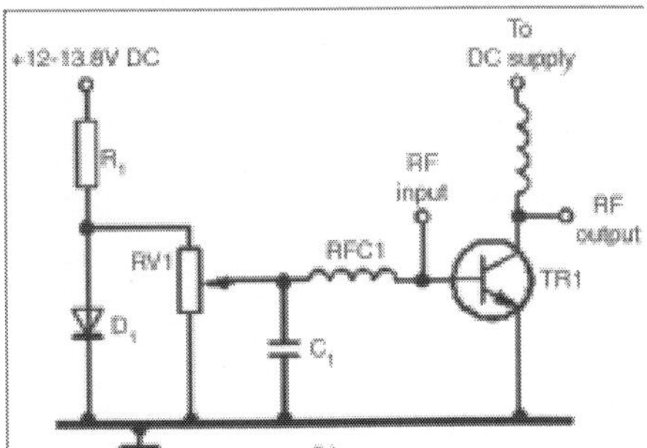

(b)

There are many problems with this simple circuit. The main one is that in order to maintain a constant voltage across D_1, the permanent standing current through D_1 must be several times higher than the maximum base current drawn by TR1 at the peak of RF drive. This requires an enormous standing current through RV1 and D_1 - one ampere or even more. Most designers fail to provide this. Some commercial circuits even use a low-current signal diode such as a lN914 for D_1! Bias regulation totally fails when peaks of RF drive create a heavy demand for base current, so this circuit is a real splatter-generator.

OPERATING CONDITIONS OF A TRANSISTOR

When a transistor is in the fully-off state (like an open switch), it is said to be *cutoff*. Conversely, when it is fully conductive between emitter and collector (passing as much current through the collector as the collector power supply and load will allow), it is said to be *saturated*. These are the two modes of operation explored thus far in using the transistor as a switch.

However, bipolar transistors don't have to be restricted to these two extreme modes of operation. As we learned in the previous section, base current "opens a gate" for a limited amount of current through the collector. If this limit for the controlled current is greater than zero but less than the maximum allowed by the power supply and load circuit, the transistor will "throttle" the collector current in a mode somewhere between cutoff and saturation. This mode of operation is called the *active* mode.

An automotive analogy for transistor operation is as follows: *cutoff* is the condition of no motive force generated by the mechanical parts of the car to make it move.

In cutoff mode, the brake is engaged (zero base current), preventing motion (collector current).

Active mode is the automobile cruising at a constant, controlled speed (constant, controlled collector current) as dictated by the driver. *Saturation* the automobile driving up a steep hill that prevents it from going as fast as the driver wishes.

In other words, a "saturated" automobile is one with the accelerator pedal pushed all the way down (base current calling for more collector current than can be provided by the power supply/load circuit).

Let's set up a circuit for SPICE simulation to demonstrate what happens when a transistor is in its active mode of operation.

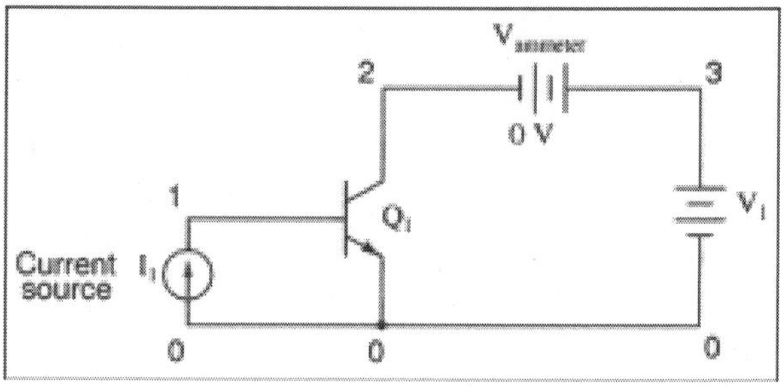


```
bipolar transistor simulation
   i1 0 1 dc 20u
   q1 2 1 0 mod1
   vammeter 3 2 dc 0
   v1 3 0 dc
   .model mod1 npn
   .dc v1 0 2 0.05
   .plot dc i(vammeter)
   .end
```

"Q" is the standard letter designation for a transistor in a schematic diagram, just as "R" is for resistor and "C" is for capacitor. In this circuit, we have an NPN transistor powered by a battery (V_1) and controlled by current through a *current source* (I_1).

A current source is a device that outputs a specific amount of current, generating as much or as little voltage across its terminals to ensure that exact amount of current through it.

Current sources are notoriously difficult to find in nature (unlike voltage sources, which by contrast attempt to maintain a constant voltage, outputting as much or as little current in the fulfillment of that task), but can be simulated with a small collection of electronic components.

As we are about to see, transistors themselves tend to mimic the constant-current behaviour of a current source in their ability to *regulate* current at a fixed value.

In the SPICE simulation, we'll set the current source at a constant value of 20 μA, then vary the voltage source (V_1) over a range of 0 to 2 volts and monitor how much current goes through it. The "dummy" battery ($V_{ammeter}$) in Figure above with its output of 0 volts serves merely to provide SPICE with a circuit element for current measurement.

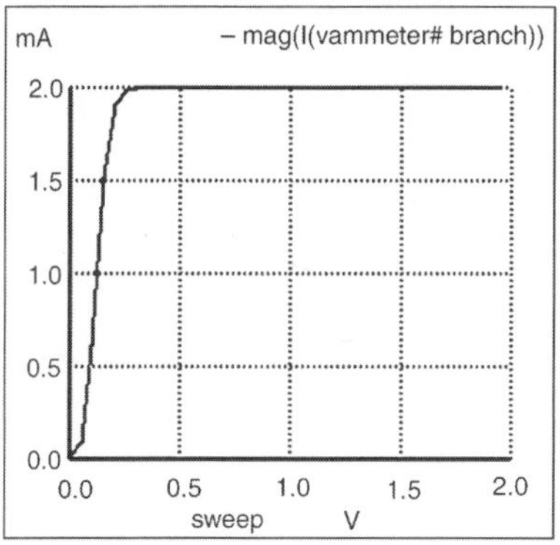

A Sweeping collector voltage 0 to 2 V with base current constant at 20 μA yields constant 2 mA collector current in the saturation region.

The constant base current of 20 μA sets a collector current limit of 2 mA, exactly 100 times as much.

Notice how flat the curve is in (Figure above for collector current over the range of battery voltage from 0 to 2 volts. The only exception to this featureless plot is at the very beginning, where the battery increases from 0 volts to 0.25 volts.

There, the collector current increases rapidly from 0 amps to its limit of 2 mA. Let's see what happens if we vary the battery voltage over a wider range, this time from 0 to 50 volts. We'll keep the base current steady at 20 μA.

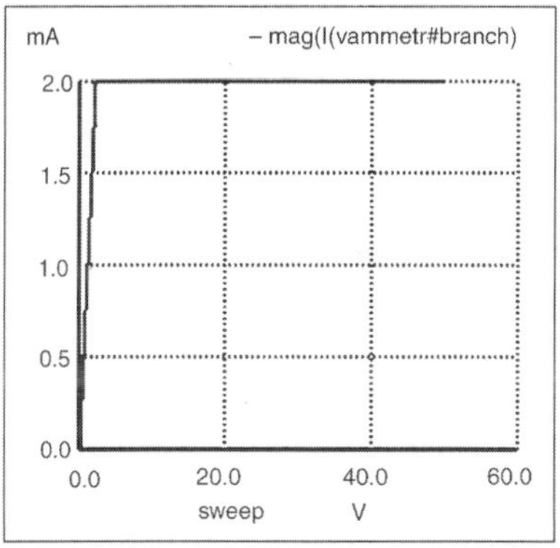


```
bipolar transistor simulation
    i1 0 1 dc 20u
    q1 2 1 0 mod1
    vammeter 3 2 dc 0
    v1 3 0 dc
    .model mod1 npn
    .dc v1 0 50 2
    .plot dc i(vammeter)
    .end
```

Sweeping collector voltage 0 to 50 V with base current constant at 20 μA yields constant 2 mA collector current.

The collector current in Figure above holds absolutely steady at 2 mA, although the battery (v_1) voltage varies all the way from 0 to 50 volts. It would appear from our simulation that collector-to-emitter voltage has little effect over collector current, except at very low levels (just above 0 volts).

The transistor is acting as a current regulator, allowing exactly 2 mA through the collector and no more. Now let's see what happens if we increase the controlling (I_1) current from 20 μA to 75 μA, once again sweeping the battery (V_1) voltage from 0 to 50 volts and graphing the collector current in Figure below.

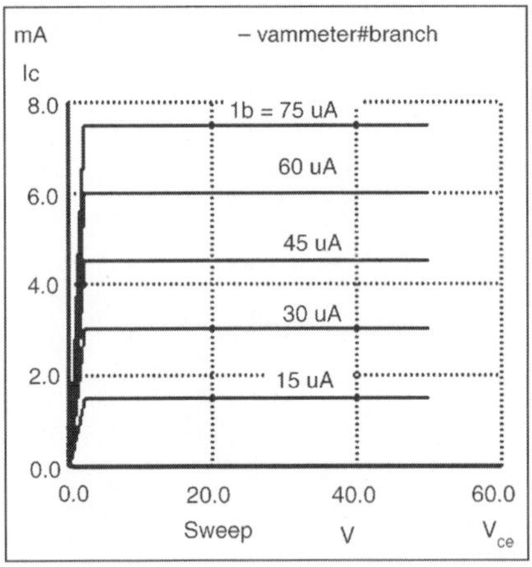


```
bipolar transistor simulation
    i1 0 1 dc 75u
    q1 2 1 0 mod1s
    vammeter 3 2 dc 0
    v1 3 0 dc
    .model mod1 npn
    .dc v1 0 50 2 i1 15u 75u 15u
    .plot dc i(vammeter)
    .end
```

Sweeping collector voltage 0 to 50 V (.dc v_1 0 50 2) with base current constant at 75 μA yields constant 7.5 mA collector current. Other curves are generated by current sweep (i1 15u 75u 15u) in DC analysis statement (.dc v_1 0 50 2 i_1 15u 75u 15u).

Not surprisingly, SPICE gives us a similar plot: a flat line, holding steady this time at 7.5 mA—exactly 100 times the base current—over the range of battery voltages from just above 0 volts to 50 volts. It appears that the base current is the deciding factor for collector current, the V_1 battery voltage being irrelevant as long as it is above a certain minimum level.

This voltage/current relationship is entirely different from what we're used to seeing across a resistor. With a resistor, current increases linearly as the voltage across it increases. Here, with a transistor, current from emitter to collector stays limited at a fixed, maximum value no matter how high the voltage across emitter and collector increases.

Often it is useful to superimpose several collector current/voltage graphs for different base currents on the same graph as in Figure below. A collection of curves like this—one curve plotted for each distinct level of base current—for a particular transistor is called the transistor's *characteristic curves*:

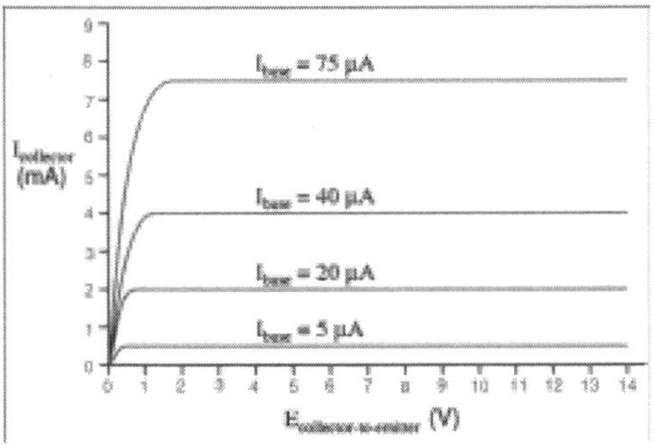

VOLTAGE COLLECTOR TO EMITTER VS COLLECTOR CURRENT FOR VARIOUS BASE CURRENTS

Each curve on the graph reflects the collector current of the transistor, plotted over a range of collector-to-emitter voltages, for a given amount of base current. Since a transistor tends to act as a current regulator, limiting collector current to a proportion set by the base current, it is useful to express this proportion as a standard transistor performance measure. Specifically, the ratio of collector current to base current is known as the *Beta* ratio (symbolized by the Greek letter b):

$$\beta = \frac{I_{\text{Collector}}}{I_{\text{base}}}.$$

Sometimes the β ratio is designated as "h_{fe}," a label used in a branch of mathematical semiconductor analysis known as "hybrid parameters" which strives to achieve precise predictions of transistor performance with detailed equations. Hybrid parameter variables are many, but each is labeled with the general letter "h" and a specific subscript. The variable "h_{fe}" is just another (standardized) way of expressing the ratio of collector current to base current, and is interchangeable with "β." The β ratio is unit less.

β for any transistor is determined by its design: it cannot be altered after manufacture. It is rare to have two transistors of the same design exactly match because of the physical variables affecting b. If a circuit design relies on equal β ratios between multiple transistors, "matched sets" of transistors may be purchased at extra cost. However, it is generally considered bad design practice to engineer circuits with such dependencies.

The β of a transistor does not remain stable for all operating conditions. For an actual transistor, the β ratio may vary by a factor of over 3 within its operating current limits. For example, a transistor with advertised β of 50 may actually test with I_c/I_b ratios as low as 30 and as high as 100, depending on the amount of collector current, the transistor's temperature, and frequency of amplified signal, among other factors. For tutorial purposes it is adequate to assume a constant β for any given transistor; realise that real life is not that simple!

Sometimes it is helpful for comprehension to "model" complex electronic components with a collection of simpler, better-understood components. The model in Figure below is used in many introductory electronics texts.

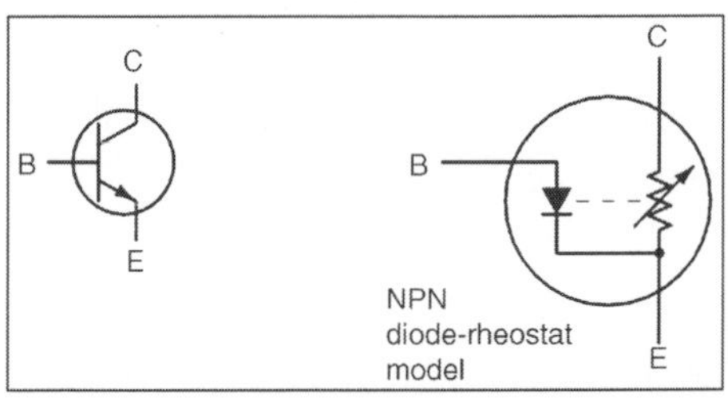

ELEMENTARY DIODE RESISTOR TRANSISTOR MODEL

This model casts the transistor as a combination of diode and rheostat (variable resistor). Current through the base-emitter diode controls the resistance of the collector-emitter rheostat (as implied by the dashed line connecting the two components), thus controlling collector current. An NPN transistor is modeled in the figure shown, but a PNP transistor would be only slightly different (only the base-emitter diode would be reversed). This model succeeds in illustrating the basic concept of transistor amplification: how the base current signal can exert control over the collector current.

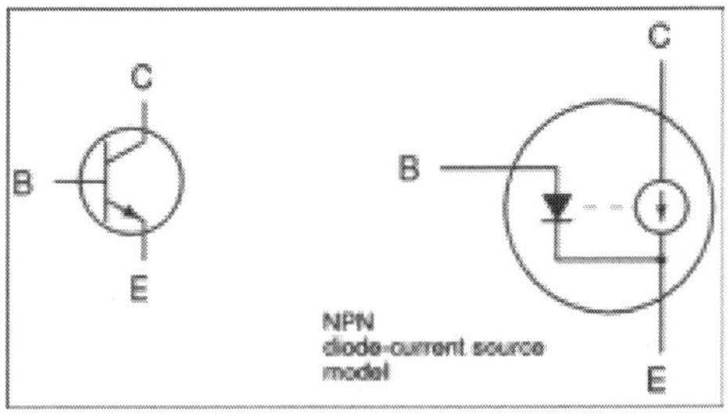

However, I don't like this model because it miscommunicates the notion of a set amount of collector-emitter resistance for a given amount of base current. If this were true, the transistor wouldn't *regulate* collector current at all like the characteristic curves show. Instead of the collector current curves flattening out after their brief rise as the collector-emitter voltage increases, the collector current would be directly proportional to collector-emitter voltage, rising steadily in a straight line on the graph.

A better transistor model, often seen in more advanced textbooks, is shown in Figure.

CURRENT SOURCE MODEL OF TRANSISTOR

It casts the transistor as a combination of diode and current source, the output of the current source being set at a multiple (β ratio) of the base current. This model is far more accurate in depicting the true input/output characteristics of a transistor: base current establishes a certain amount of collector *current*, rather than a certain amount of collector-emitter *resistance* as the first model implies. Also, this model is favoured when performing network analysis on transistor circuits, the current source being a well-understood theoretical component.

Unfortunately, using a current source to model the transistor's current-controlling behaviour can be misleading: in no way will the transistor ever act as a *source* of electrical energy. The current source does not model the fact that its source of energy is a external power supply, similar to an amplifier.

FIELD EFFECT TRANSISTORS (FET)

The field-effect transistor (FET) is a transistor that relies on an electric field to control the shape and hence the conductivity of a channel of one type of charge carrier in a semiconductor material. FETs are sometimes called *unipolar transistors* to contrast their single-carrier-type operation with the dual-carrier-type operation of bipolar (junction) transistors (BJT). The *concept* of the FET predates the BJT, though it was not physically implemented until *after* BJTs due to the limitations of semiconductor materials and the relative ease of manufacturing BJTs compared to FETs at the time.

The field-effect transistor (FET) controls the current between two points but does so differently than the bipolar transistor. The FET operates by the effects of an electric field on the flow of electrons through a single type of semiconductor material. This is why the FET is sometimes called a unipolar transistor. Also, unlike bipolar semiconductors that can be arranged in many configurations to provide diodes, transistors, photoelectric devices. temperature sensitive devices and so on, the field effect is usually only used to make transistors, although FETs are also available as special-purpose diodes, for use as constant current sources.

Current moves within the FET in a channel, from the source connection to the drain connection. A gate terminal generates an electric field that controls the current. The channel is made of either N-type or P-type semiconductor material; an FET is specified as either an N-channel or P-channel device. Majority carriers flow from source to drain. In N-channel devices, electrons flow so the drain potential must be higher than that of the Source (VDS > O)- In P-channel devices, the flow of holes requires that VDS < 0. The polarity of the electric field that controls current in the channel is determined by the majority carriers of the channel, ordinarily positive for P-channel FETs and negative for N-channel FETS.

Variations of FET technology are based on different ways of generating the electric field. In all of these, however, electrons at the gate are used only for their charge in order to create an electric field around the channel, and there is a minimal flow of electrons through the gate. This leads to a very high de input resistance in devices that use FETs for their input circuitry. There may be quite a bit of capacitance between the gate and the other FET terminals, however. The input impedance may be quite low at RF.

The current through an FET only has to pass through a single type of semiconductor material. There is very little resistance in the absence of an electric field (no bias voltage). The drain-source resistance (rDS ON) is between a few hundred ohms to less than an ohm. The output impedance of devices made with FETs is generally quite low. If a gate bias voltage is added to operate the transistor near cut off, the circuit output impedance may be much higher.

FET devices are constructed on a substrate of doped semiconductor material. The channel is formed within the substrate and has the opposite polarity (a P-channel FET has N-type substrate). Most FETS are constructed with silicon. In order to achieve a higher gain-bandwidth product, other materials have been used. Gallium Arsenide (GaAs) has electron mobility and drift velocities that are far higher than the standard doped silicon, Amplifiers designed with GaAs FET devices have much higher frequency response and lower noise factor at VHF and UHF than those made with standard FETS.

The common transistor is called a junction transistor, and it was the key device which led to the solid state electronics revolution. In application, the junction transistor has the disadvantage of a low input impedance because the base of the transistor is the signal input and the base-emitter diode is forward biased.

Another device achieved transistor action with the input diode junction reversed biased, and this device is called a "field effect transistor" or a "junction field effect transistor", JFET. With the reverse biased input junction, it has a very high input impedance. Having a high input impedance minimizes the interference with or "loading" of the signal source when a measurement is made.

CONSTRUCTION OF A JEFT

All FETs have *gate*, *drain*, and *source* terminals that correspond roughly to the *base*, *collector*, and *emitter* of BJTs. Most FETs also have a fourth terminal called the *body*, *base*, *bulk*, or *substrate*.

This fourth terminal serves to bias the transistor into operation; it is rare to make non-trivial use of the body terminal in circuit designs, but its presence is important when setting up the physical layout of an integrated circuit. The size of the gate, length L in the diagram, is the distance between source and drain. The *width* is the extension of the transistor, in the diagram perpendicular to the cross section. Typically the width is much larger than the length of the gate. A gate length of 1 μm limits the upper frequency to about 5 GHz, 0.2 μm to about 30 GHz.

The names of the terminals refer to their functions. The gate terminal may be thought of as controlling the opening and closing of a physical gate. This gate permits electrons to flow through or blocks their passage by creating or eliminating a channel between the source and drain. Electrons flow from the source terminal towards the drain terminal if influenced by an applied voltage.

The body simply refers to the bulk of the semiconductor in which the gate, source and drain lie. Usually the body terminal is connected to the highest or lowest voltage within the circuit, depending on type. The body terminal and the source terminal are sometimes connected together since the source is also sometimes connected to the highest or lowest voltage within the circuit, however there are several uses of FETs which do not have such a configuration, such as transmission gates and cascode circuits

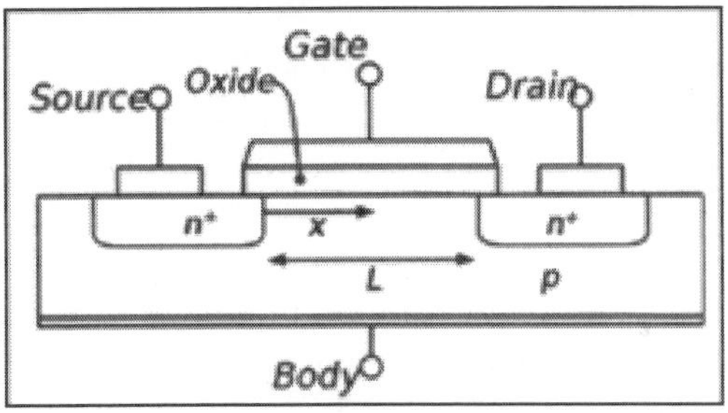

The FET controls the flow of electrons (or electron holes) from the source to drain by affecting the size and shape of a "conductive channel" created and influenced by voltage (or lack of voltage) applied across the gate and source terminals (For ease of discussion, this assumes body and source are connected). This conductive channel is the "stream" through which electrons flow from source to drain.

In an n-channel *depletion-mode* device, a negative gate-to-source voltage causes a *depletion region* to expand in width and encroach on the channel from the sides, narrowing the channel. If the depletion region expands to completely close the channel, the resistance of the channel from source to drain becomes large, and the FET is effectively turned off like a switch. Likewise a positive gate-to-source voltage increases the channel size and allows electrons to flow easily.

Conversely, in an n-channel *enhancement-mode* device, a positive gate-to-source voltage is necessary to create a conductive channel, since one does not exist naturally within the transistor. The positive voltage attracts free-floating electrons within the body towards the gate, forming a conductive channel. But first, enough electrons must be attracted near the gate to counter the dopant ions added to the body of the FET; this forms a region free of mobile carriers called a depletion region, and the phenomenon is referred to as the *threshold voltage* of the FET. Further gate-to-source voltage increase will attract even more electrons towards the gate which are able to create a conductive channel from source to drain; this process is called *inversion*.

For either enhancement- or depletion-mode devices, at drain-to-source voltages much less than gate-to-source voltages, changing the gate voltage will alter the channel resistance, and drain current will be proportional to drain voltage (referenced to source voltage). In this mode the FET operates like a variable resistor and the FET is said to be operating in a *linear mode* or *ohmic mode*.

If drain-to-source voltage is increased, this creates a significant asymmetrical change in the shape of the channel due to a gradient of voltage potential from source to drain. The shape of the inversion region becomes "pinched-off" near the drain end of the channel. If drain-to-source voltage is increased further, the pinch-off point of the channel begins to move away from the drain towards the source.

The FET is said to be in *saturation mode*;[7] some authors refer to it as *active mode*, for a better analogy with bipolar transistor operating regions. The saturation mode, or the region between ohmic and saturation, is used when amplification is needed. The in-between region is sometimes considered to be part of the ohmic or linear region, even where drain current is not approximately linear with drain voltage.

Even though the conductive channel formed by gate-to-source voltage no longer connects source to drain during saturation mode, carriers are not blocked from flowing. Considering again an n-channel device, a depletion region exists in the p-type body, surrounding the conductive channel and drain and source regions. The electrons which comprise the channel are free to move out of the channel through the depletion region if attracted to the drain by drain-to-source voltage.

The depletion region is free of carriers and has a resistance similar to silicon. Any increase of the drain-to-source voltage will increase the distance from drain to the pinch-off point, increasing resistance due to the depletion region proportionally to the applied drain-to-source voltage.

This proportional change causes the drain-to-source current to remain relatively fixed independent of changes to the drain-to-source voltage and quite unlike the linear mode operation.

Thus in saturation mode, the FET behaves as a constant-current source rather than as a resistor and can be used most effectively as a voltage amplifier. In this case, the gate-to-source voltage determines the level of constant current through the channel.

BIASING OF FET

Biasing in electronics is the method of establishing predetermined voltages and/or currents at various points of an electronic circuit to set an appropriate operating point. The operating point of a device, also known as bias point, quiescent point, or simply Q-point, is the point on the output characteristics that shows the direct current (DC), collector-emitter voltage (V_{CE}), and the collector current (I_C) with no input signal applied. The term is normally used in connection with devices such as transistors.

For bipolar junction transistors the bias point is chosen to keep the transistor operating in the active mode, using a variety of circuit techniques, establishing the Q-point DC voltage and current.

A small signal is then applied on top of the Q-point bias voltage, thereby either modulating or switching the current, depending on the purpose of the circuit.

The quiescent point of operation is typically near the middle of DC load line. The process of obtaining certain DC collector current at a certain DC collector voltage by setting up operating point is called biasing.

After establishing the operating point, when input signal is applied, the output signal should not move the transistor either to saturation or to cut-off.

However, this unwanted shift still might occur, due to the following reasons:

- Parameters of transistors depend on junction temperature. As junction temperature increases, leakage current due to minority charge carriers (I_{CBO}) increases. As I_{CBO} increases, I_{CEO} also increases, causing an increase in collector current I_C. This produces heat at the collector junction. This process repeats, and, finally, Q-point may shift into the saturation region. Sometimes, the excess heat produced at the junction may even burn the transistor. This is known as thermal runaway.
- When a transistor is replaced by another of the same type, the Q-point may shift, due to changes in parameters of the transistor, such as *current gain* (β) which varies slightly for each unique transistor.

To avoid a shift of Q-point, bias-stabilization is necessary. Various biasing circuits can be used for this purpose.

THE JUNCTION FIELD EFFECT TRANSISTOR

We saw previously that a bipolar junction transistor is constructed using two PN-junctions in the main current carrying path between the Emitter and the Collector terminals.

The Junction Field Effect Transistor (JUGFET or JFET) has no PN-junctions but instead has a narrow piece of high-resistivity semiconductor material forming a "Channel" of either N-type or P-type silicon for the majority carriers to flow through with two ohmic electrical connections at either end commonly called the Drain and the Source respectively.

There are two basic configurations of junction field effect transistor, the N-channel JFET and the P-channel JFET. The N-channel JFET's channel is doped with donor impurities meaning that the flow of current through the channel is negative (hence the term N-channel) in the form of electrons. Likewise, the P-channel JFET's channel is doped with acceptor impurities meaning that the flow of current through the channel is positive (hence the term P-channel) in the form of holes. N-channel JFET's have a greater channel conductivity (lower resistance) than their equivalent P-channel types, since electrons have a higher mobility through a conductor compared to holes. This makes the N-channel JFET's a more efficient conductor compared to their P-channel counterparts.

We have said previously that there are two ohmic electrical connections at either end of the channel called the Drain and the Source. But within this channel there is a third electrical connection which is called the Gate terminal and this can also be a P-type or N-type material forming a PN-junction with the main channel.

The relationship between the connections of a junction field effect transistor and a bipolar junction transistor are compared below.

COMPARISON OF CONNECTIONS BETWEEN A JFET AND A BJT

Bipolar Transistor	**Field Effect Transistor**
Emitter - (E) >> Source - (S)	
Base - (B) >> Gate - (G)	
Collector - (C) >> Drain - (D)	

The symbols and basic construction for both configurations of JFETs are shown below.

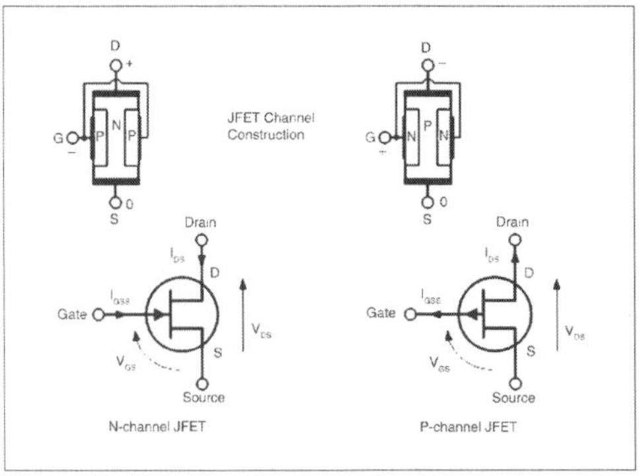

The semiconductor "channel" of the Junction Field Effect Transistor is a resistive path through which a voltage V_{DS} causes a current $I_D$to flow. The JFET can conduct current equally well in either direction. A voltage gradient is thus formed down the length of the channel with this voltage becoming less positive as we go from the Drain terminal to the Source terminal. The PN-junction therefore has a high reverse bias at the Drain terminal and a lower reverse bias at the Source terminal. This bias causes a "depletion layer" to be formed within the channel and whose width increases with the bias.

The magnitude of the current flowing through the channel between the Drain and the Source terminals is controlled by a voltage applied to the Gate terminal, which is a reverse-biased. In an N-channel JFET this Gate voltage is negative while for a P-channel JFET the Gate voltage is positive. The main difference between the JFET and a BJT device is that when the JFET junction is reverse-biased the Gate current is practically zero, whereas the Base current of the BJT is always some value greater than zero.

Bias Arrangement for an N-channel JFET and Corresponding Circuit Symbols

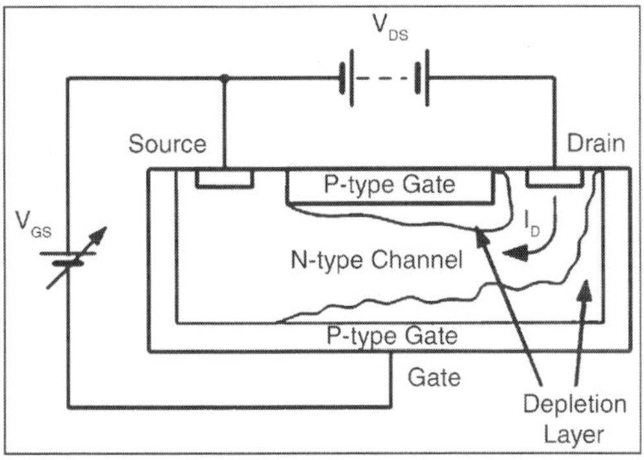

The cross sectional diagram above shows an N-type semiconductor channel with a P-type region called the Gate diffused into the N-type channel forming a reverse biased PN-junction and it is this junction which forms the depletion region around the Gate area when no external voltages are applied. JFETs are therefore known as depletion mode devices.

This depletion region produces a potential gradient which is of varying thickness around the PN-junction and restrict the current flow through the channel by reducing its effective width and thus increasing the overall resistance of the channel itself. The most-depleted portion of the depletion region is in between the Gate and the Drain, while the least-depleted area is between the Gate and the Source. Then the JFET's channel conducts with zero bias voltage applied (*i.e.,* the depletion region has near zero width).

With no external Gate voltage ($V_G = 0$), and a small voltage (V_{DS}) applied between the Drain and the Source, maximum saturation current (I_{DSS}) will flow through the channel from the Drain to the Source restricted only by the small depletion region around the junctions.

If a small negative voltage ($-V_{GS}$) is now applied to the Gate the size of the depletion region begins to increase reducing the overall effective area of the channel and thus reducing the current flowing through it, a sort of "squeesing" effect takes place.

So by applying a reverse bias voltage increases the width of the depletion region which in turn reduces the conduction of the channel. Since the PN-junction is reverse biased, little current will flow into the gate connection. As the Gate voltage ($-V_{GS}$) is made more negative, the width of the channel decreases until no more current flows between the Drain and the Source and the FET is said to be "pinched-off" (similar to the cut-off region for a BJT). The voltage at which the channel closes is called the "pinch-off voltage", (V_P).

JFET Channel Pinched-off

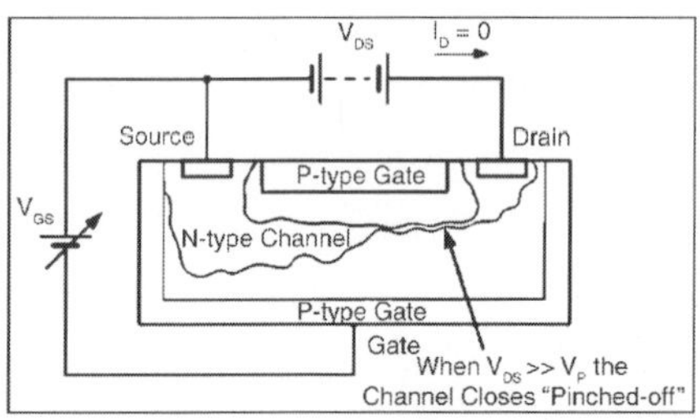

In this pinch-off region the Gate voltage, V_{GS} controls the channel current and V_{DS} has little or no effect.

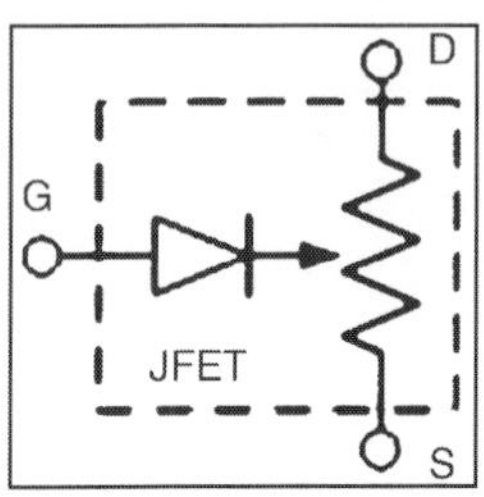

JFET Model

The result is that the FET acts more like a voltage controlled resistor which has zero resistance when $V_{GS} = 0$ and maximum "ON" resistance (R_{DS}) when the Gate voltage is very negative. Under normal operating conditions, the JFET gate is always negatively biased relative to the source. It is essential that the Gate voltage is never positive since if it is all the channel current will flow to the Gate and not to the Source, the result is damage to the JFET. Then to close the channel:

- No Gate voltage (V_{GS}) and V_{DS} is increased from zero.
- No V_{DS} and Gate control is decreased negatively from zero.
- V_{DS} and V_{GS} varying.

The P-channel Junction Field Effect Transistor operates the same as the N-channel above, with the following exceptions:

- Channel current is positive due to holes,
- The polarity of the biasing voltage needs to be reversed.

The output characteristics of an N-channel JFET with the gate short-circuited to the source is given as

Output Characteristic V-I Curves of a Typical Junction FET.

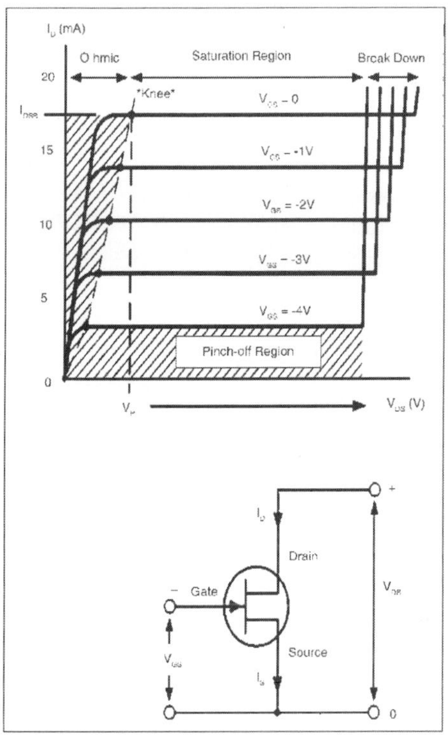

The voltage V_{GS} applied to the Gate controls the current flowing between the Drain and the Source terminals. V_{GS} refers to the voltage applied between the Gate and the Source while V_{DS} refers to the voltage applied between the Drain and the Source. Because a Junction Field Effect Transistor is a voltage controlled device, "NO current flows into the gate!" then the Source current (I_S) flowing out of the device equals the Drain current flowing into it and therefore ($I_D = I_S$). The characteristics curves example shown above, shows the four different regions of operation for a JFET and these are given as:

- *Ohmic Region*: When $V_{GS} = 0$ the depletion layer of the channel is very small and the JFET acts like a voltage controlled resistor.
- *Cut-off Region*: This is also known as the pinch-off region were the Gate voltage, V_{GS} is sufficient to cause the JFET to act as an open circuit as the channel resistance is at maximum.

- *Saturation or Active Region*: The JFET becomes a good conductor and is controlled by the Gate-Source voltage, (V_{GS}) while the Drain-Source voltage, (V_{DS}) has little or no effect.
- *Breakdown Region*: The voltage between the Drain and the Source, (V_{DS}) is high enough to causes the JFET's resistive channel to break down and pass uncontrolled maximum current.

The characteristics curves for a P-channel junction field effect transistor are the same as those above, except that the Drain current I_D decreases with an increasing positive Gate-Source voltage,V_{GS}.

The Drain current is zero when $V_{GS} = V_P$. For normal operation, V_{GS} is biased to be somewhere between V_P and 0. Then we can calculate the Drain current, I_D for any given bias point in the saturation or active region as follows:

Drain current in the active region.

$$I_D = I_{DSS}\left[1 - \frac{V_{GS}}{V_P}\right]^2$$

Note that the value of the Drain current will be between zero (pinch-off) and I_{DSS} (maximum current). By knowing the Drain current I_D and the Drain-Source voltage V_{DS} the resistance of the channel (I_D) is given as:

Drain-Source channel resistance.

$$R_{DS} = \frac{\Delta V_{DS}}{\Delta I_D} = \frac{1}{gm}$$

Where: g_m is the "transconductance gain" since the JFET is a voltage controlled device and which represents the rate of change of the Drain current with respect to the change in Gate-Source voltage.

JFET Amplifier

Just like the bipolar junction transistor, JFET's can be used to make single stage class A amplifier circuits with the JFET common source amplifier and characteristics being very similar to the BJT common emitter circuit. The main advantage JFET amplifiers have over BJT amplifiers is their high input impedance which is controlled by the Gate biasing resistive network formed by $R1$ and $R2$ as shown.

Biasing of JFET Amplifier

This common source (CS) amplifier circuit is biased in class A mode by the voltage divider network formed by $R1$ and $R2$. The voltage across the Source resistor R_S is generally set at one fourth V_{DD}, ($V_{DD}/4$). The required Gate voltage can then be calculated using this R_S value. Since the Gate current is zero, ($I_G = 0$) we can set the required DC quiescent voltage by the proper selection of resistors $R1$ and $R2$.

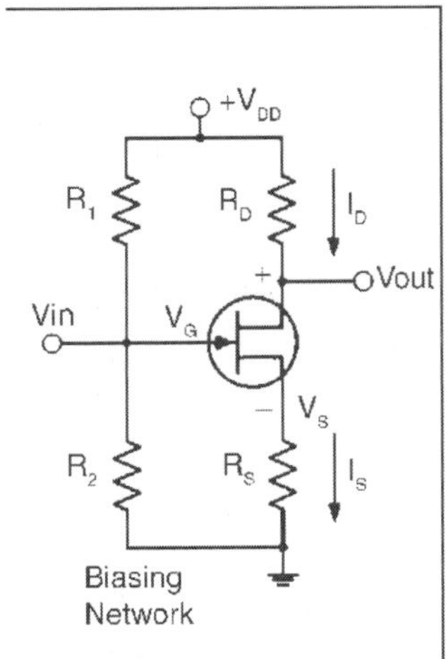

$$V_S = I_D R_S = \frac{V_{DD}}{4}$$

$$V_S = V_G - V_{GS}$$

$$V_G = \left(\frac{R2}{R1 + R2}\right) V_{DD}$$

$$I_D = \frac{V_S}{R_S} = \frac{V_G - V_{GS}}{R_S}$$

The control of the Drain current by a negative Gate potential makes the Junction Field Effect Transistor useful as a switch and it is essential that the Gate voltage is never positive for an N-channel JFET as the channel current will flow to the Gate and not the Drain resulting in damage to the JFET.

The principals of operation for a P-channel JFET are the same as for the N-channel JFET, except that the polarity of the voltages need to be reversed.

In the next tutorial about Transistors, we will look at another type of Field Effect Transistor called a MOSFET whose Gate connection is completely isolated from the main current carrying channel.

OPERATING POINT IN PRESENCE OF SELF HEATING

For this situation, the total heat flow $\dot{Q}$ is made up of the heat flow in the two parallel paths, $\dot{Q} = \dot{Q}_1 + \dot{Q}_2$,, with the total resistance given by,

$$\frac{1}{R} = \frac{1}{R_1} + \frac{1}{R_2}.$$

More complex configurations can also be examined; for example, a brick wall with insulation on both sides.

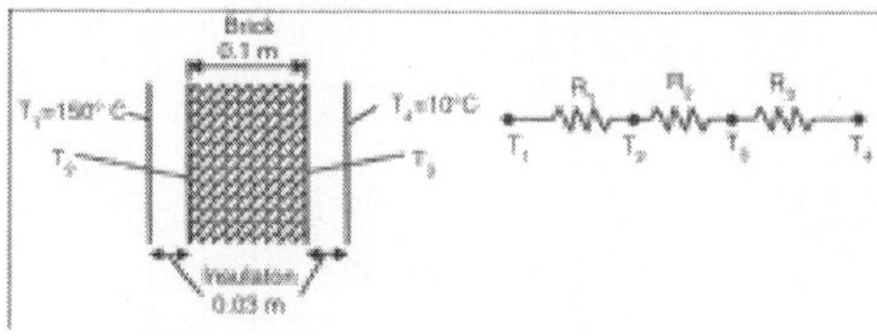

Fig. Heat Transfer Through an Insulated Wall

The overall thermal resistance is given by

$$R = R_1 + R_2 + R_3 = \frac{L_1}{k_1 A_1} + \frac{L_2}{k_2 A_2} + \frac{L_3}{k_3 A_3}.$$

Some representative values for the brick and insulation thermal conductivity are:

$$k_{brik} = k_2 \qquad = 0.7 W/m-K$$

$$k_{insulation} = k_1 = k_3 \qquad = 0.07 W/m-K$$

Using these values, and noting that $A_1 = A_2 = A_3 = A$, we obtain,

$$Ar_1 = Ar_3 = \frac{L_1}{k_1} = \frac{0.03}{0.07 W/m-k} = 0.42 m^2 k/W$$

$$Ar_2 = \frac{L_2}{k_2} = \frac{0.1m}{0.7 W/m-k} = 0.14 m^2 k/W.$$

This is a series circuit so,

$$\dot{q} = \frac{\dot{Q}}{A} = \text{Constant throughout} = \frac{T_1 - T_4}{RA} = \frac{140}{0.98 m^2 K/W} = 142 W/m^2.$$

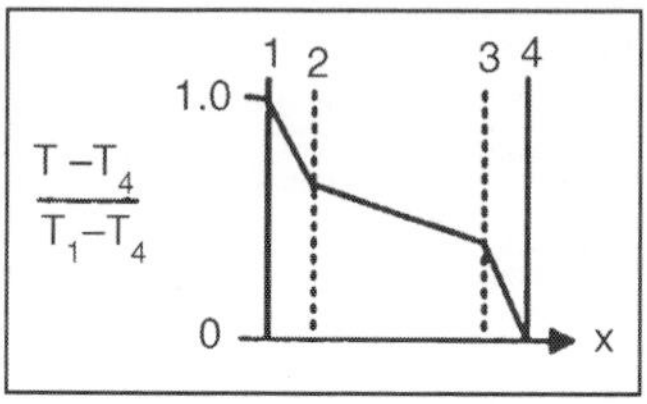

Fig. Temperature Distribution Through an Insulated Wall

The temperature is continuous in the wall and the intermediate temperatures can be found from applying the resistance equation across each slab, since $\dot{Q}$ is constant across the slab. For example, to find T_2:

$$\dot{q} = \frac{T_1 - T_2}{R_1 A} = 142 W/m^2.$$

This yields

$$T_1 - T_2 = 60k$$

or

$$T_2 = 90°.$$

The same procedure gives T_3 = 90°. As sketched in Figure, the larger drop is across the insulating layer even though the brick layer is much thicker.

BASE RESISTOR/FIXED BIAS CIRCUIT

The first biasing method, called *Base Current Bias* or sometimes FIXED BIAS, was used in the picture below. As you recall, it consisted basically of a resistor (RB) connected between the collector supply voltage and the base. Unfortunately, this simple arrangement is quite thermally unstable. If the temperature of the transistor rises for any reason (due to a rise in ambient temperature or due to current flow through it), collector current will increase.

This increase in current also causes the dc operating point, sometimes called the quiescent or static point, to move away from its desired position (level).

This reaction to temperature is undesirable because it affects amplifier gain (the number of times of amplification) and could result in distortion, as you will see later in this discussion.

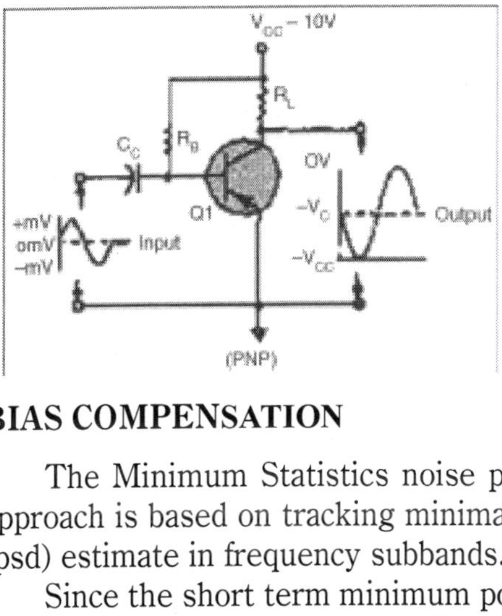

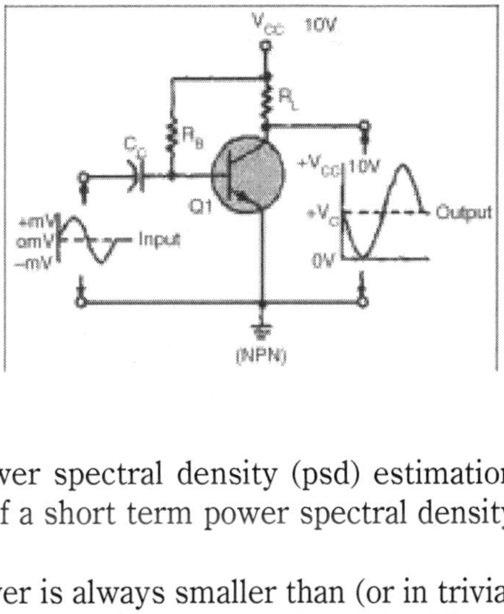

BIAS COMPENSATION

The Minimum Statistics noise power spectral density (psd) estimation approach is based on tracking minima of a short term power spectral density (psd) estimate in frequency subbands.

Since the short term minimum power is always smaller than (or in trivial cases equal to) the mean power, the minimum noise power estimate is a biased estimate of the mean power.

For an accurate noise power estimate this bias must be compensated.In this paper we review bias compensation methods for moving average and first-order recursive smoothed psd estimates. While for some cases exact expressions for the bias are available, approximations are required in general.

We present approximations which allow an efficient computation and compensation of the bias. We discuss factors that influence the bias and show that the method is to some extent robust to variations of the signal statistics. Besides different smoothing methods, we discuss the effect of overlapping spectral analysis windows and of signal correlation.

BIASING WITH A FEEDBACK RESISTOR

INTRODUCTION

If some percentage of an amplifier's output signal is connected to the input, so that the amplifier amplifies part of its own output signal, we have what is known as *feedback*. Feedback comes in two varieties:*positive* (also called *regenerative*), and *negative* (also called *degenerative*). Positive feedback reinforces the direction of an amplifier's output voltage change, while negative feedback does just the opposite.

A familiar example of feedback happens in public-address ("PA") systems where someone holds the microphone too close to a speaker: a high-pitched "whine" or "howl" ensues, because the audio amplifier system is detecting and amplifying its own noise. Specifically, this is an example of *positive* or *regenerative*feedback, as any sound detected by the microphone is amplified and turned into a louder sound by the speaker, which is then detected by the microphone again, and so on... the result being a noise of steadily increasing volume until the system becomes "saturated" and cannot produce any more volume.

One might wonder what possible benefit feedback is to an amplifier circuit, given such an annoying example as PA system "howl." If we introduce positive, or regenerative, feedback into an amplifier circuit, it has the tendency of creating and sustaining oscillations, the frequency of which determined by the values of components handling the feedback signal from output to input. This is one way to make an *oscillator* circuit to produce AC from a DC power supply. Oscillators are very useful circuits, and so feedback has a definite, practical application for us. "Phase shift oscillator", for a practical application of positive feedback.

Negative feedback, on the other hand, has a "dampening" effect on an amplifier: if the output signal happens to increase in magnitude, the feedback signal introduces a decreasing influence into the input of the amplifier, thus opposing the change in output signal. While positive feedback drives an amplifier circuit towards a point of instability (oscillations), negative feedback drives it the opposite direction: towards a point of stability.

An amplifier circuit equipped with some amount of negative feedback is not only more stable, but it distorts the input waveform less and is generally capable of amplifying a wider range of frequencies. The tradeoff for these advantages (there just *has* to be a disadvantage to negative feedback, right?) is decreased gain. If a portion of an amplifier's output signal is "fed back" to the input to oppose any changes in the output, it will require a greater input signal amplitude to drive the amplifier's output to the same amplitude as before. This constitutes a decreased gain. However, the advantages of stability, lower distortion, and greater bandwidth are worth the tradeoff in reduced gain for many applications.

Let's examine a simple amplifier circuit and see how we might introduce negative feedback into it, starting with Figure below.

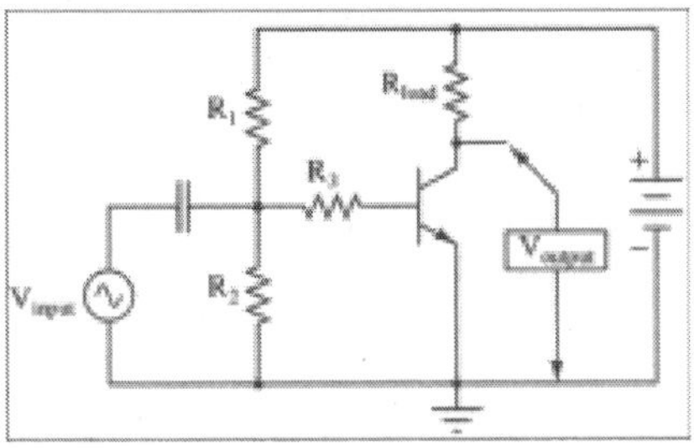

COMMON-EMITTER AMPLIFIER WITHOUT FEEDBACK.

The amplifier configuration shown here is a common-emitter, with a resistor bias network formed by R_1 and R_2. The capacitor couples V_{input} to the amplifier so that the signal source doesn't have a DC voltage imposed on it by the R_1/R_2 divider network. Resistor R_3 serves the purpose of controlling voltage gain. We could omit it for maximum voltage gain, but since base resistors like this are common in common-emitter amplifier circuits, we'll keep it in this schematic. Like all common-emitter amplifiers, this one *inverts* the input signal as it is amplified. In other words, a positive-going input voltage causes the output voltage to decrease, or move towards negative, and vice versa. The oscilloscope waveforms are shown in Figure below.

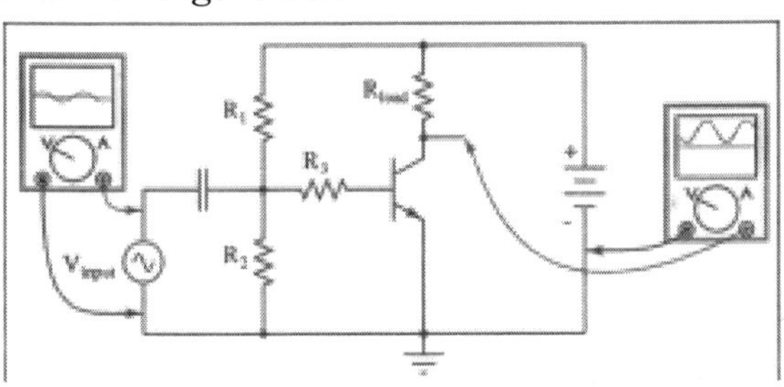

COMMON-EMITTER AMPLIFIER, NO FEEDBACK, WITH REFERENCE WAVEFORMS FOR COMPARISON

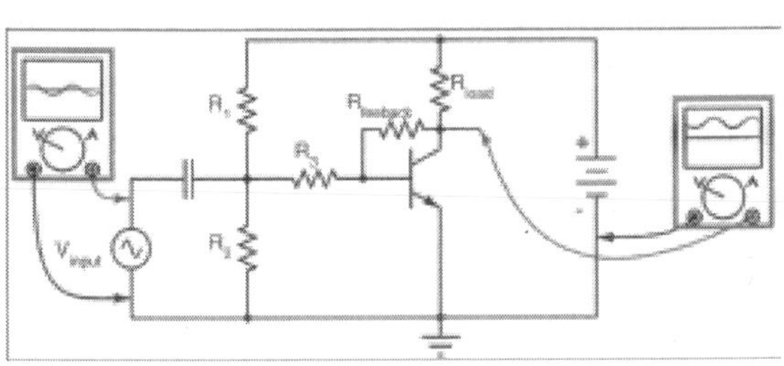

Because the output is an inverted, or mirror-image, reproduction of the input signal, any connection between the output (collector) wire and the input (base) wire of the transistor in Figure will result in *negative* feedback.

NEGATIVE FEEDBACK, COLLECTOR FEEDBACK, DECREASES THE OUTPUT SIGNAL

The resistances of R_1, R_2, R_3, and $R_{feedback}$ function together as a signal-mixing network so that the voltage seen at the base of the transistor (with respect to ground) is a weighted average of the input voltage and the feedback voltage, resulting in signal of reduced amplitude going into the transistor. So, the amplifier circuit in Figure above will have reduced voltage gain, but improved linearity (reduced distortion) and increased bandwidth.

EMITTER RESISTANCE BIASING (OR SELF BIAS)

A better method of biasing is obtained by inserting the bias resistor directly between the base and collector, as shown in the next illustration below. By tying the collector to the base in this manner, feedback voltage can be fed from the collector to the base to develop forward bias. This arrangement is called SELF-BIAS.

Now, if an increase of temperature causes an increase in collector current, the collector voltage (VC) will fall because of the increase of voltage produced across the load resistor (R_L).

This drop in VC will be fed back to the base and will result in a decrease in the base current. The decrease in base current will oppose the original increase in collector current and tend to stabilize it. The exact opposite effect is produced when the collector current decreases.

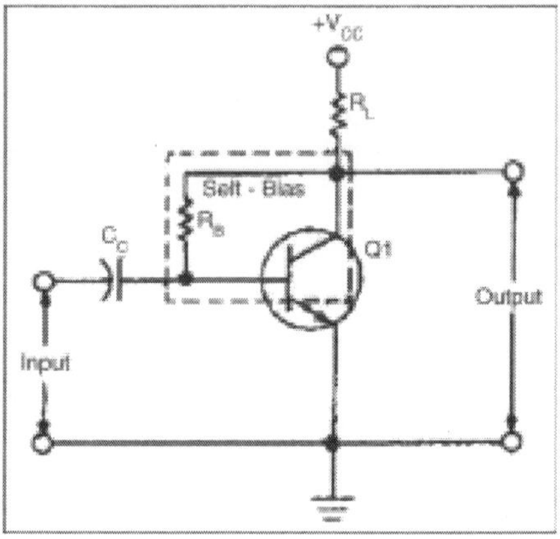

Self-bias has two small drawbacks:

- It is only partially effective and, therefore, is only used where moderate changes in ambient temperature are expected;
- It reduces amplification since the signal on the collector also affects the base voltage.

This is because the collector and base signals for this particular amplifier configuration are 180 degrees out of phase (opposite in polarity) and the part of the collector signal that is fed back to the base cancels some of the input signal.

This process of returning a part of the output back to its input is known as *Degeneration* or *Negative Feedback*. Sometimes degeneration is desired to prevent amplitude distortion (an output signal that fails to follow the input exactly) and self-bias may be used for this purpose.

VOLTAGE OR POTENTIAL DIVIDER BIASING

This is the most widely used biasing scheme in general electronics. For a single stage amplifier this circuit offers the best resilience against changes in temperature and device characteristics. The disadvantage is that a couple of extra resistors are required, but this is outweighed by the advantage of excellent stability. The circuits below:

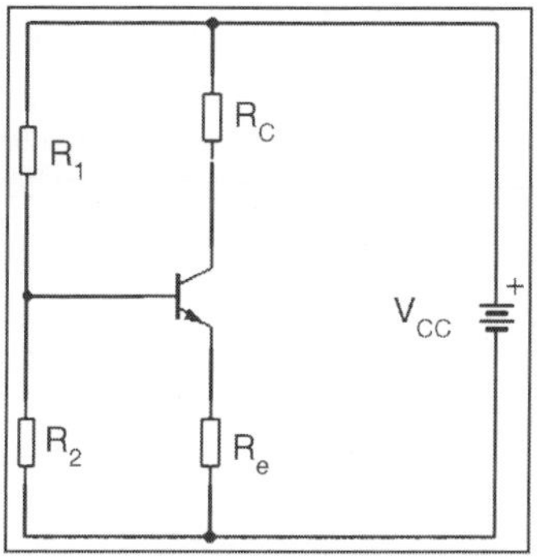

Here R_1 and R_2 form a potential divider, which will fix the base potential of the transistor. The current through this bias chain is usually set at 10 times greater than the base current required by the transistor. The base emitter voltage drop of the transistor is approximated as 0.6 volt. There will also be a voltage drop across the emitter resistor, Re, this is generally set to about 10per cent of the supply voltage. The inclusion of this resistor also helps to stabilize the bias: If the temperature increases, then extra collector current will flow. If Ic increases, then so will Ie as Ie= Ib + Ic. The extra current flow through Re increases the voltage drop across this resistor reducing the effective base emitter voltage and therefore stabilizing the collector current. The equations follow:

$$R_c = V_c/I_c$$

$$Ie = Ib + Ic \text{ as } Ic >> Ib \text{ then } Ie \sim Ic$$

$$Ve = 10\text{per cent} * Vcc$$

$$Re = V_e/I_e$$

$$V_b = V_e + 0.6$$

$$R_2 = V_b/10 * I_b$$

$$R_1 = V_{cc} - V_b/10 * I_b$$

AN EXAMPLE

Using the values of the previous examples a direct comparison of stability can be demonstrated. The values are;

V_{cc}=20V, V_c=10V, I_c = 1mA, transistor is BC107A with h_{FE}=195

$$R_c = Vc/I_c = 10/1m = 10k$$

$$V_e = 10\text{per cent} * 20 = 2V$$

$$R_e = V_e/I_e = 2/1 = 2k$$

$$V_b = 2 + 0.6 = 2.6V$$

$$I_b = I_c/h_{FE} = 1/195 = 0.005128mA$$

$$R_2 = V_b/10 * Ib = 2.6/0.05128 = 50.7k \text{ (use 47K)}$$

$$R_1 = V_{cc} - V_b/10 * I_b = (20-2.6)/0.05128 = 339.3k \text{ (use 330K)}$$

Using these values and plotting the change in quiescent conditions for V_c and I_c over a temperature range of–50 to +50 celcius is displayed below:

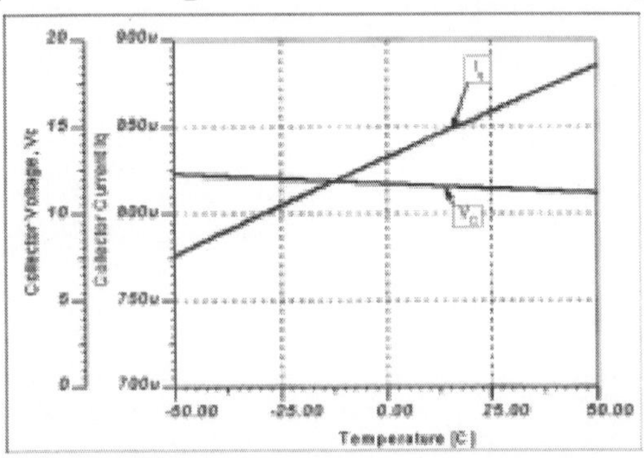

As shown above, this bias circuit gives the best stability against variations in V_C and I_C over a very wide temperature range. As the resistor values were approximated the quiescent voltage point is slightly different from the intended value. Using a load line gives a slightly more accurate design.

TWO BATTERY BIAS STABILISATION

Transistors are inevitable parts of Electronic circuits. The success of a circuit design lies in the selection of proper transistor type and calculation of voltage and current flowing through it. A small variation in the voltage or current level in the transistor will affects the working of the whole circuit. Here explains how a transistor works.

The Fig. explains how voltage and current are flowing through a bipolar transistor. Input voltage to the circuit is 12 volt DC. The base of T_1 is connected to a potential divider R_1–R_2. If they have equal values, half supply voltage will be available at the base of T_1. Here the value of R_1 is 3.2 Ohms.

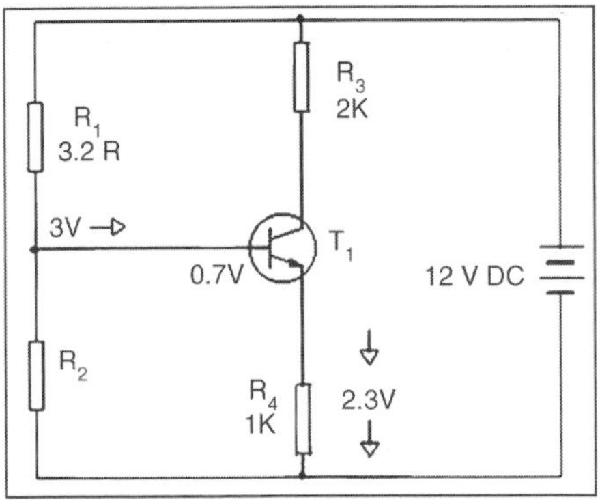

If the value of R_1 is three times greater than R_2, then three quarter of 12V drops by R_1 and allow one quarter to pass through R_2. Therefore the base voltage of T_1 will be 12/4 = 3 V.

Thus the voltage provided by R1 to the base of T_1 is 3 volts. The emitter voltage of T_1 will be 0.7 volts less than 3 volts since T1 drops 0.7 volts for its biasing. Thus the emitter voltage appears as 3–0.7 = 2.3 volts.

If the value of the emitter resistor R_4 is 1K, then if 2.3 volt passes through it, emitter current will be 2.3V/1 = 2.3 mA. Collector current also remains same. If the value of the load resistor R_3 is 2K, two times higher than that of R_4, then the voltage drop across it will be 2 × 2.3V = 4.6 volts. There fore the collector voltage of T1 remains as 12–4.6 = 7.4 volts.

LOAD CURRENT

In the circuit shown in Fig volt DC supply is provided. T1 is a general purpose NPN transtior like BC 548. A potential divider comprising R_1 and R_2 bias the base of T1. Minimum base voltage necessary for biasing T1 is 0.7 volts. The potential divider R_1-R_2 drops 6-0.7 = 6.3 volts.

If the load takes 4 volts,then the collector voltage will be 2 volts. 6-4=2 volts.

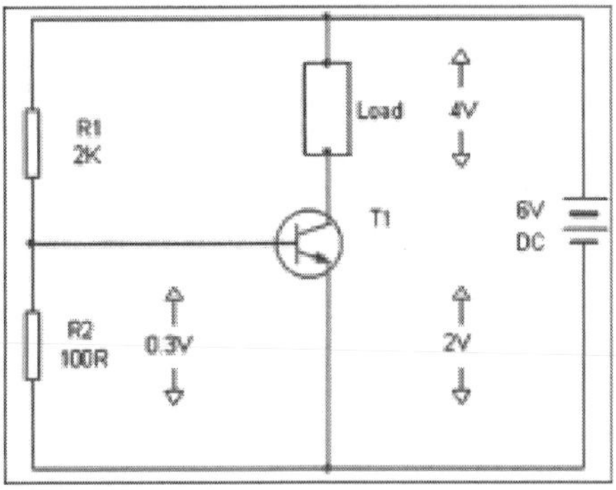

Value of the collector current depends on the base voltage. When the base voltage increases, collector current also increases. This results in more volts in the load. In short, 0.1 volt increase in base voltage causes 1 Volt increases in the load.

CURRENT IN THE TRANSISTOR AMPLIFIER

Normally when a High volt is present at the collector and Low volt in the base, Base-Emitter junction of T1 will be reverse biased.If the collector remains open, collector voltage will be 0 and hence the base current will be 100 mA. If collector of T1 is connected to the Vcc, 98 mA current flows through the collector and 2mA to the base.That is

$$\text{Emitter current} = \text{Base current} + \text{Collector current}$$

$$= 2\text{mA} + 98\text{mA} = 100\text{mA}$$

$$\text{Collector current} = \text{Emitter current–Base current}$$

$$= 100\text{mA–}2\text{ mA} = 98\text{ mA}$$

$$\text{Base current} = \text{Emitter current–Collector current}$$

$$= 100\text{mA–}98\text{mA} = 2\text{ mA}$$

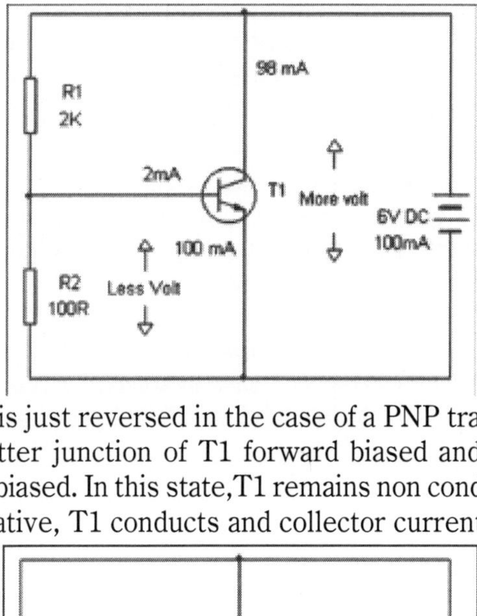

The condition is just reversed in the case of a PNP transistor as shown in Fig. The base-emitter junction of T1 forward biased and the base-collector junction is reverse biased. In this state,T1 remains non conducting. If we makes the base more negative, T1 conducts and collector current appears.

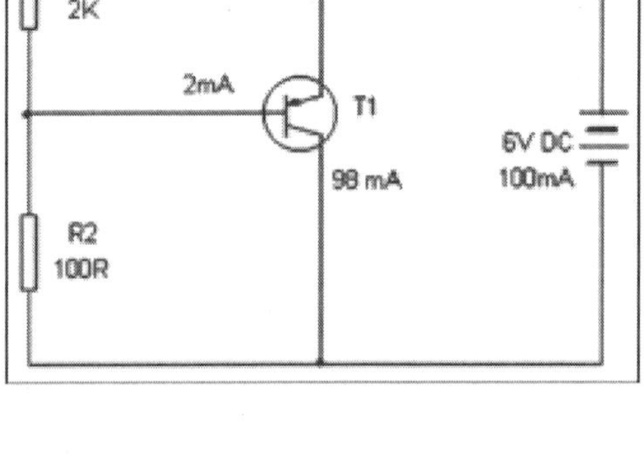

THERMAL RESISTANCE

There is an electrical analogy with conduction heat transfer that can be exploited in problem solving. The analog of $\dot{Q}$ is current, and the analog of the temperature difference, $T_1 - T_2$, is voltage difference. From this perspective the slab is a pure resistance to heat transfer and we can define,

$$\dot{Q} = \frac{T_1 - T_2}{R}$$

where $R = L / kA$, the thermal resistance. The thermal resistance R increases as L increases, as A decreases, and as k decreases.

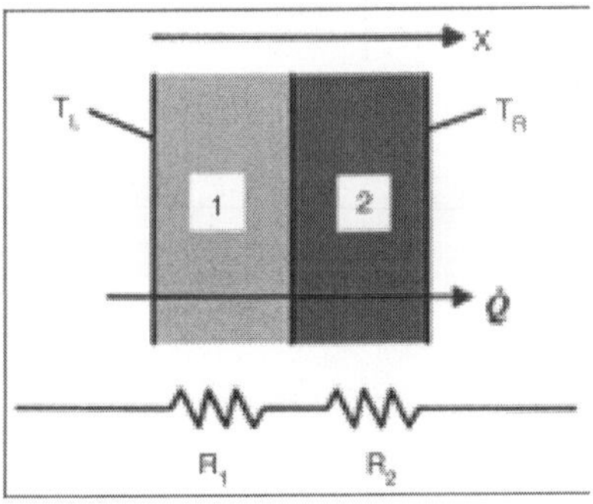

Fig. Heat Transfer Across a Composite Slab (Series Thermal Resistance)

The concept of a thermal resistance circuit allows ready analysis of problems such as a composite slab (composite planar heat transfer surface). In the composite slab shown in Figure, the heat flux is constant with x. The resistances are in series and sum to $R = R_1 + R_2$. If T_L is the temperature at the left, and T_R is the temperature at the right, the heat transfer rate is given by,

$$\dot{Q} = \frac{T_L - T_R}{R} = \frac{T_L - T_R}{R_1 + R_2}.$$

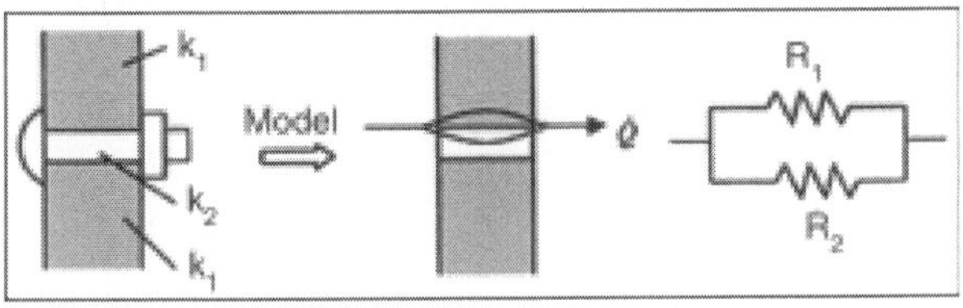

Fig. Heat Transfer for a Wall with Dissimilar Materials (Parallel Thermal Resistance)

Another example is a wall with a dissimilar material such as a bolt in an insulating layer. In this case, the heat transfer resistances are in parallel. Figure shows the physical configuration, the heat transfer paths and the thermal resistance circuit.

2

Circuit Analysis

INTRODUCTION

Circuit analysis begins with a circuit diagram or a black box and ends with an identification of the sequential circuit implemented by the device - normally a truth table.

The steps are:

- Identify the inputs and the outputs.
- Express each output as a Boolean function of the inputs and the present state Q(T).
- Identify the circuit if possible.

CIRCUIT FOR ANALYSIS

We first study the analysis of digital circuits, then we study their design. There are a number of steps in the analysis of a circuit. Where to begin depends on what one has.

When given a circuit diagram, the following steps are used to begin the analysis:

- Determine the inputs and outputs of the circuit. Assign variables to represent these.
- Determine the inputs and outputs of the flip-flops.
- Construct the Next State and Output Tables.
- Construct the State Diagram.
- If possible, identify the circuit. There are no good rules for this step.

Consider the following circuit. We want to discover what it does.

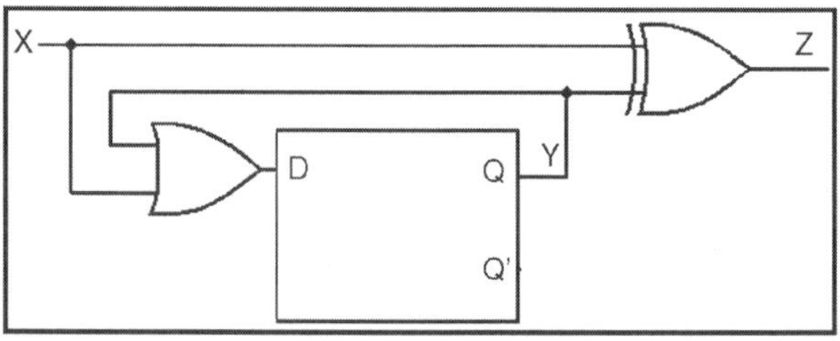

Fig. Circuit to Be Analysed

Step 1: Identify and Label the Inputs, Outputs, and Internal States

We use the following variables in the analysis of this circuit with a single flip-flop. X denotes the input, Y denotes the output of the flip-flop (Y' also), and Z the output of the circuit. If we had more than one flip-flop, we would label the flip-flop with a number beginning at 0 and use that as a subscript, so flip-flop 0 would have output Y0, etc.

Step 2: Determine the Inputs and Outputs of the Flip-flops

The next step is to determine the equations for Z, the output, and D, the input to the flip-flop. By inspection, we determine the following for the equations:

$$Z = X \oplus Y$$

$$D = X + Y$$

Step 3: Construct the Next State and Output Tables.

We begin this state by recalling the characteristic table of each flip-flop that is used in the design. Here we have only one flip-flip, a D with a very simple characteristic table that is better represented as an equation: Q(t+1) = D - the next state is what you put in now.

Noting that Q(t) = Y (the state of a flip-flop is also its output) we construct the following Next-State diagram for the flip-flop, based on the characteristic table of a D flip-flop and the equation we derived for the D input: D = X + Y. One simple caution here is that the input to a flip-flop is a function of the present state only, having nothing to do with the next state (as we have no crystal balls). Thus Y = Q(t). Here is the present state (PS)/next state (NS) diagram for the circuit.

X	Q(t) = Y	D = X + Y	Q(t+1)
0	0	0	0
0	1	1	1
1	0	1	1
1	1	1	1

The output table is similarly constructed, using the equation

$$Z = X \oplus Y = Z = X \oplus Q(t).$$

Again, note that the output is not a function of the next state.

X	Q(t)	Z
0	0	0
0	1	1
1	0	1
1	1	0

These two tables are combined to form the transition/output table.

X	Q(t) = Y	D = X + Y	Q(t+1)/Z
0	0	0	0/0
0	1	1	1/1
1	0	1	1/1
1	1	1	1/0

At this point, we should produce a state table in the standard format. This involves assigning labels to each of the two states, currently identified only as Q(t) = 0 and Q(t) = 1. For lack of anything more imaginative, we label the states 0 and 1.

Table. State Table for Circuit to be Analysed.

Present State	Next State/Output	
	X = 0	X = 1
0	0/0	1/1
1	1/1	1/0

Step 4: Construct the State Diagram

The final step in the process may be the creation of the state diagram.

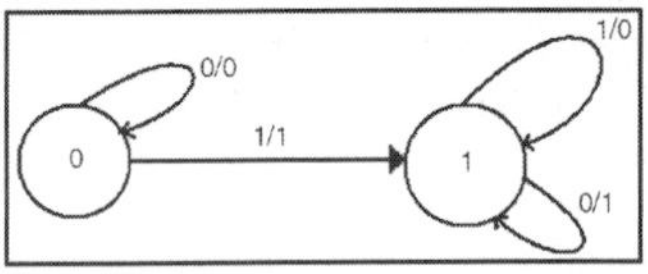

At this point, we have a complete description of the circuit. It may be possible to proceed from this diagram to obtain an understanding of what the circuit does.

Step 5: If Possible, Identify the Circuit.

This circuit is a 2-state device, with memory represented by one bit. The circuit stays in state 0 from the start until a 1 is input at which time it transitions to state 1 and remains there. Note the relation of the output to the input, depending on the state of the machine.

Input	Q(t)	Output	
0	0	0	For Q(t) = 0, the output is X
1	0	1	
0	1	1	For Q(t) = 1, the output is X'.
1	1	0	

A verbal description of the circuit is then that it copies its input to the output until the first 1 is encountered in the input stream. After that event, all input is output as complemented.

What this circuit does is take the two's-complement of a binary integer, presented to the circuit least-significant bit first. It is easy to prove that such a strategy produces the two's complement of a number. First, consider a number ending in 1, say $X_nX_{n-1}\ldots.\ X_2X_{1 1}$.

The one's complement of this number is $X_n'X_{n-1}'\ \ldots.\ X_2'X_{1'0}$. Adding 1 to this produces the number $X_n'X_{n-1}'\ \ldots.\ X_2'X_{1'1}$, in which the least significant 1 is copied and the remaining bits are complemented. The proof is completed by supposing that the number terminates in a one or more zeroes; *i.e.*, its least significant bits are 10 .... 0, where the count of zeroes is not important.

The one's-complement of this number will end with 01 .... 1, where each zero in the original has turned to a 1. But 01 .... 1 + 1 = 10 .... 0, and up to the least significant 1 the two's-complement is a copy of the bits in the integer itself.

Note that there is no carry out of the addition that produced the right-most 1, so the remainder of the integer is formed by the one's-complement. Thus we have a complete description of the circuit.

SOURCE OF ELECTRIC POWER

The source of electric power in our houses, offices, class rooms etc is the AC source. A single source works as voltage and current source. Systems like Radio, TV, Music etc have inbuilt dc powers supply that converts ac in to dc to supply power to these and we enjoy programme of our choices.

The value of voltage (v) or current (i) of ac source varies sinusoidally with time t and can be written as

$$v(t) = V_{max}\sin\omega t \quad or \quad v(t) = V_{max}\cos\omega t$$

$$i(t) = I_{max}\sin\omega t \quad or \quad i(t) = I_{max}\cos\omega t$$

V_{max} and I_{max} the peak values of voltage and current

and $\omega = 2\pi f$ is the anugar frequency of the sourse.

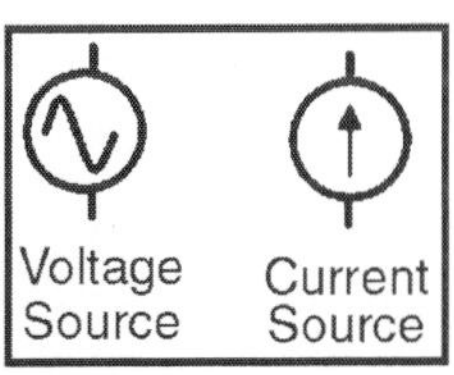

For simplicity we write $v(t)$ as v and $i(t)$ as i.

The variation of output of voltage/current source is represented below

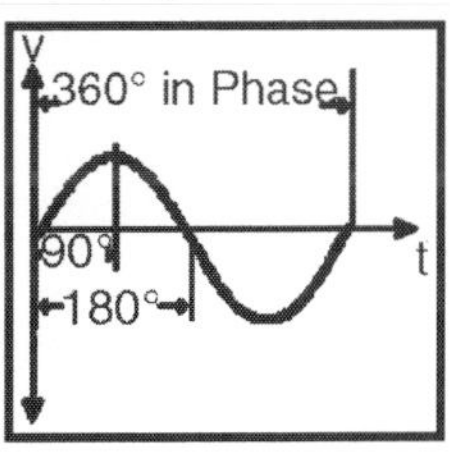

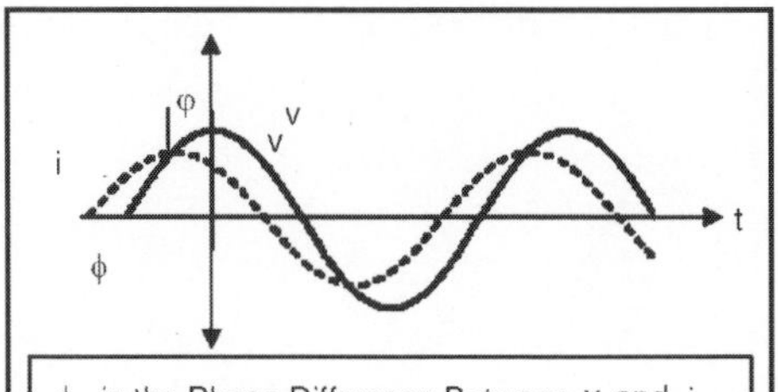

ϕ is the Phase Difference Between v and i

AC SOURCE AND RESISTOR(R)

We shall study the phase difference between the current in the resistor and the voltage across the R.

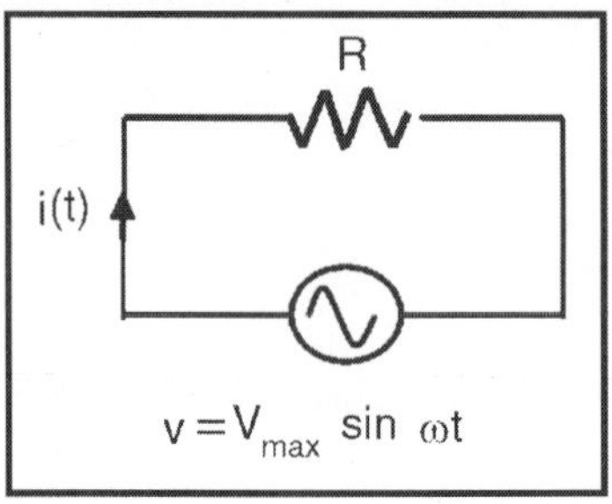

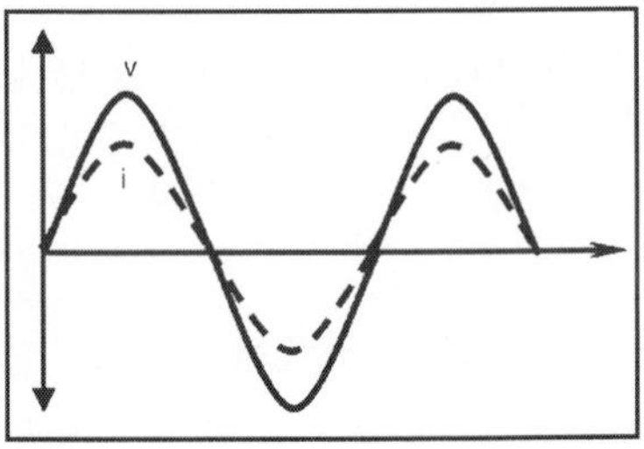

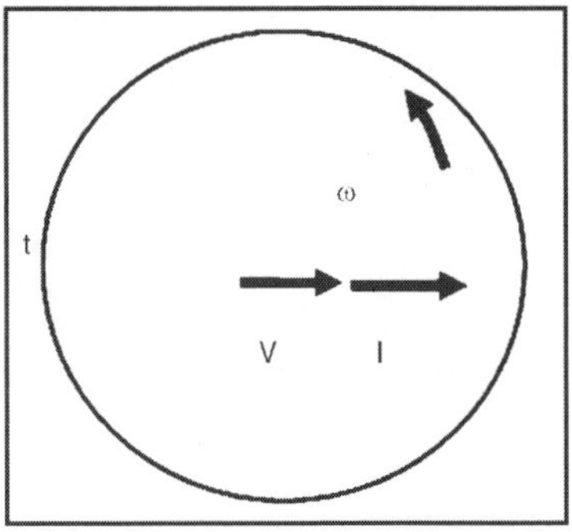

$$v = V_{\max} \sin \omega t$$

Let instantaneous value of potential from source is given by,

$$v = V_{\max} \sin \omega t$$

Instantaneous value of current from source can be obtained as,

$$i = \frac{v}{R} = \frac{V_{max} \sin \omega t}{R}$$

$$= I_{max} \sin \omega t$$

$$Where \quad \frac{V_{max}}{R} = I_{max} \quad \text{(max. value of current)}$$

The ratio,

$$V_{max}/I_{max}$$

is called the *reactance* due to resistor, X_R.

$X_R = R$ is called resistance

Above Eqs; indicate that v and i in the resistor are in the phase.

The average values of v and i in one cycle are zero. Therefore, average value of i and v are given by ttheir *rms* values. The effective or *rms* value of voltage in one cycle is given by,

$$V_{rms} = \sqrt{\frac{\int_0^T v^2 dt}{T}} = \frac{V_{\max}}{\sqrt{2}} = 0.707\, V_{\max}$$

Similarly *rms* value of current can be obtained as,

$$I_{rms} = \frac{V_{max}}{\sqrt{2}}$$

AC Source and inductance (L):

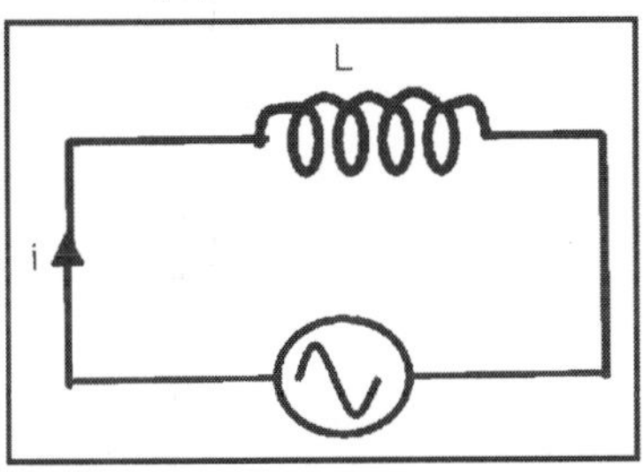

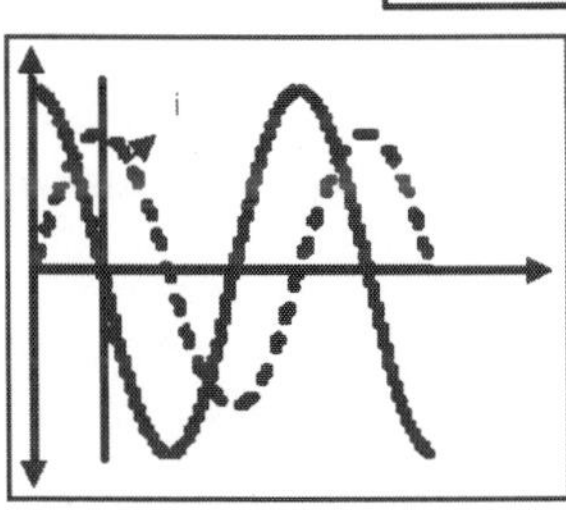

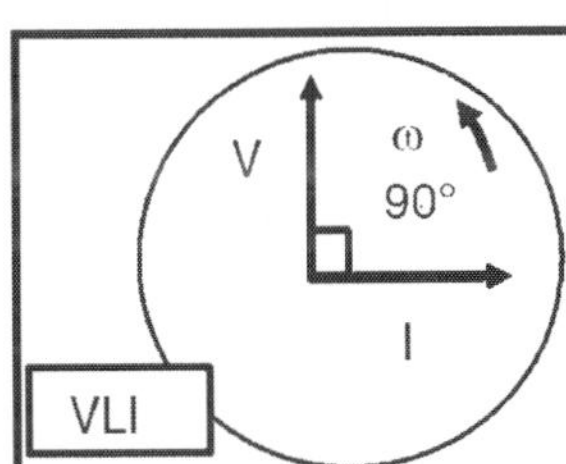

$$v = V_{max} \sin \omega t$$

Assume the resistance of the coil is negligible.

Let,

$$i = I_{max} \sin wt,$$

the changing current sets up a back- *emf* in the coil, given by,

$$\varepsilon = -L\ \frac{di}{dt}$$

To maintain the current, the applied voltage must be equal to the back-*emf i.e.*,

$$v = L\frac{di}{dt}$$

substituting the value of $i(t)$ in above Eq;

$$X_C = \frac{V_{max}}{I_{max}} = \frac{1}{\omega C}$$

we get,

$$v = LI_{max}\omega\cos\omega t \quad or$$

where,

$$V_{max} = \omega L I_{max}$$

$$\frac{V_{max}}{I_{max}} = \omega L$$

is called reactance due to inductor and is written as X_L.

Reactance of,

$$L, X_L = w L$$

The voltage across the L leads the current through the L by 90° or the current through the L lags the voltage across the L by 90°.

Problem

An inductor has a 54.0-Ω reactance at 60.0 Hz. What is the maximum current if this inductor is connected to a 50.0-Hz source that produces a 100-V rms voltage?

Sol

At 50.0 Hz, S,

$$X_L = 2\pi(50.0\text{ Hz})L = 2\pi(50.0\text{ Hz})\left(\frac{X_L|_{60.0\text{ Hz}}}{2\pi(60.0\text{ Hz})}\right) = \frac{50.0}{60.0}(54.0\ \Omega)$$

$= 45.0\ \Omega$

$$I_{max} = \frac{\Delta V_{max}}{X_L} = \frac{\sqrt{2}(\Delta V_{rms})}{X_L} = \frac{\sqrt{2}(100\text{ V})}{45.0\ \Omega} = \boxed{3.14\text{ A}}$$

AC SOURCE AND CAPACITOR C

The charge (+, -) will be flowing from terminals of the supply to the plates of capacitor in the circuit and deposit on plates that results potential difference across the plates of capacitor. The dc current does not pass through a capacitor since it offers infinite reactance to dc current.

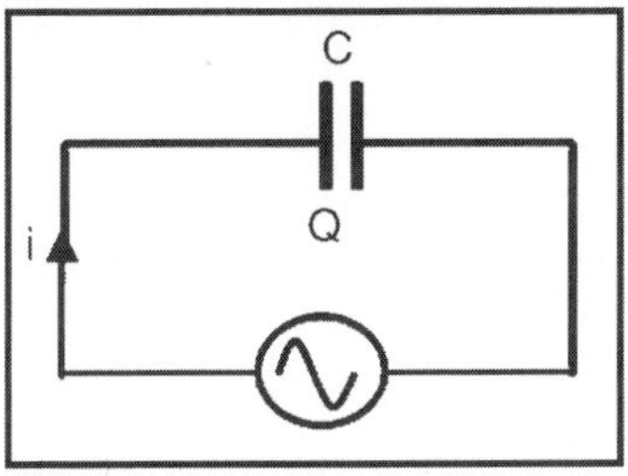

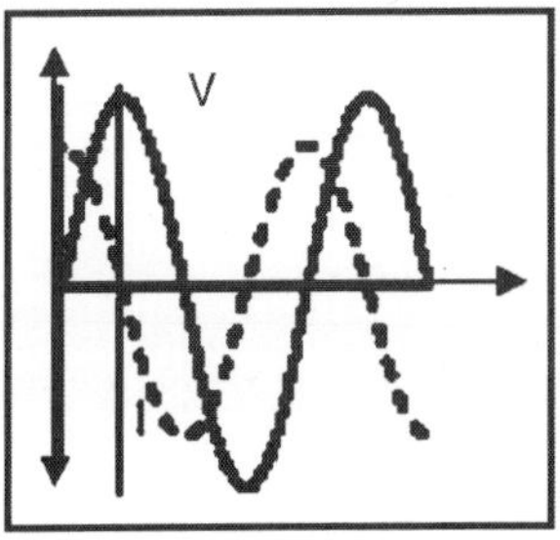

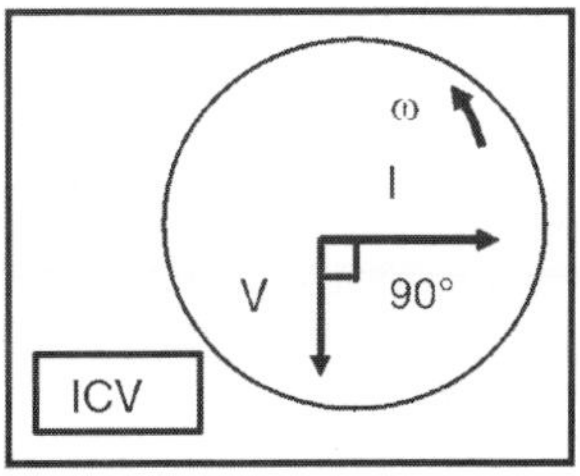

$$v = V_{\max} \sin \omega t$$

Given,

$$v = V_{\max} \sin \omega t$$

The Charge Q in the capacitor with time t is given by,

$$Q = C\,v = C\,V_{max} \sin \omega t$$

The current in the circuit is,

$$i = \frac{dQ}{dt} = \frac{d(CV_{\max} \sin \omega t)}{dt} = \omega CV_{\max} \cos \omega t$$

$$i(t) = \omega CV_{\max} \sin\left(\omega t + \frac{\pi}{2}\right)$$

From above Eq; we can infer that current leads the voltage by 90°. Or the voltage across the C lags the current through C by 90°.

The $i(t)$ can be written as,

$$i(t) = I_{\max} \cos \omega t$$

From above Eq;

$$I_{max} = \omega\, CV_{max}$$

Hence reactance of capacitor,

$$X_C = \frac{V_{\max}}{I_{\max}} = \frac{1}{\omega C}$$

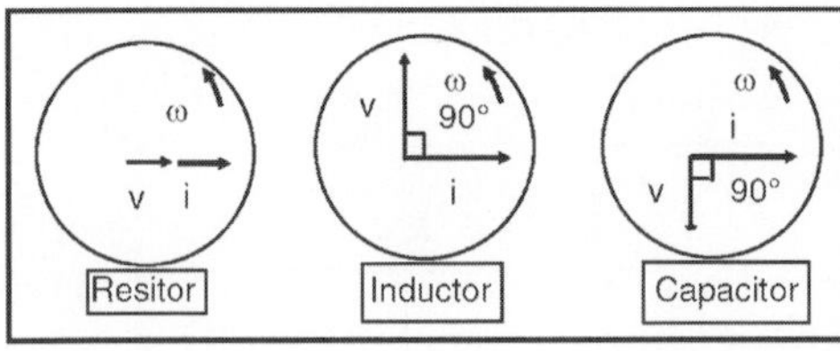

Fig. Simple Diagram to Understand the Phase Difference Between v and i.

Problem

What maximum current is delivered by an AC source with $V_{max} = 48.0$ V and $f = 90.0$ Hz when connected across a 3.70-μF capacitor?

Sol:

$$I_{max} = (\Delta V_{max})\omega C = (48.0\ \text{V})(2\pi)(90.0\ \text{s}^{-1})(3.70\times10^{-6}\ \text{F})$$

$$= \boxed{100\ \text{mA}}$$

RLC SERIES CIRCUIT

There are many ways of analysing AC circuits but we will use a simple method using a phasor diagram. This is like a vector diagram and voltages and currents are represented by lines with arrows drawn on them.

We have to find the impedance of the circuit, current in the circuit, resonating frequency and phase difference between v and i in the ckt.

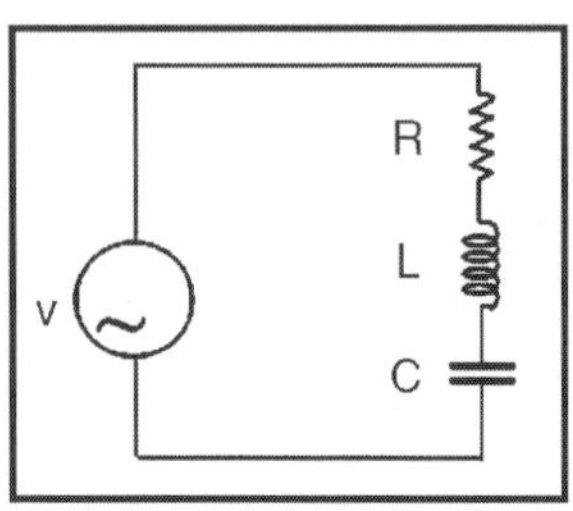

Fig. Phasor

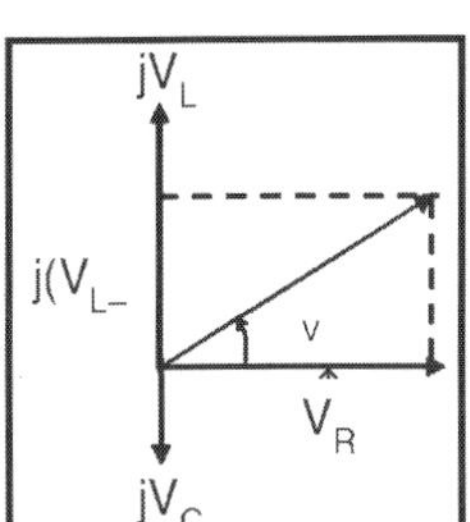

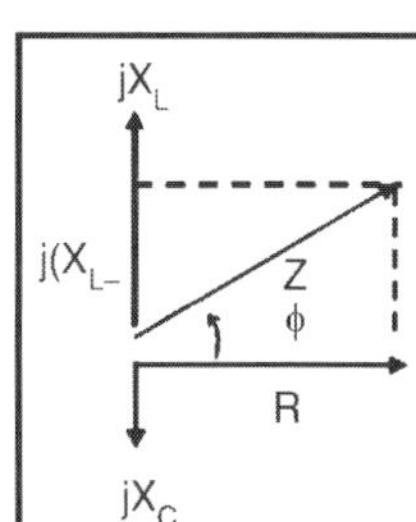

- *The impedance Z:* Z of the circuit $Z = R + j\,(X_L - X_C)$,

$$Z = \sqrt{R^2 + (X_L - X_C)^2}$$

- *Resultant voltage:*

$$V = \sqrt{V_R^2 + (V_L - V_C)^2}$$

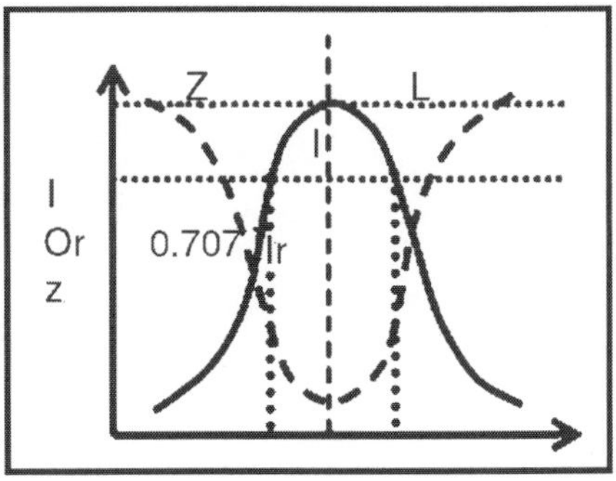

- *Current I:* The current,

$$I = \frac{V}{Z} = \frac{V}{\sqrt{R^2 + (X_L - X_C)^2}}$$

- *Phase differencef:* The phase difference between v and i.

$$\phi = \tan^{-1}\left(\frac{X_L - X_C}{R}\right) = \tan^{-1}\left(\frac{V_L - V_C}{V_R}\right)$$

- *Resonating Frequency fx:* At resonance, $X_L = X_C$, there fore resonating frequency is given by

$$f_r = \frac{1}{2\pi\sqrt{LC}}$$

and

$$Z = R$$

At resonance the impedance and resultant voltage are minimum and current is maximum.

POWER IN AN AC CIRCUIT

The average power dissipated in any element can also be written as,

$$P_{av} = \int_o^T \frac{v \times i}{T} dt$$

Where v is the voltage across the element and i is the current flowing through it. For the inductor and the capacitor the average power dissipated is given by,

$$P_{av} = \int_o^T \frac{V_{max} I_{max} \sin \omega t \cos \omega t}{T} dt$$

The value of integral,

$$\int_0^T \frac{\sin \omega t \ \cos \omega t\, dt}{T} dt = 0$$

Therefore, power dissipated (loss) in pure Inductor L and capacitor C are Zero. This is the power supplied by the source. If there is a resistor in the circuit, the power is lost as heat. In L and C the average power supplied by the source is zero. In RLC circuit there is a phase difference f between *v(t)* and *i(t)*.

Let

$$v = V_{max} \sin wt$$

Then

$$i = I_{max} \sin (wt \pm f)$$

$$= I_{max} (\sin wt \cos f \pm \cos wt \sin f)$$

Power dissipated in RLC circuit:

$$P_{av} = \frac{1}{T}\int_0^T V_{max} I_{max} \left(\sin^2 \omega t \cos\phi \, dt \pm \sin\omega t \cos\omega t \sin\phi dt\right)$$

The integral of the first factor (time average) is ½ cos φ and the integral of the second factor (time average) is 0. Therefore, average power P_{av} delivered by the source,

$$P_{av} = \frac{V_{max}\ I_{max}}{2} \cos\phi$$

$$= V_{rms}\ I_{rms} \cos f = I_{rms}{}^2 R$$

PROBLEM

A resonating circuit shown in Fig. has, C=300 pF, L= 220 μH R= 20Ω. If supply voltage is v(t)=10 sin*wt* V. Calculate,

- Resonating frequency
- Impedance of ckt Z_s at ω=300 k rad/sec
- Impedance at resonance,
- Voltage across each component at resonance
- Q factor and bandwidth of ckt.

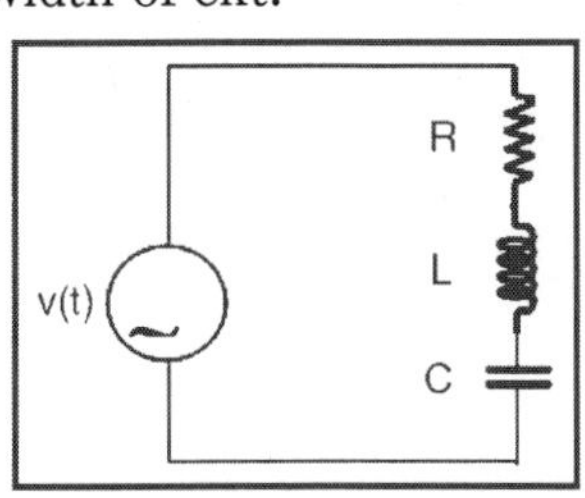

Sol:

$$f_r = \frac{1}{2\pi\sqrt{LC}}$$

$$= \frac{1}{2\pi\sqrt{220\times10^{-6}\ 300\times10^{-12}}} = 620kHz$$

impedance at w=300 k rad/sec

$$Z_s = \left[R^2 + \left(X_C - X_L\right)^2\right]^{1/2}$$

$$= \sqrt{20^2 + (0.6-6.6)10^6} = 2.5k\Omega$$

- Impedance at resonance

$$Z = R = 20W$$

- Currsent at resonance

$$I_r = \frac{V}{R} = \frac{10}{20} = 0.5A$$

- Potentail across R, L, and C

$$V_R = IR = 10\,V,$$

$$V_L = I\omega_r L = 428V$$

$$V_C = I\frac{1}{\omega_r C} = 428V$$

- Quality factor Q

$$Q = \frac{X_L}{R} = \frac{\omega_r L}{R} = \frac{2\times\pi\times620\times10^3\times220\times10^{-6}}{20} = 42.8$$

PROBLEM

A series RLC circuit has a resistance of 45 Ω and an impedance of 75 Ω. What average power is delivered to the circuit when V_{rms} = 210 V.

Sol

$$P = (V_{rms})I_{rms}\cos\phi$$

$$(X_L - X_C) = \sqrt{(75.0\ \Omega)^2 - (45.0\ \Omega)^2} = 60.0\ \Omega$$

$$\phi = \tan^{-1}\left(\frac{X_L - X_C}{R}\right) = \tan^{-1}\left(\frac{60.0\ \Omega}{45.0\ \Omega}\right) = 53.1^\circ$$

$$I_{rms} = \frac{V_{rms}}{Z} = \frac{210\ \text{V}}{75.0\ \Omega} = 2.80\ \text{A}$$

$$\text{P} = (V_{rms})I_{rms}\cos\phi = (210\ \text{V})(2.80\ \text{A})\cos(53.1¢^a)$$
$$= 353\ \text{W}$$

A network of passive elements and sources is a circuit.

Analysis: To determine currents or voltages in various elements (effects) due to various sources (cause).

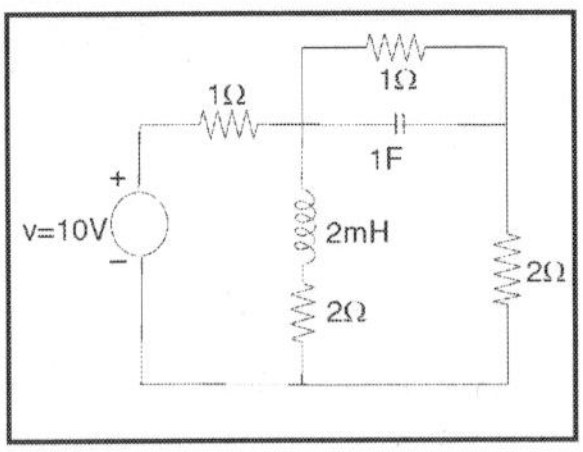

In circuit 4.1 v=10V all the time. Expected that all current (voltage) in (across) the elements is constant.

INDUCTOR

$$v = L\frac{di}{dt}$$

$$\frac{di}{dt} = 0$$

Implies i = constant.

Hence $v = 0$

Inductors act as a short cicuit for DC inputs. This would not be the case if I put a switch across a source.

Capacitor: $i = C\dfrac{dv}{dt}$

as $\dfrac{dv}{dt} = 0$ (expected), i=0

Thus capacitor acts as open circuit for DC analysis.

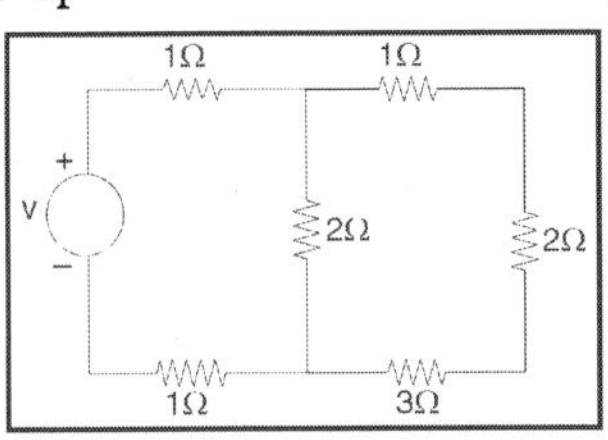

The resultant circuit will be as shown in Fig..

Analysis

To find currents in all branches, voltage across all branches.

We can use Kirchoff's law (voltage and current). For as many independent equations as number of unknown variables. Solve the simultaneous equations, and get the result.

- V_{ab} = Voltage drop from'a'to'b'.Therefore,

$$V_{ab} = -V_{ba}$$

- I_{ab} = Current in branch ab in the direction from 'a' to 'b'. Here

$$I_{ab} = -I_{ba}$$

Hence

$$V_{fa} = -v$$

$$V_{ab} = I_{ab} \times 1 = I_{ab}$$

$$V_{be} = I_{be} 2$$

$$V_{ef} = I_{ef} 1$$

$$V_{bc} = I_{bc} 1$$

$$V_{cd} = I_{cd} 2$$

$$V_{de} = I_{de} 3$$

$$V_{ab} + V_{be} + V_{ef} = v$$

$$I_{fa} = I_{ab}$$

$$I_{ef} = I_{fa}$$

$$I_{bc} = I_{cd}$$

$$I_{cd} = I_{de}$$

$$V_{ab} + V_{bc} + V_{cd} + V_{de} + V_{ef} = v$$

$$I_{bc} + V_{be} = I_{ab}$$

Note: In a circuit with N nodes, the number of branches B will alwasy be$^{(N-1)+}$ ^{L}, where L are maximum number of independent closed paths possible in the circuit.

Hence, we can always form B equations using Ohm's law1, $^{(N-1)}$ equations using KCL, L equations using KVL. Hence in total 2B equations can be formed, which are sufficient to solve for 2B variables (voltage and current in each branch).

SIMPLIFYING

Loop Currents Method

We do away with branch currents and define loop currents. The branch currents can be written in terms of loop currents once all the loop currents passing through the branch and their directions are known.

The branch voltages can always be written using Ohm's law and branch current written in terms of loop currents. So now our objective is to find loop currents.

For this we choose maximum number of independent loops Figure and apply KVL in them.

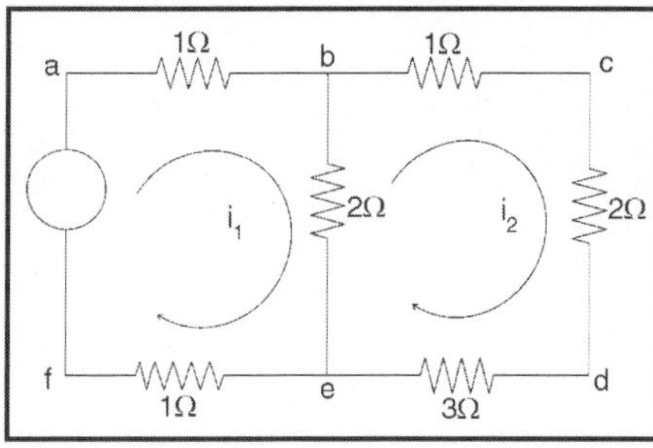

If ^{i1} and ^{i2} are known, voltages across all elements can befound.

Make Two Independent Equations:

For loop abef:

$$v - i_1 \times 1 - (i_1 - i_2) \times 2 - i_1 \times 1 = 0$$

$$v = i_1(1 + 2 + 1) - 2i_2$$

For loop bcde:

$$-(i_2 - i_1)2 - 1i_2 - 2i_2 - 3i_3 = 0$$

$$0 = -2i_2 + i_2(2 + 1 + 2 + 3)$$

Use any technique to solve these (such as using matrices).

We get:

$$i_1 = \frac{2}{7} \, Amperes$$

$$i_2 = \frac{2}{7} \, Amperes$$

NODAL VOLTAGE METHOD

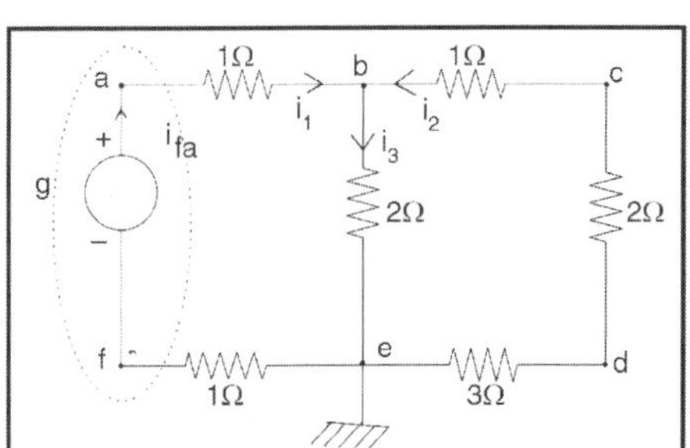

INDEPENDENT NODES

One of the nodes in circuit need to be considered as reference node. Hence its node potential is zero. For other nodes, nodal voltage is potential differetial w.r.t. to reference node.

The nodes are called independent nodes. In general for N node network, N - 1 nodes will be independent.

At Node B:

$$i_1 + i_2 - i_3 = 0.$$

Similarly, Other Equations are:

$$\frac{E_a - E_b}{1} - \frac{E_c - E_b}{1} - \frac{E_b}{2} = 0$$

$$\frac{E_a - E_b}{1} = i_{fa}$$

$$E_a = E_f + v$$

$$\frac{-E_f}{1} = i_{fa}$$

$$\frac{E_c - E_b}{1} = \frac{E_d - E_c}{2}$$

$$\frac{-E_d}{3} = \frac{E_d - E_c}{2}$$

These are six equations, in six unknowns. Thus can be solved for a unique solution. One can make a supernode and use KCL combining the nodes nodes a and f. We also make extra equations for potential difference between two nodes.

With supernode, no. of equations is equal to no. of independent nodes whose voltage w.r.t. reference needs to be determined.

CURRENT SOURCES IN LOOP ANALYSIS

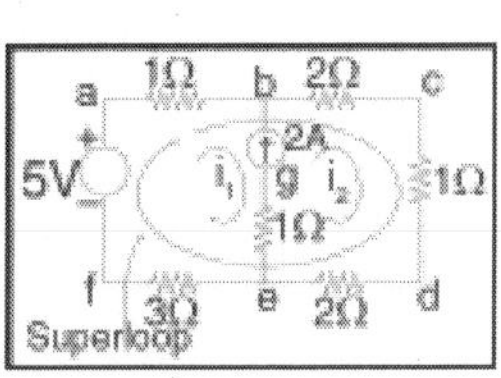

Fig. Current sources in Loop current analysis

*Using KVL for Loop 1 in Fig*ure

$$5 - i_1 - v_{bg} - (i_1 - i_2)1 - i_1 3 = 0$$

The other equations are:

$$i_1 - i_2 = 2$$

$$v_{bg} - i_2 2 - i_2 1 - i_2 2 - (i_2 - i_1) = 0$$

The first and the third equation can be combined for taking care of V_{bg}. This can be done by making superloop for writing KVL.

GRAPH

For analysing circuits efficiently.

Loop Current Method

One can form a spanning tree from graph such that current sources are in links (Those elements which do not form part of the tree). Each link when added to the tree gives a loop. All voltage sources should be kept in branches of tree.

For example, refer to the following figures.

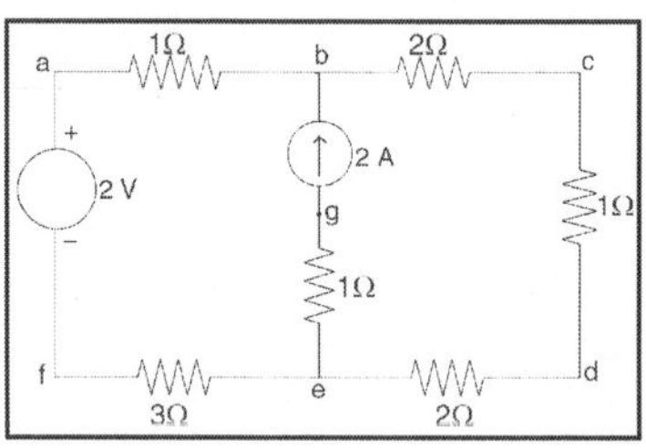

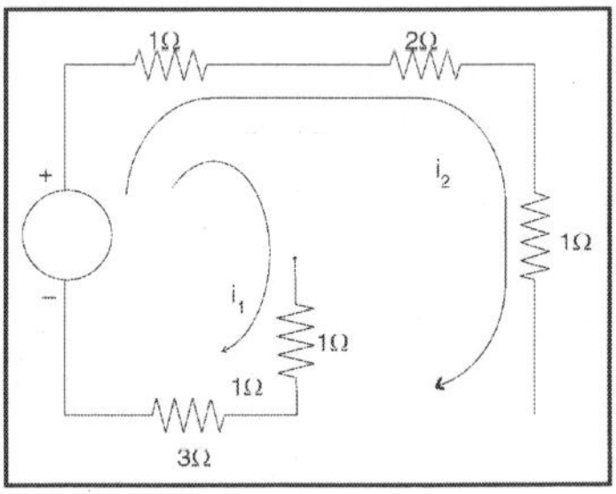

Node Voltages

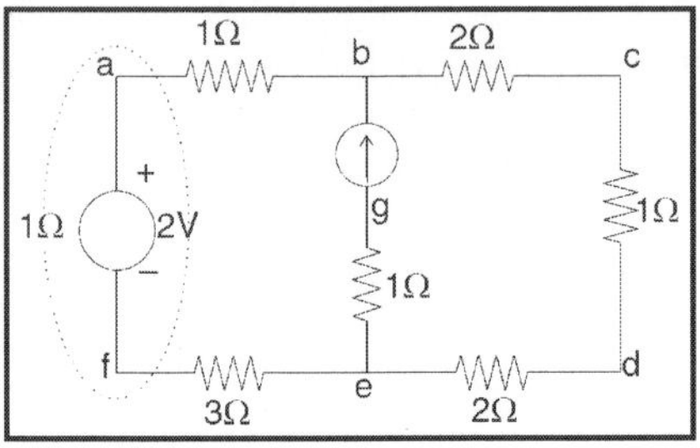

In the above figure $V_{af} = 2V$. There are five unknown node voltages in the above circuit, namely,

$$V_{ab}, V_{fe}, V_{bc}, V_{cd} \text{ and } V_{ge}.$$

Correspondingly, we have five equations. Note that we can merge af into one supernode,

$$V_{ab} + V_{fe}/3 = 0$$

$$V_{ab} + V_{bc}/2 + 2 = 0$$

$$V_{bc}/2 - V_{cd} = 0$$

$$V_{cd} - V_{de}/2 = 0$$

$$V_{ge} + 2 = 0$$

The second equation follows from looking at node b, while the third one from doing the same at node c.

3

Synchronous Circuit

We saw a circuit using one J-K flip-flop that counted backward in a two-bit binary sequence, from 11 to 10 to 01 to 00.

Since it would be desirable to have a circuit that could count forward and not just backward, it would be worthwhile to examine a forward count sequence again and look for more patterns that might indicate how to build such a circuit.

Since we know that binary count sequences follow a pattern of octave (factor of 2) frequency division, and that J-K flip-flop multivibrators set up for the "toggle" mode are capable of performing this type of frequency division, we can envision a circuit made up of several J-K flip-flops, cascaded to produce four bits of output.

The main problem facing us is to determine how to connect these flip-flops together so that they toggle at the right times to produce the proper binary sequence.

Examine the following binary count sequence, paying attention to patterns preceding the "toggling" of a bit between 0 and 1:

0 0 0 0
0 0 0 1
0 0 1 0
0 0 1 1
0 1 0 0
0 1 0 1
0 1 1 0
0 1 1 1
1 0 0 0
1 0 0 1
1 0 1 0
1 0 1 1
1 1 0 0
1 1 0 1
1 1 1 0
1 1 1 1

Note that each bit in this four-bit sequence toggles when the bit before it (the bit having a lesser significance, or place-weight), toggles in a particular direction: from 1 to 0. Small arrows indicate those points in the sequence where a bit toggles, the head of the arrow pointing to the previous bit transitioning from a "high" (1) state to a "low" (0) state:

0 0 0 0
0 0 0 1
→
0 0 1 0
0 0 1 1
→→
0 1 0 0
0 1 0 1
→
0 1 1 0
0 1 1 1
→→→
1 0 0 0
1 0 0 1
→
1 0 1 0
1 0 1 1
→→
1 1 0 0
1 1 0 1
→
1 1 1 0
1 1 1 1

way so that each succeeding bit toggles when the bit before it transitions from 1 to 0. The Q outputs of each flip-flop will serve as the respective binary bits of the final, four-bit count:

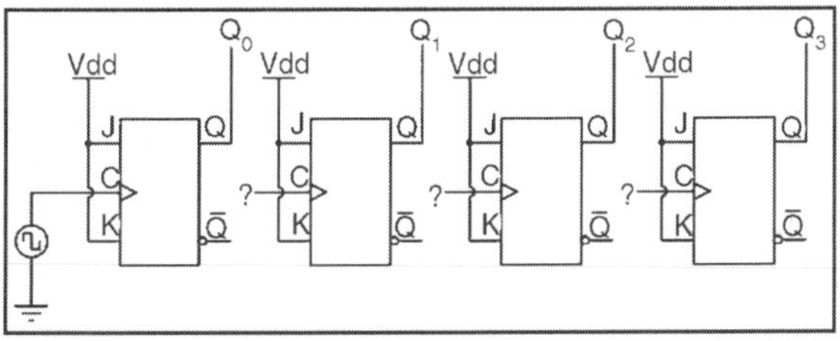

If we used flip-flops with negative-edge triggering (bubble symbols on the clock inputs), we could simply connect the clock input of each flip-flop to the Q output of the flip-flop before it, so that when the bit before it changes from a 1 to a 0, the "falling edge" of that signal would "clock" the next flip-flop to toggle the next bit:

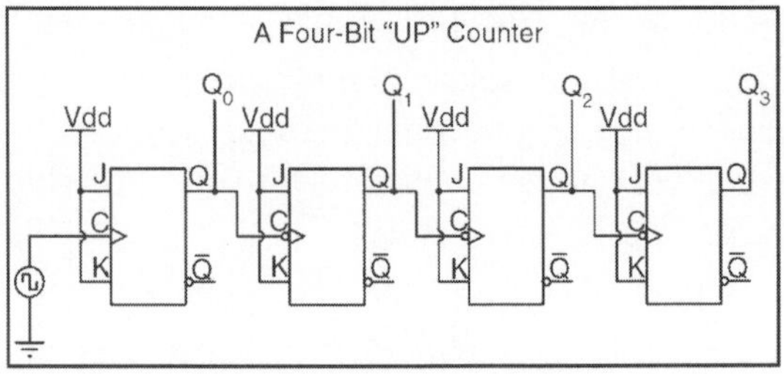

This circuit would yield the following output waveforms, when "clocked" by a repetitive source of pulses from an oscillator:

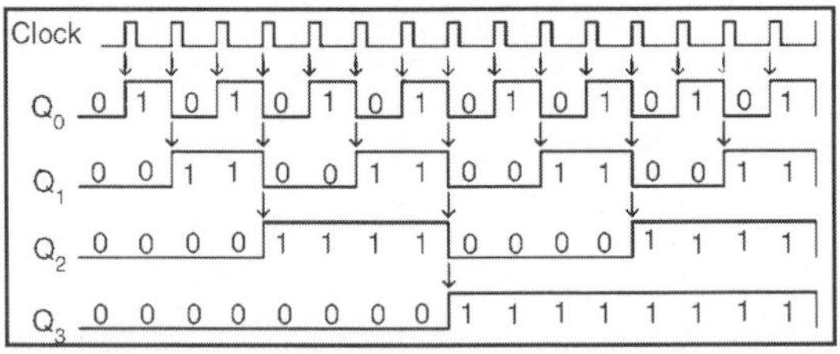

The first flip-flop (the one with the Q0 output), has a positive-edge triggered clock input, so it toggles with each rising edge of the clock signal. Notice how the clock signal in this example has a duty cycle less than 50per cent. I've shown the signal in this manner for the purpose of demonstrating how the clock signal need not be symmetrical to obtain reliable, "clean" output bits in our four-bit binary sequence. In the very first flip-flop circuit shown in this chapter, I used the clock signal itself as one of the output bits.

This is a bad practice in counter design, though, because it necessitates the use of a square wave signal with a 50per cent duty cycle ("high" time = "low" time) in order to obtain a count sequence where each and every step pauses for the same amount of time. Using one J-K flip-flop for each output bit, however, relieves us of the necessity of having a symmetrical clock signal, allowing the use of practically any variety of high/low waveform to increment the count sequence.

As indicated by all the other arrows in the pulse diagram, each succeeding output bit is toggled by the action of the preceding bit transitioning from "high" (1) to "low" (0). This is the pattern necessary to generate an "up" count sequence. A less obvious solution for generating an "up" sequence using positive-edge triggered flip-flops is to "clock" each flip-flop using the Q' output of the preceding flip-flop rather than the Q output.

Since the Q' output will always be the exact opposite state of the Q output on a J-K flip-flop (no invalid states with this type of flip-flop), a high-to-low transition on the Q output will be accompanied by a low-to-high transition on the Q' output. In other words, each time the Q output of a flip-flop transitions from 1 to 0, the Q' output of the same flip-flop will transition from 0 to 1, providing the positive-going clock pulse we would need to toggle a positive-edge triggered flip-flop at the right moment:

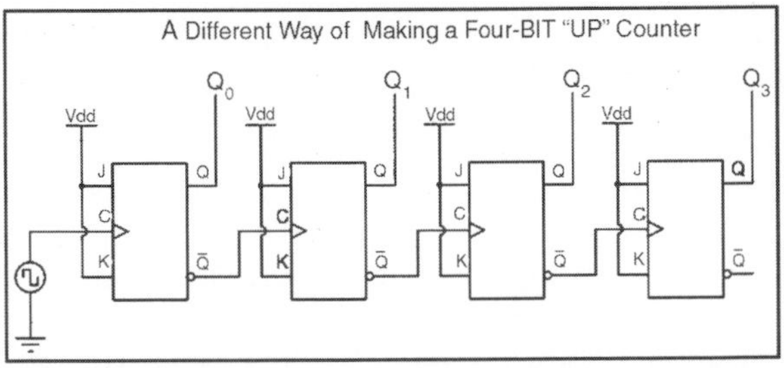

One way we could expand the capabilities of either of these two counter circuits is to regard the Q' outputs as another set of four binary bits. If we examine the pulse diagram for such a circuit, we see that the Q' outputs generate a down-counting sequence, while the Q outputs generate an up-counting sequence:

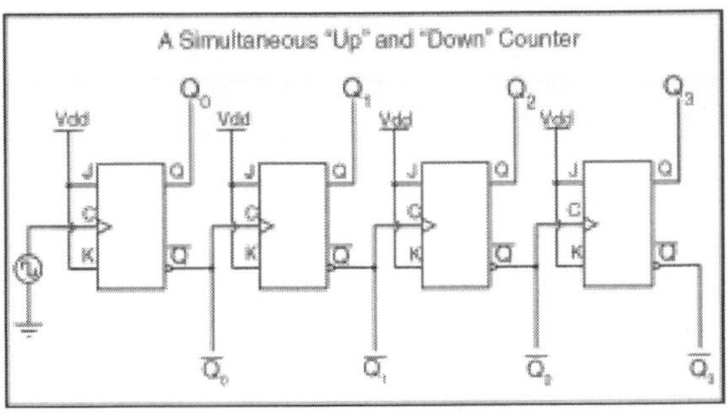

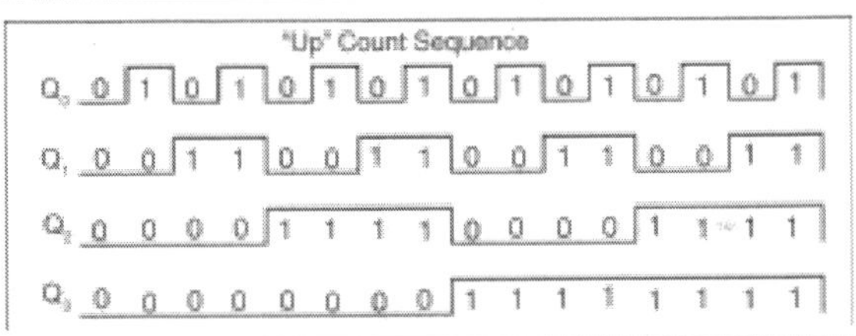

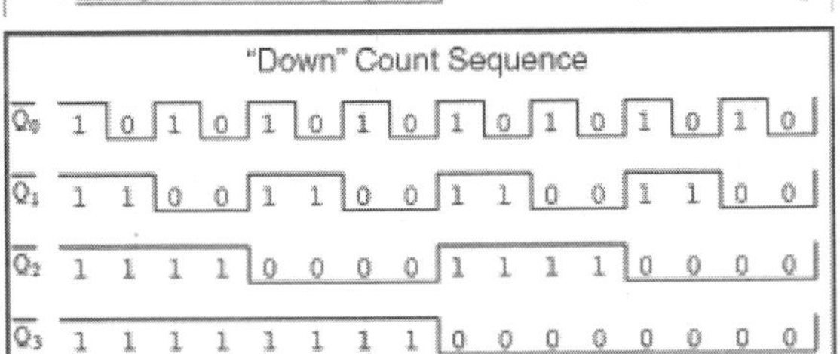

Unfortunately, all of the counter circuits shown thusfar share a common problem: the ripple effect. This effect is seen in certain types of binary adder and data conversion circuits, and is due to accumulative propagation delays between cascaded gates. When the Q output of a flip-flop transitions from 1 to 0, it commands the next flip-flop to toggle. If the next flip-flop toggle is a transition from 1 to 0, it will command the flip-flop after it to toggle as well, and so on.

However, since there is always some small amount of propagation delay between the command to toggle (the clock pulse) and the actual toggle response

(Q and Q' outputs changing states), any subsequent flip-flops to be toggled will toggle some time after the first flip-flop has toggled. Thus, when multiple bits toggle in a binary count sequence, they will not all toggle at exactly the same time:

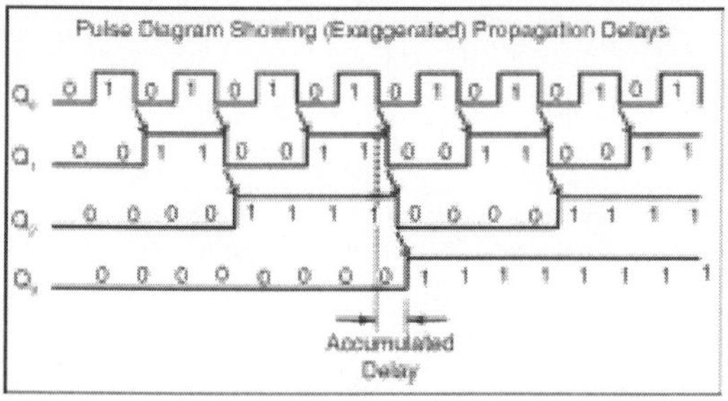

As you can see, the more bits that toggle with a given clock pulse, the more severe the accumulated delay time from LSB to MSB. When a clock pulse occurs at such a transition point (say, on the transition from 0111 to 1000), the output bits will "ripple" in sequence from LSB to MSB, as each succeeding bit toggles and commands the next bit to toggle as well, with a small amount of propagation delay between each bit toggle. If we take a close-up look at this effect during the transition from 0111 to 1000, we can see that there will be false output counts generated in the brief time period that the "ripple" effect takes place:

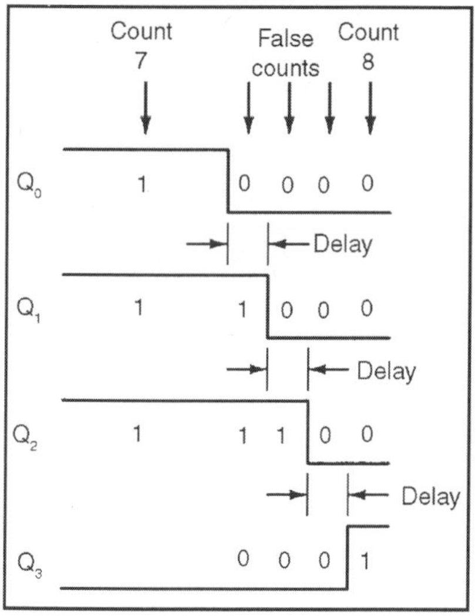

Instead of cleanly transitioning from a "0111" output to a "1000" output, the counter circuit will very quickly ripple from 0111 to 0110 to 0100 to 0000 to 1000, or from 7 to 6 to 4 to 0 and then to 8. This behaviour earns the counter circuit the name ofripple counter, or asynchronous counter. In many applications, this effect is tolerable, since the ripple happens very, very quickly (the width of the delays has been exaggerated here as an aid to understanding the effects). If all we wanted to do was drive a set of light-emitting diodes (LEDs)

with the counter's outputs, for example, this brief ripple would be of no consequence at all. However, if we wished to use this counter to drive the "select" inputs of a multiplexer, index a memory pointer in a microprocessor (computer) circuit, or perform some other task where false outputs could cause spurious errors, it would not be acceptable. There is a way to use this type of counter circuit in applications sensitive to false, ripple-generated outputs, and it involves a principle known as strobing.

Most decoder and multiplexer circuits are equipped with at least one input called the "enable." The output(s) of such a circuit will be active only when the enable input is made active. We can use this enable input to strobe the circuit receiving the ripple counter's output so that it is disabled (and thus not responding to the counter output) during the brief period of time in which the counter outputs might be rippling, and enabled only when sufficient time has passed since the last clock pulse that all rippling will have ceased. In most cases, the strobing signal can be the same clock pulse that drives the counter circuit:

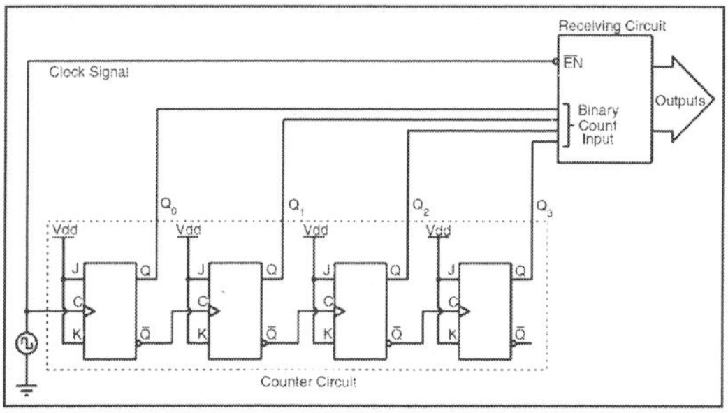

With an active-low Enable input, the receiving circuit will respond to the binary count of the four-bit counter circuit only when the clock signal is "low." As soon as the clock pulse goes "high," the receiving circuit stops responding to the counter circuit's output. Since the counter circuit is positive-edge triggered (as determined by the first flip-flop clock input), all the counting action takes place on the low-to-high transition of the clock signal, meaning that the receiving circuit will become disabled just before any toggling occurs on the counter circuit's four output bits.

The receiving circuit will not become enabled until the clock signal returns to a low state, which should be a long enough time after all rippling has ceased to be "safe" to allow the new count to have effect on the receiving circuit. The crucial parameter here is the clock signal's "high" time: it must be at least as long as the maximum expected ripple period of the counter circuit. If not, the clock signal will prematurely enable the receiving circuit, while some rippling is still taking place. Another disadvantage of the asynchronous, or ripple, counter circuit is limited speed. While all gate circuits are limited in terms of maximum signal frequency, the design of asynchronous counter circuits compounds this problem by making propagation delays additive. Thus, even if strobing is used in the receiving circuit, an asynchronous counter circuit cannot be clocked at any frequency higher than that which allows the greatest possible accumulated propagation delay to elapse well before the next pulse.

The solution to this problem is a counter circuit that avoids ripple altogether. Such a counter circuit would eliminate the need to design a "strobing" feature into whatever digital circuits use the counter output as an input, and would also enjoy a much greater operating speed than its asynchronous equivalent.

ASYNCHRONOUS (RIPPLE) COUNTERS

A two-bit asynchronous counter is shown below. The external clock is connected to the clock input of the first flip-flop (FF0) only.

So, FF0 changes state at the falling edge of each clock pulse, but FF1 changes only when triggered by the falling edge of the Q output of FF0.

Because of the inherent propagation delay through a flip-flop, the transition of the input clock pulse and a transition of the Q output of FF0 can never occur at exactly the same time. Therefore, the flip-flops cannot be triggered simultaneously, producing an asynchronous operation.

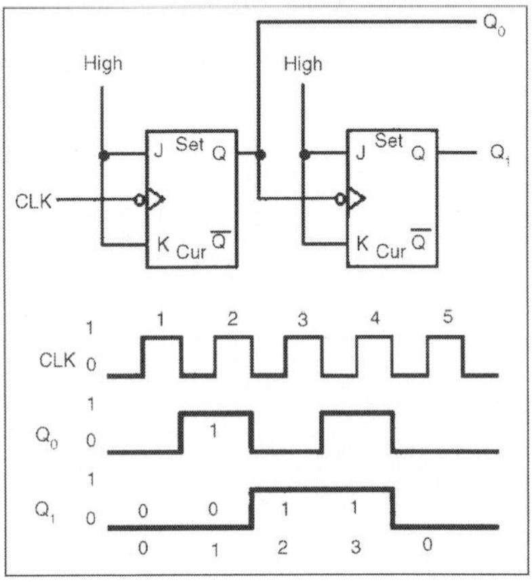

Note that for simplicity, the transitions of Q0, Q1 and CLK in the timing diagram above are shown as simultaneous even though this is an asynchronous counter.

Actually, there is some small delay between the CLK, Q0 and Q1 transitions. 0. Usually, all the CLEAR inputs are connected together, so that a single pulse can clear all the flip-flops before counting starts. The clock pulse fed into FF0 is rippled through the other counters after propagation delays, like a ripple on water, hence the name Ripple Counter.

The 2-bit ripple counter circuit above has four different states, each one corresponding to a count value. Similarly, a counter with n flip-flops can have 2 to the power n states. The number of states in a counter is known as its mod (modulo) number. Thus a 2-bit counter is amod-4 counter. A mod-n counter may also described as a divide-by-n counter. This is because the most significant flip-flop (the furthest flip-flop from the original clock pulse) produces one pulse for every n pulses at the clock input of the least significant flip-flop (the one triggers by the clock pulse). Thus, the above counter is an example of a divide-by-4 counter.

The following is a three-bit asynchronous binary counter 0 and its timing diagram for one cycle. It works exactly the same way as a two-bit asynchronous binary counter mentioned above, except it has eight states due to the third flip-flop.

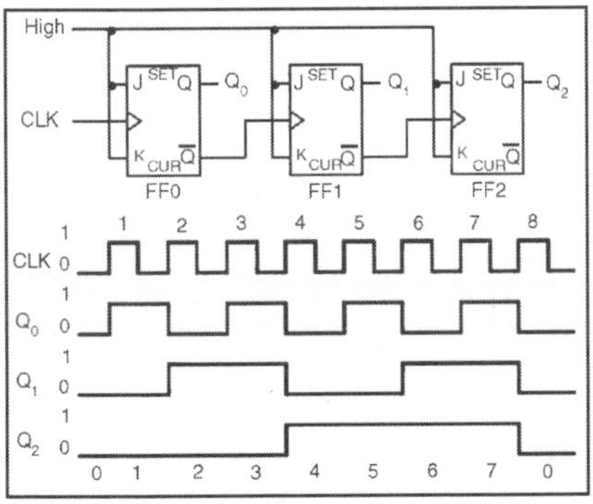

Since we know that binary count sequences follow a pattern of octave (factor of 2) frequency division, and that J-K flip-flop multivibrators set up for the"toggle" mode are capable of performing this type of frequency division, we can envision a circuit made up of several J-K flip-flops, cascaded to produce four bits of output.

The main problem facing us is to determine how to connect these flip-flops together so that they toggle at the right times to produce the proper binary sequence. Examine the following binary count sequence, paying attention to patterns preceding the"toggling" of a bit between 0 and 1:

0 0 0 0
0 0 0 1
0 0 1 0
0 0 1 1
0 1 0 0
0 1 0 1
0 1 1 0
0 1 1 1
1 0 0 0
1 0 0 1
1 0 1 0
1 0 1 1
1 1 0 0
1 1 0 1
1 1 1 0
1 1 1 1

Asynchronous Decade Counters

The binary counters previously introduced have two to the power n states. But counters with states less than this number are also possible.

They are designed to have the number of states in their sequences, which are called truncated sequences. These sequences are achieved by forcing the counter to recycle before going through all of its normal states.

A common modulus for counters with truncated sequences is ten. A counter with ten states in its sequence is called a decade counter. The circuit below is an implementation of a decade counter.

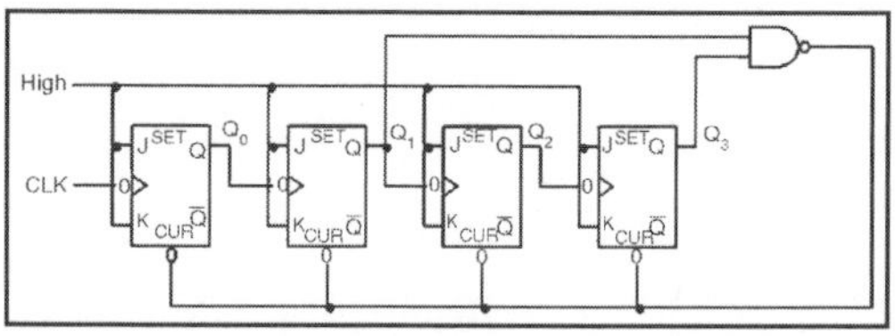

Once the counter counts to ten (1010), all the flip-flops are being cleared. Notice that only Q1 and Q3 are used to decode the count of ten. This is called partial decoding, as none of the other states (zero to nine) have both Q1 and Q3 HIGH at the same time. The sequence of the decade counter is shown in the table below:

Clock Pulse	Q3	Q2	Q1	Q0
0	0	0	0	0
1	0	0	0	1
2	0	0	1	0
3	0	0	1	1
4	0	1	0	0
5	0	1	0	1
6	0	1	1	0
7	0	1	1	1
8	1	0	0	0
9	1	0	0	1

Asynchronous Up-Down Counters

In certain applications a counter must be able to count both up and down. The circuit below is a 3-bit up-down counter. It counts up or down depending on the status of the control signals UP and DOWN.

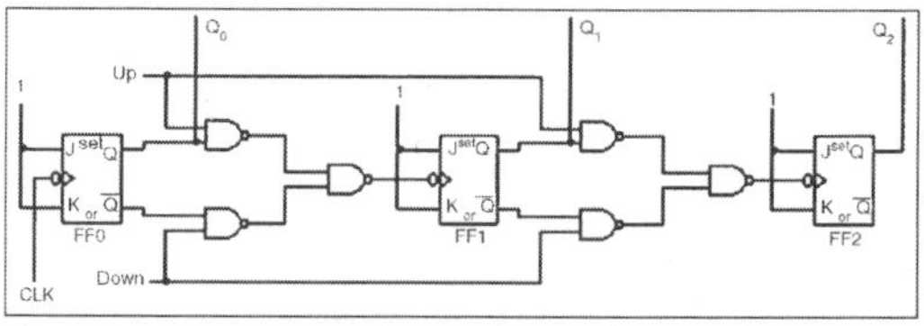

When the UP input is at 1 and the DOWN input is at 0, the NAND network between FF0 and FF1 will gate the non-inverted output (Q) of FF0 into the clock input of FF1. Similarly, Q of FF1 will be gated through the other NAND network into the clock input of FF2. Thus the counter will count up.

When the control input UP is at 0 and DOWN is at 1, the inverted outputs of FF0 and FF1 are gated into the clock inputs of FF1 and FF2 respectively. If the flip-flops are initially reset to 0's, then the counter will go through the following sequence as input pulses are applied.

FF2	FF1	FF0
0	0	0
1	1	1
1	1	0
1	0	1
1	0	0
0	1	1
0	1	0
0	0	1

Notice that an asynchronous up-down counter is slower than an up counter or a down counter because of the additional propagation delay introduced by the NAND networks.

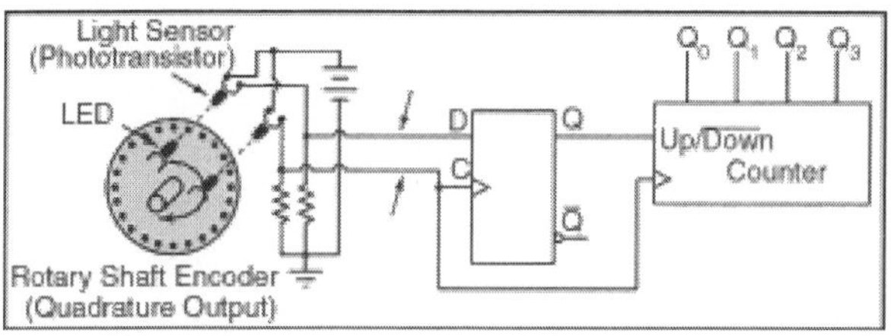

When the encoder rotates clockwise, the "D" input signal square-wave will lead the "C" input square-wave, meaning that the "D" input will already be "high" when the "C" transitions from "low" to "high," thus setting the D-type flip-flop (making the Q output "high") with every clock pulse. A "high" Q output places the counter into the "Up" count mode, and any clock pulses received by the clock from the encoder (from either LED) will increment it. Conversely, when the encoder reverses rotation, the "D" input will lag behind the "C" input waveform, meaning that it will be "low" when the "C" waveform transitions from "low" to "high," forcing the D-type flip-flop into the reset state (making the Q output "low") with every clock pulse. This "low" signal commands the counter circuit to decrement with every clock pulse from the encoder. This circuit, or something very much like it, is at the heart of every position-measuring circuit based on a pulse encoder sensor. Such applications are very common in robotics, CNC machine tool control, and other applications involving the measurement of reversible, mechanical motion.

SYNCHRONOUS COUNTERS

In synchronous counters, the clock inputs of all the flip-flops are connected together and are triggered by the input pulses. Thus, all the flip-flops change state simultaneously (in parallel).

The circuit below is a 3-bit synchronous counter. The J and K inputs of FF0 are connected to HIGH. FF1 has its J and K inputs connected to the output of FF0, and the J and K inputs of FF2 are connected to the output of an AND gate that is fed by the outputs of FF0 and FF1.

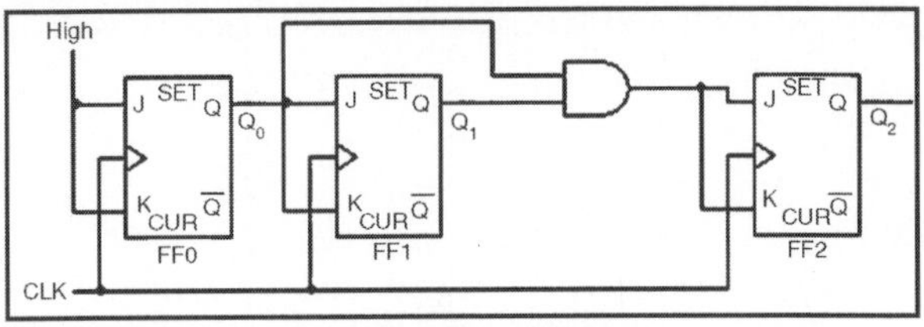

Pay attention to what happens after the 3rd clock pulse. Both outputs of FF0 and FF1 are HIGH. The positive edge of the 4th clock pulse will cause FF2 to change its state due to the AND gate.

FF2	FF1	FF0
0	0	0
0	0	1
0	1	0
0	1	1
1	0	0
1	0	1
1	1	0
1	1	1

The count sequence for the 3-bit counter is shown on the right. The most important advantage of synchronous counters is that there is no cumulative time delay because all flip-flops are triggered in parallel. Thus, the maximum operating frequency for this counter will be significantly higher than for the corresponding ripple counter.

A synchronous counter, in contrast to an asynchronous counter, is one whose output bits change state simultaneously, with no ripple. The only way we can build such a counter circuit from J-K flip-flops is to connect all the clock inputs together, so that each and every flip-flop receives the exact same clock pulse at the exact same time:

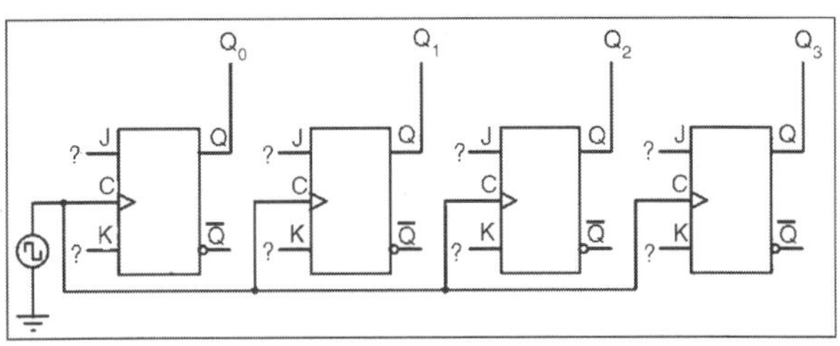

Now, the question is, what do we do with the J and K inputs? We know that we still have to maintain the same divide-by-two frequency pattern in order to count in a binary sequence, and that this pattern is best achieved utilizing the"toggle" mode of the flip-flop, so the fact that the J and K inputs must both be (at times)"high" is clear. However, if we simply connect all the J and K inputs to the positive rail of the power supply as we did in the asynchronous circuit, this would clearly not work because all the flip-flops would toggle at the same time: with each and every clock pulse!

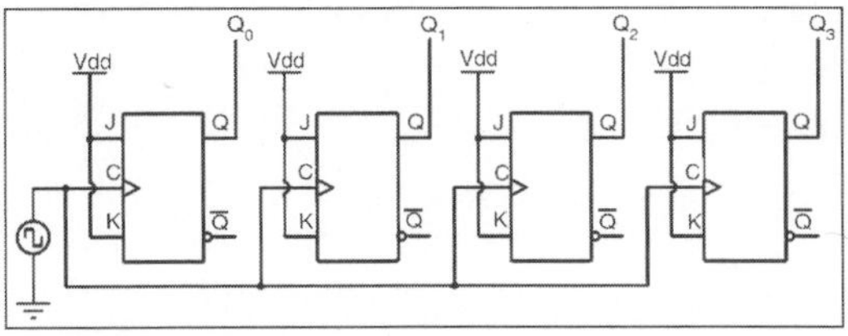

Fig. This Circuit will Not Function as a Counter!

Let's examine the four-bit binary counting sequence again, and see if there are any other patterns that predict the toggling of a bit. Asynchronous counter circuit design is based on the fact that each bit toggle happens at the same time that the preceding bit toggles from a"high" to a"low" (from 1 to 0). Since we cannot clock the toggling of a bit based on the toggling of a previous bit in a synchronous counter circuit (to do so would create a ripple effect) we must find some other pattern in the counting sequence that can be used to trigger a bit toggle: Examining the four-bit binary count sequence, another predictive pattern can be seen. Notice that just before a bit toggles, all preceding bits are"high:"

```
0000
0001
0010
0011
0100
0101
0110
0111
1000
1001
1010
1011
1100
1101
1110
1111
```

This pattern is also something we can exploit in designing a counter circuit. If we enable each J-K flip-flop to toggle based on whether or not all preceding flip-flop outputs (Q) are"high," we can obtain the same counting sequence as the asynchronous circuit without the ripple effect, since each flip-flop in this circuit will be clocked at exactly the same time:

The result is a four-bit synchronous"up" counter. Each of the higher-order flip-flops are made ready to toggle (both J and K inputs"high") if the Q outputs of all previous flip-flops are"high." Otherwise, the J and K inputs for that flip-flop will both be"low," placing it into the"latch" mode where it will maintain its present output state at the next clock pulse. Since the first (LSB) flip-flop needs to toggle at every clock pulse, its J and K inputs are connected to Vcc or Vdd, where they will be"high" all the time. The next flip-flop need only"recognize" that the first flip-flop's Q output is high to be made ready to toggle, so no AND gate is needed. However, the remaining flip-flops should be made ready to toggle only when all lower-order output bits are"high," thus the need for AND gates.

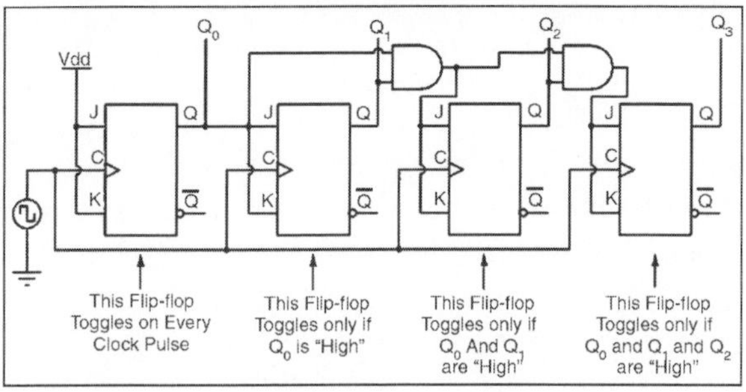

Fig. A Four-bit Synchronous "up" Counter

To make a synchronous"down" counter, we need to build the circuit to recognize the appropriate bit patterns predicting each toggle state while counting down. Not surprisingly, when we examine the four-bit binary count sequence, we see that all preceding bits are"low" prior to a toggle (following the sequence from bottom to top):

```
0000
0001
0010
0011
0100
0101
0110
0111
1000
1001
1010
1011
1100
1101
1110
1111
```

Since each J-K flip-flop comes equipped with a Q' output as well as a Q output, we can use the Q' outputs to enable the toggle mode on each succeeding flip-flop, being that each Q' will be"high" every time that the respective Q is"low:"

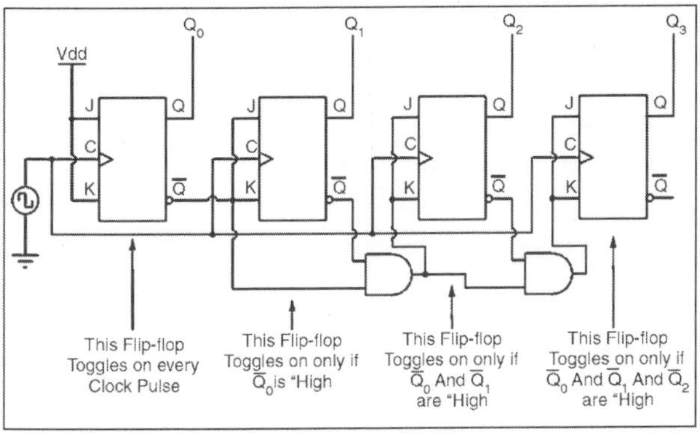

Fig. A four-bit Synchronous "down" Counter

Taking this idea one step further, we can build a counter circuit with selectable between"up" and"down" count modes by having dual lines of and gates detecting the appropriate bit conditions for an"up" and a"down" counting sequence, respectively, then use or gates to combine the and gate outputs to the J and K inputs of each succeeding flip-flop:

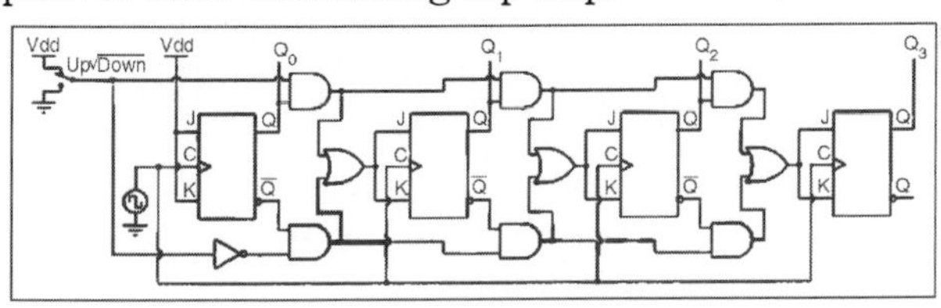

Fig. A Four-bit Synchronous "up/down" Counter

This circuit isn't as complex as it might first appear. The Up/Down control input line simply enables either the upper string or lower string of AND gates to pass the Q/Q' outputs to the succeeding stages of flip-flops.

If the Up/Down control line is"high," the top and gates become enabled, and the circuit functions exactly the same as the first ("up") synchronous counter circuit shown in this section. If the Up/Down control line is made"low," the bottom and gates become enabled, and the circuit functions identically to the second ("down" counter) circuit shown in this section.

To illustrate, here is a diagram showing the circuit in the"up" counting mode (all disabled circuitry shown in grey rather than black):

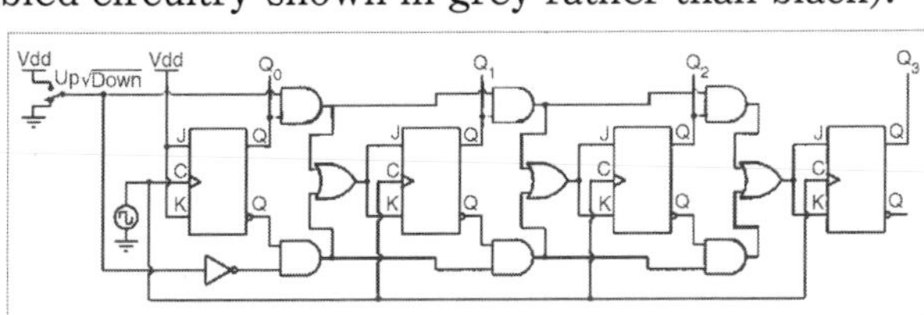

Fig. Counter in "up" Counting Mode

Here, shown in the"down" counting mode, with the same grey colouring representing disabled circuitry.

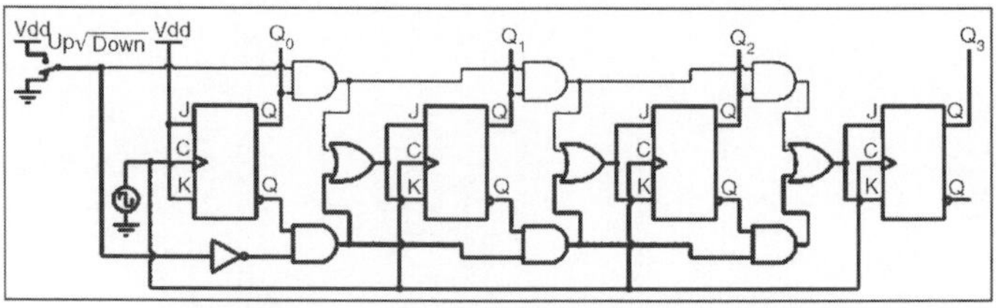

Fig. Counter in "Down" Counting Mode

Up/down counter circuits are very useful devices. A common application is in machine motion control, where devices called rotary shaft encoders convert mechanical rotation into a series of electrical pulses, these pulses"clocking" a counter circuit to track total motion:

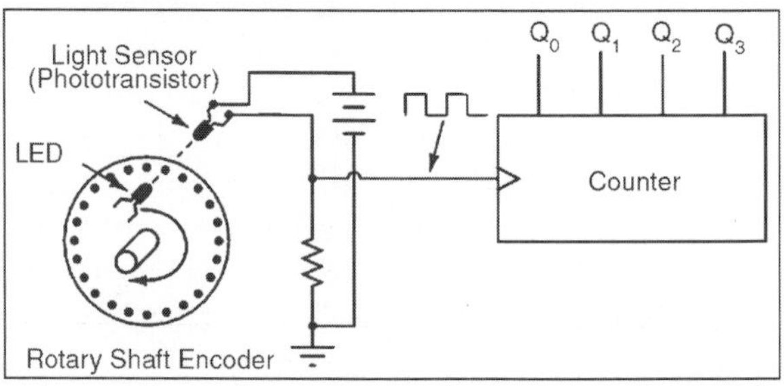

As the machine moves, it turns the encoder shaft, making and breaking the light beam between LED and phototransistor, thereby generating clock pulses to increment the counter circuit.

Thus, the counter integrates, or accumulates, total motion of the shaft, serving as an electronic indication of how far the machine has moved. If all we care about is tracking total motion, and do not care to account for changes in the direction of motion, this arrangement will suffice.

However, if we wish the counter to increment with one direction of motion and decrement with the reverse direction of motion, we must use an up/down counter, and an encoder/decoding circuit having the ability to discriminate between different directions.

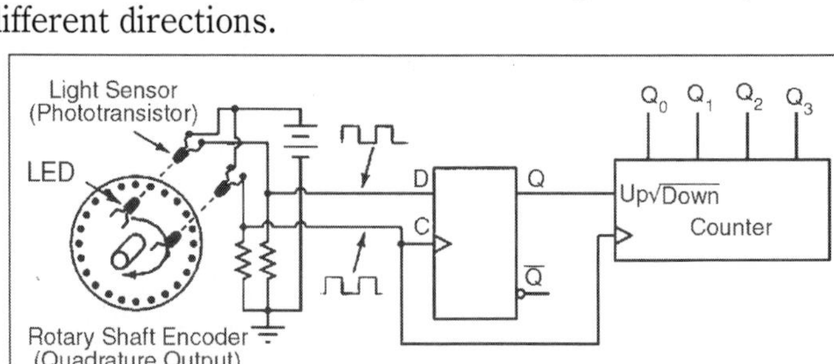

If we re-design the encoder to have two sets of LED/phototransistor pairs, those pairs aligned such that their square-wave output signals are 90o out of phase with each other, we have what is known as a quadrature output encoder (the word"quadrature" simply refers to a 90o angular separation).

A phase detection circuit may be made from a D-type flip-flop, to distinguish a clockwise pulse sequence from a counter-clockwise pulse sequence:

When the encoder rotates clockwise, the"D" input signal square-wave will lead the"C" input square-wave, meaning that the"D" input will already be"high" when the"C" transitions from"low" to"high," thus setting the D-type flip-flop (making the Q output"high") with every clock pulse.

A"high" Q output places the counter into the"Up" count mode, and any clock pulses received by the clock from the encoder (from either LED) will increment it. Conversely, when the encoder reverses rotation, the"D" input will lag behind the"C" input waveform, meaning that it will be"low" when the"C" waveform transitions from"low" to"high," forcing the D-type flip-flop into the reset state (making the Q output"low") with every clock pulse.

This"low" signal commands the counter circuit to decrement with every clock pulse from the encoder.

Synchronous Decade Counters

Similar to an asynchronous decade counter, a synchronous decade counter counts from 0 to 9 and then recycles to 0 again. This is done by forcing the 1010 state back to the 0000 state. This so called truncated sequence can be constructed by the following circuit.

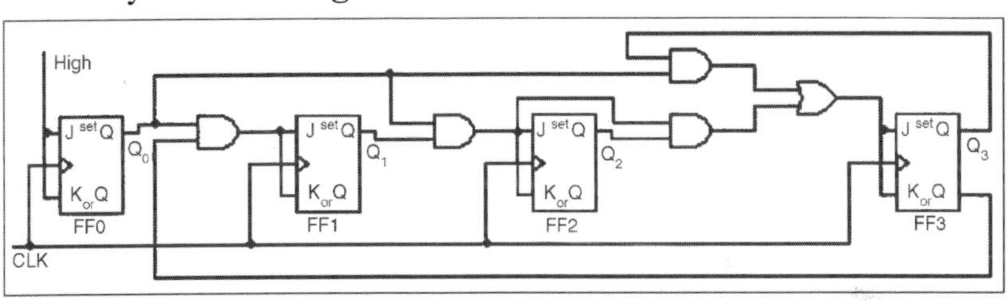

Clock Pulse	Q3	Q2	Q1	Q0
0	0	0	0	0
1	0	0	0	1
2	0	0	1	0
3	0	0	1	1
4	0	1	0	0
5	0	1	0	1
6	0	1	1	0
7	0	1	1	1
8	1	0	0	0
9	1	0	0	1

From the sequence on the left, we notice that:

- Q0 toggles on each clock pulse.
- Q1 changes on the next clock pulse each time Q0=1 and Q3=0.
- Q2 changes on the next clock pulse each time Q0=Q1=1.
- Q3 changes on the next clock pulse each time Q0=1, Q1=1 and Q2=1 (count 7), or when Q0=1 and Q3=1 (count 9).

Synchronous Up-Down Counters

A circuit of a 3-bit synchronous up-down counter and a table of its sequence are shown below. Similar to an asynchronous up-down counter, a synchronous up-down counter also has an up-down control input. It is used to control the direction of the counter through a certain sequence.

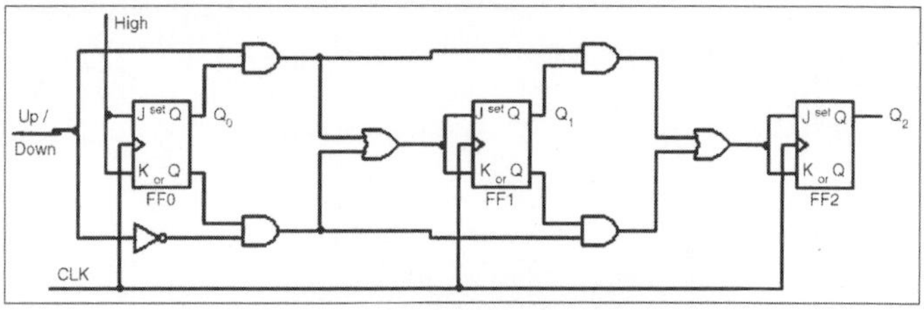

Up Down

Q2	Q1	Q0
0	0	0
0	0	1
0	1	0
0	1	1
1	0	0
1	0	1
1	1	0
1	1	1

An examination of the sequence table shows:

- For both the UP and DOWN sequences, Q0 toggles on each clock pulse.
- For the UP sequence, Q1 changes state on the next clock pulse when Q0=1.
- For the DOWN sequence, Q1 changes state on the next clock pulse when Q0=0.
- For the UP sequence, Q2 changes state on the next clock pulse when Q0=Q1=1.
- For the DOWN sequence, Q2 changes state on the next clock pulse when Q0=Q1=0.

4

Rectifier Circuits

Now we come to the most popular application of the diode: rectification. Simply defined, rectification is the conversion of alternating current (AC) to direct current (DC). This involves a device that only allows one-way flow of electrons. As we have seen, this is exactly what a semiconductor diode does. The simplest kind of rectifier circuit is the half-wave rectifier. It only allows one half of an AC waveform to pass through to the load.

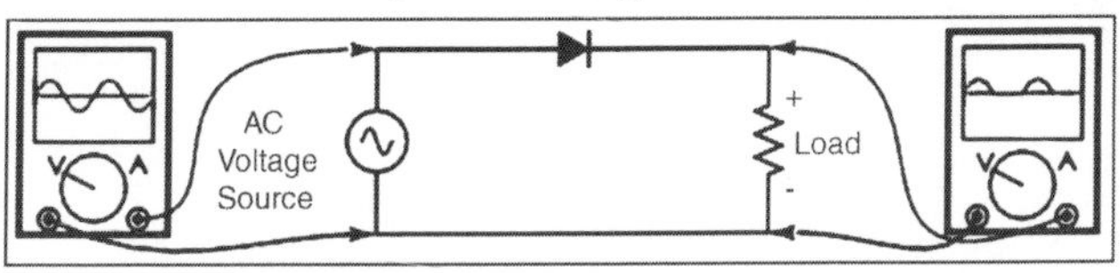

Fig. Half-wave rectifier circuit.

For most power applications, half-wave rectification is insufficient for the task. The harmonic content of the rectifier's output waveform is very large and consequently difficult to filter. Furthermore, the AC power source only supplies power to the load one half every full cycle, meaning that half of its capacity is unused. Half-wave rectification is, however, a very simple way to reduce power to a resistive load. Some two-position lamp dimmer switches apply full AC power to the lamp filament for "full" brightness and then half-wave rectify it for a lesser light output.

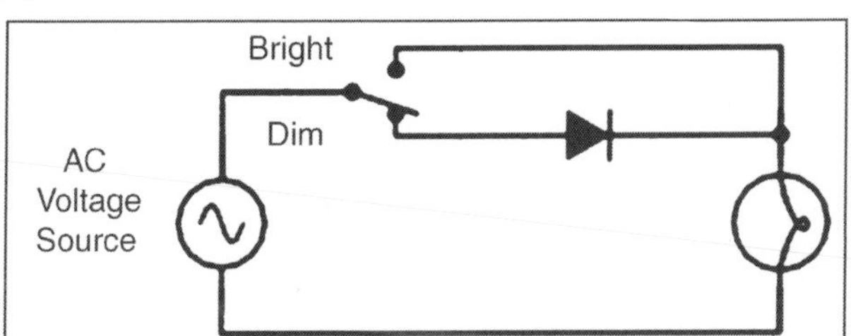

Fig. Half-wave rectifier application: Two level lamp dimmer.

In the "Dim" switch position, the incandescent lamp receives approximately one-half the power it would normally receive operating on full-wave AC. Because the half-wave rectified power pulses far more rapidly than the filament has time to heat up and cool down, the lamp does not blink. Instead, its filament merely operates at a lesser temperature than normal, providing less light output.

This principle of "pulsing" power rapidly to a slow-responding load device to control the electrical power sent to it is common in the world of industrial electronics. Since the controlling device (the diode, in this case) is either fully conducting or fully non-conducting at any given time, it dissipates little heat energy while controlling load power, making this method of power control very energy-efficient. This circuit is perhaps the crudest possible method of pulsing power to a load, but it suffices as a proof-of-concept application.

If we need to rectify AC power to obtain the full use of both half-cycles of the sine wave, a different rectifier circuit configuration must be used. Such a circuit is called a *full-wave* rectifier. One kind of full-wave rectifier, called the *centre-tap* design, uses a transformer with a centre-tapped secondary winding and two diodes.

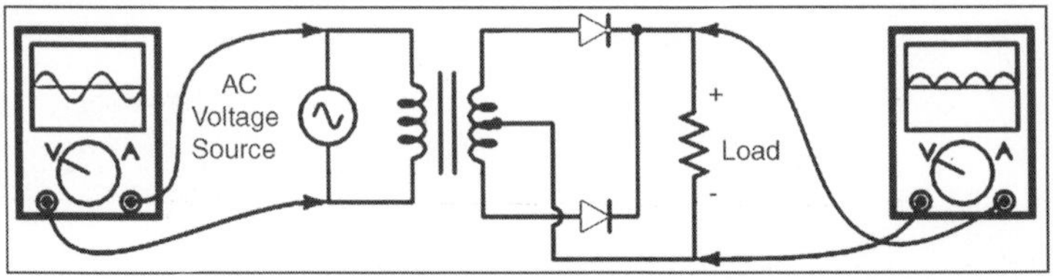

Fig. Full-wave rectifier, centre-tapped design.

This circuit's operation is easily understood one half-cycle at a time. Consider the first half-cycle, when the source voltage polarity is positive (+) on top and negative (–) on bottom. At this time, only the top diode is conducting; the bottom diode is blocking current, and the load "sees" the first half of the sine wave, positive on top and negative on bottom. Only the top half of the transformer's secondary winding carries current during this half-cycle.

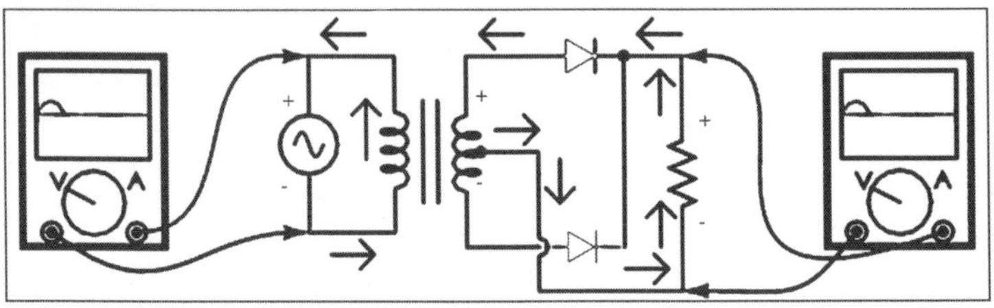

FULL-WAVE CENTRE-TAP RECTIFIER

Top half of secondary winding conducts during positive half-cycle of input, delivering positive half-cycle to load.

During the next half-cycle, the AC polarity reverses. Now, the other diode and the other half of the transformer's secondary winding carry current while the portions of the circuit formerly carrying current during the last half-cycle sit idle. The load still "sees" half of a sine wave, of the same polarity as before: positive on top and negative on bottom.

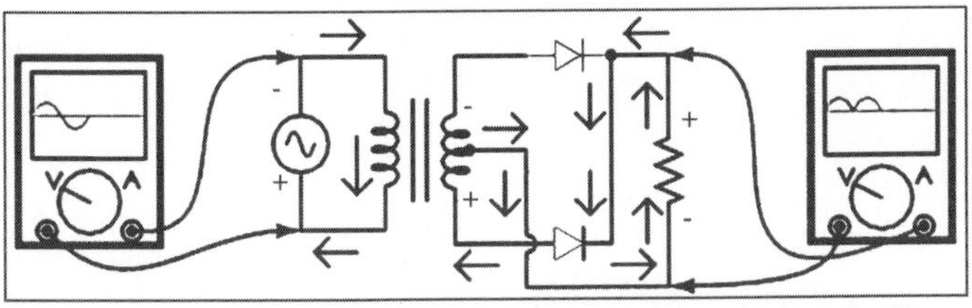

Full-wave centre-tap rectifier: During negative input half-cycle, bottom half of secondary winding conducts, delivering a positive half-cycle to the load. One disadvantage of this full-wave rectifier design is the necessity of a transformer with a centre-tapped secondary winding. If the circuit in question is one of high power, the size and expense of a suitable transformer is significant. Consequently, the centre-tap rectifier design is only seen in low-power applications.

The full-wave centre-tapped rectifier polarity at the load may be reversed by changing the direction of the diodes. Furthermore, the reversed diodes can be paralleled with an existing positive-output rectifier. The result is dual-polarity full-wave centre-tapped rectifier. Note that the connectivity of the diodes themselves is the same configuration as a bridge.

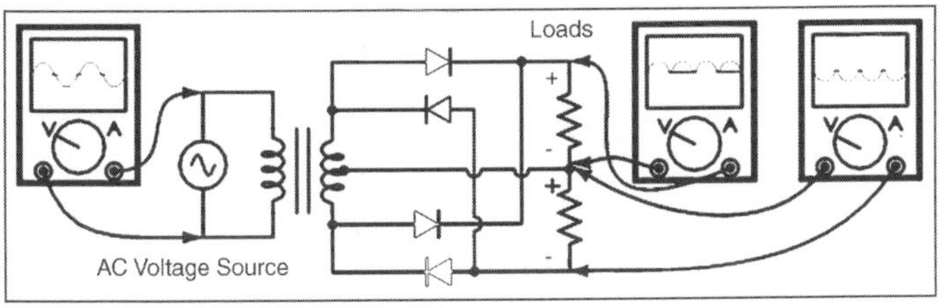

Dual Polarity Full-Wave Centre Tap Rectifier

Another, more popular full-wave rectifier design exists, and it is built around a four-diode bridge configuration. For obvious reasons, this design is called a full-wave bridge.

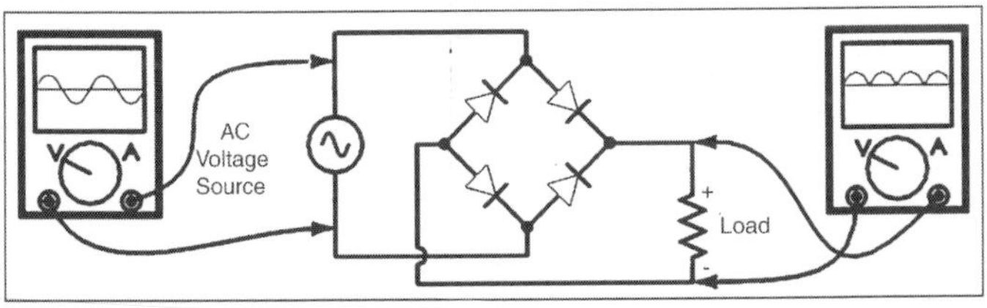

Fig. Full-wave bridge rectifier.

Current directions for the full-wave bridge rectifier circuit for positive half-cycle for negative half-cycles of the AC source waveform. Note that regardless of the polarity of the input, the current flows in the same direction through the load. That is, the negative half-cycle of source is a positive half-cycle at the load. The current flow is through two diodes in series for both polarities. Thus, two diode drops of the source voltage are lost ($0.7 \cdot 2 = 1.4$ V for Si) in the diodes.

This is a disadvantage compared with a full-wave centre-tap design. This disadvantage is only a problem in very low voltage power supplies.

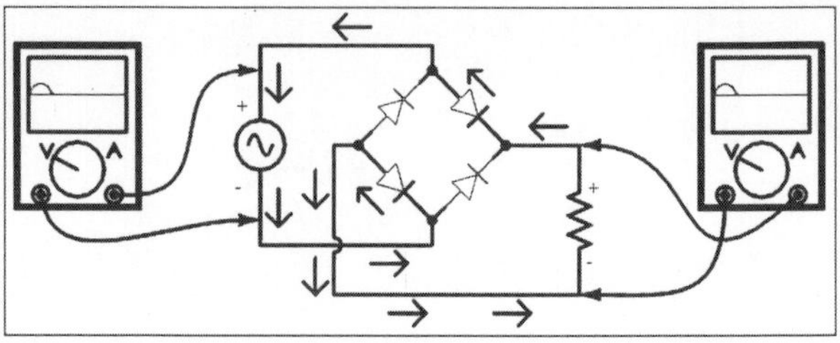

Fig. Full-wave bridge rectifier: Electron flow for positive half-cycles.

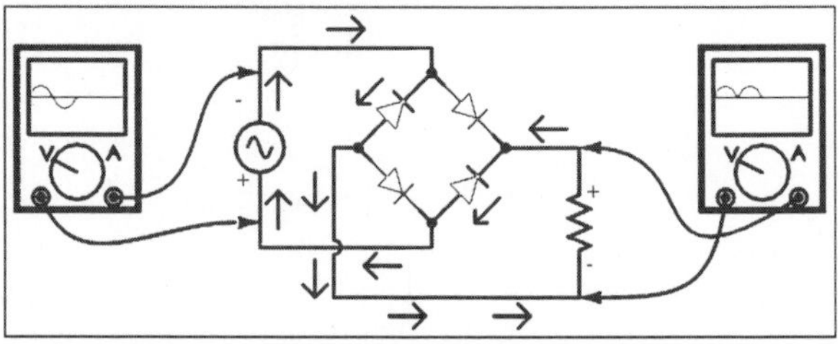

Fig. Full-wave bridge rectifier: Electron flow for negative half=cycles.

Remembering the proper layout of diodes in a full-wave bridge rectifier circuit can often be frustrating to the new student of electronics. I've found that an alternative representation of this circuit is easier both to remember and to comprehend. It's the exact same circuit, except all diodes are drawn in a horizontal attitude, all "pointing" the same direction.

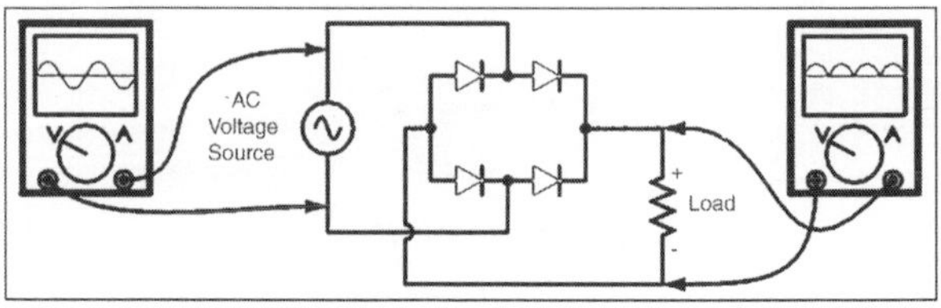

Alternative layout style for Full-wave bridge rectifier.

One advantage of remembering this layout for a bridge rectifier circuit is that it expands easily into a polyphase version.

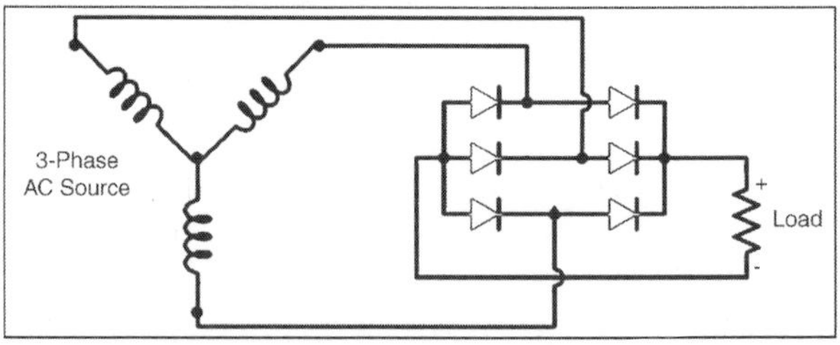

Fig. Three-phase full-wave bridge rectifier circuit.

Each three-phase line connects between a pair of diodes: one to route power to the positive (+) side of the load, and the other to route power to the negative

(–) side of the load. Polyphase systems with more than three phases are easily accommodated into a bridge rectifier scheme. Take for instance the six-phase bridge rectifier circuit.

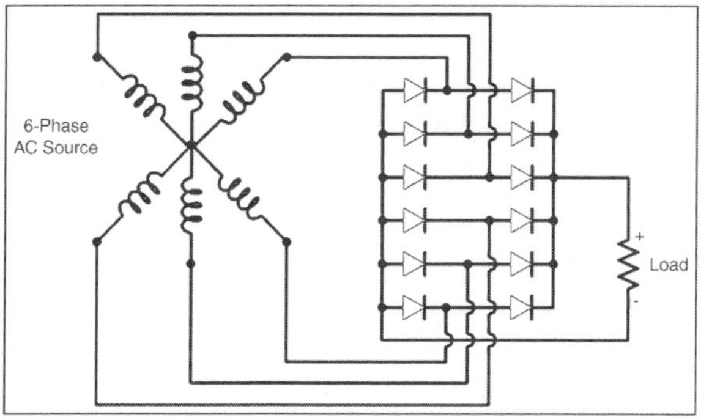

Fig. Six-phase full-wave bridge rectifier circuit.

When polyphase AC is rectified, the phase-shifted pulses overlap each other to produce a DC output that is much "smoother" (has less AC content) than that produced by the rectification of single-phase AC. This is a decided advantage in high-power rectifier circuits, where the sheer physical size of filtering components would be prohibitive but low-noise DC power must be obtained. The full-wave rectification of three-phase AC.

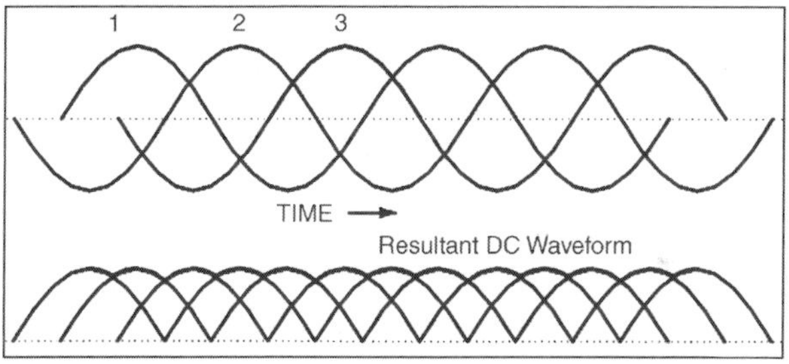

Fig. Three-phase AC and 3-phase full-wave rectifier output.

In any case of rectification -- single-phase or polyphase -- the amount of AC voltage mixed with the rectifier's DC output is called ripple voltage. In most cases, since "pure" DC is the desired goal, ripple voltage is undesirable. If the power levels are not too great, filtering networks may be employed to reduce the amount of ripple in the output voltage.

Sometimes, the method of rectification is referred to by counting the number of DC "pulses" output for every 360o of electrical "rotation." A single-phase, half-wave rectifier circuit, then, would be called a *1-pulse* rectifier, because it produces a single pulse during the time of one complete cycle (360o) of the AC waveform. A single-phase, full-wave rectifier (regardless of design, centre-tap or bridge) would be called a *2-pulse* rectifier, because it outputs two pulses of DC during one AC cycle's worth of time. A three-phase full-wave rectifier would be called a *6-pulse* unit.

Modern electrical engineering convention further describes the function of a rectifier circuit by using a three-field notation of phases, ways, and number ofpulses. A single-phase, half-wave rectifier circuit is given the somewhat cryptic designation of 1Ph1W1P (1 phase, 1 way, 1 pulse), meaning that the AC supply voltage is single-phase, that current on each phase of the AC supply lines moves in only one direction (way), and that there is a single pulse of DC produced for every 360o of electrical rotation.

A single-phase, full-wave, centre-tap rectifier circuit would be designated as 1Ph1W2P in this notational system: 1 phase, 1 way or direction of current in each winding half, and 2 pulses or output voltage per cycle.

A single-phase, full-wave, bridge rectifier would be designated as 1Ph2W2P: the same as for the centre-tap design, except current can go both ways through the AC lines instead of just one way.

The three-phase bridge rectifier circuit shown earlier would be called a 3Ph2W6P rectifier. Is it possible to obtain more pulses than twice the number of phases in a rectifier circuit? The answer to this question is yes: especially in polyphase circuits.

Through the creative use of transformers, sets of full-wave rectifiers may be paralleled in such a way that more than six pulses of DC are produced for three phases of AC. A 30° phase shift is introduced from primary to secondary of a three-phase transformer when the winding configurations are not of the same type.

In other words, a transformer connected either Y-Δ or Δ-Y will exhibit this 30° phase shift, while a transformer connected Y-Y or Δ-Δ will not. This phenomenon may be exploited by having one transformer connected Y-Y feed a bridge rectifier, and have another transformer connected Y-Δ feed a second bridge rectifier, then parallel the DC outputs of both rectifiers.

Since the ripple voltage waveforms of the two rectifiers' outputs are phase-shifted 30° from one another, their superposition results in less ripple than either rectifier output considered separately: 12 pulses per 360° instead of just six:

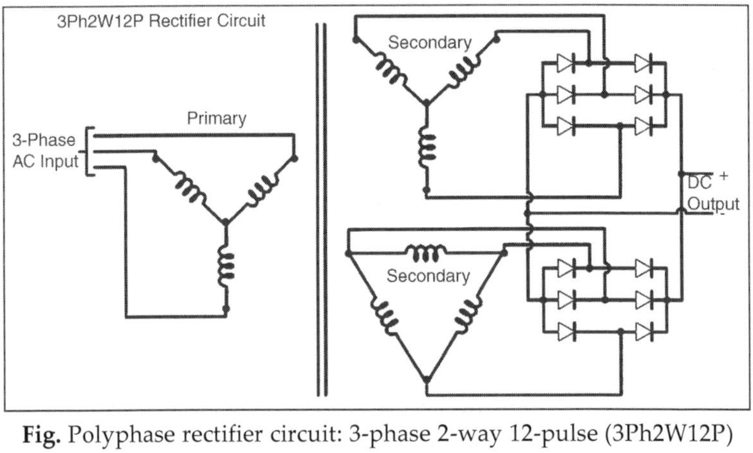

Fig. Polyphase rectifier circuit: 3-phase 2-way 12-pulse (3Ph2W12P)

RESISTOR IN SERIES OF ZENER DIODES

If we connect a diode and resistor in series with a DC voltage source so that the diode is forward-biased, the voltage drop across the diode will remain fairly constant over a wide range of power supply voltages.

According to the "diode equation" here, the current through a forward-biased *PN* junction is proportional to *e* raised to the power of the forward voltage drop. Because this is an exponential function, current rises quite rapidly for modest increases in voltage drop.

Another way of considering this is to say that voltage dropped across a forward-biased diode changes little for large variations in diode current. In the circuit, diode current is limited by the voltage of the power supply, the series resistor, and the diode's voltage drop, which as we know doesn't vary much from 0.7 volts.

If the power supply voltage were to be increased, the resistor's voltage drop would increase almost the same amount, and the diode's voltage drop just a little. Conversely, a decrease in power supply voltage would result in an almost equal decrease in resistor voltage drop, with just a little decrease in diode voltage drop. In a word, we could summarise this behaviour by saying that the diode is regulating the voltage drop at approximately 0.7 volts.

Voltage regulation is a useful diode property to exploit. Suppose we were building some kind of circuit which could not tolerate variations in power supply voltage, but needed to be powered by a chemical battery, whose voltage changes over its lifetime. We could form a circuit as shown and connect the circuit requiring steady voltage across the diode, where it would receive an unchanging 0.7 volts.

This would certainly work, but most practical circuits of any kind require a power supply voltage in excess of 0.7 volts to properly function. One way we could increase our voltage regulation point would be to connect multiple diodes in series, so that their individual forward voltage drops of 0.7 volts each would add to create a larger total.

For instance, if we had ten diodes in series, the regulated voltage would be ten times 0.7, or 7 volts.

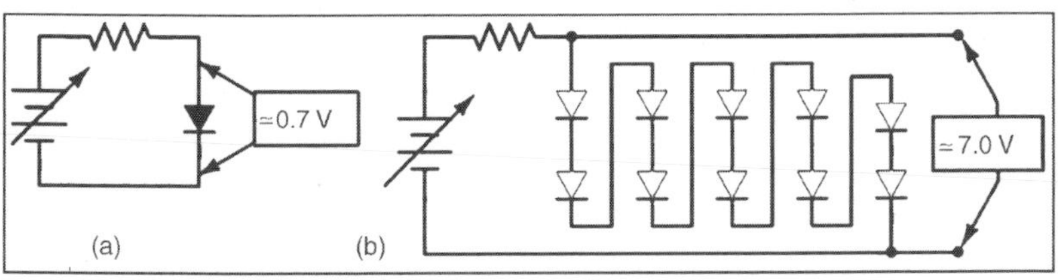

Fig. Forward biased *Si* reference: (a) single diode, 0.7V, (b) 10-diodes in series 7.0V.

So long as the battery voltage never sagged below 7 volts, there would always be about 7 volts dropped across the ten-diode "stack."

If larger regulated voltages are required, we could either use more diodes in series (an inelegant option, in my opinion), or try a fundamentally different approach.

We know that diode forward voltage is a fairly constant figure under a wide range of conditions, but so is reverse breakdown voltage, and breakdown voltage is typically much, much greater than forward voltage. If we reversed the polarity of the diode in our single-diode regulator circuit and increased the power supply voltage to the point where the diode "broke down" (could no longer withstand the reverse-bias voltage impressed across it), the diode would similarly regulate the voltage at that breakdown point, not allowing it to increase further.

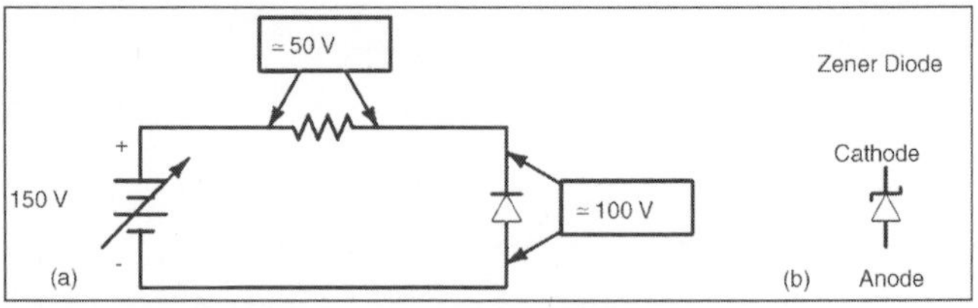

Fig. (a) Reverse biased *Si* small-signal diode breaks down at about 100V. (b) Symbol for Zener diode.

Unfortunately, when normal rectifying diodes "break down," they usually do so destructively. However, it is possible to build a special type of diode that can handle breakdown without failing completely. This type of diode is called a zener diode, and its symbol looks.

When forward-biased, zener diodes behave much the same as standard rectifying diodes: they have a forward voltage drop which follows the "diode equation" and is about 0.7 volts. In reverse-bias mode, they do not conduct until the applied voltage reaches or exceeds the so-called zener voltage, at which point the diode is able to conduct substantial current, and in doing so will try to limit the voltage dropped across it to that zener voltage point. So long as the power dissipated by this reverse current does not exceed the diode's thermal limits, the diode will not be harmed.

Zener diodes are manufactured with zener voltages ranging anywhere from a few volts to hundreds of volts. This zener voltage changes slightly with temperature, and like common carbon-composition resistor values, may be anywhere from 5 percent to 10 percent in error from the manufacturer's specifications. However, this stability and accuracy is generally good enough for the zener diode to be used as a voltage regulator device in common power supply circuit.

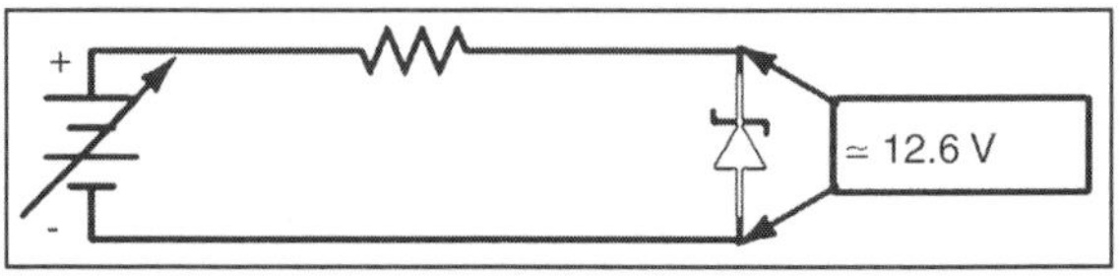

Fig. Zener diode regulator circuit, Zener voltage = 12.6V).

Please take note of the zener diode's orientation in the above circuit: the diode is reverse-biased, and intentionally so. If we had oriented the diode in the "normal" way, so as to be forward-biased, it would only drop 0.7 volts, just like a regular rectifying diode. If we want to exploit this diode's reverse breakdown properties, we must operate it in its reverse-bias mode. So long as the power supply voltage remains above the zener voltage (12.6 volts, in this example), the voltage dropped across the zener diode will remain at approximately 12.6 volts. Like any semiconductor device, the zener diode is sensitive to temperature. Excessive temperature will destroy a zener diode, and because it both drops voltage and conducts current, it produces its own heat in accordance with Joule's Law (P=IE). Therefore, one must be careful to design the regulator circuit in such a way that the diode's power dissipation rating is not exceeded. Interestingly enough, when zener diodes fail due to excessive power dissipation, they usually fail shorted rather than open. A diode failed in this manner is readily detected: it drops almost zero voltage when biased either way, like a piece of wire.

Let's examine a zener diode regulating circuit mathematically, determining all voltages, currents, and power dissipations. Taking the same form of circuit shown earlier, we'll perform calculations assuming a zener voltage of 12.6 volts, a power supply voltage of 45 volts, and a series resistor value of 1000 Ω (we'll regard the zener voltage to be exactly 12.6 volts so as to avoid having to qualify all figures as "approximate".

If the zener diode's voltage is 12.6 volts and the power supply's voltage is 45 volts, there will be 32.4 volts dropped across the resistor (45 volts - 12.6 volts = 32.4 volts). 32.4 volts dropped across 1000 Ω gives 32.4 mA of current in the circuit.

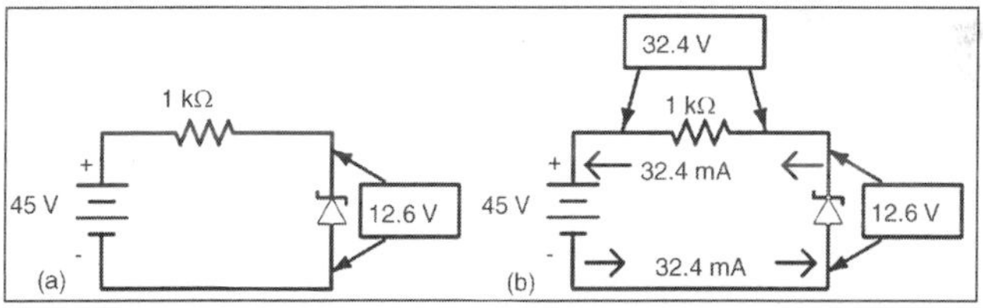

Fig. (a) Zener Voltage regulator with 1000 O resistor.
(b) Calculation of voltage drops and current.

Power is calculated by multiplying current by voltage (P = IE), so we can calculate power dissipations for both the resistor and the zener diode quite easily:

$$P_{resistor} = (32.4 \text{ mA})\ (32.4 \text{ V})$$

$$P_{resistor} = 1.0498 \text{ W}$$

$$P_{diode} = (32.4 \text{ mA})\ (12.6 \text{ V})$$

$$P_{diode} = 408.24 \text{ mW}.$$

A zener diode with a power rating of 0.5 watt would be adequate, as would a resistor rated for 1.5 or 2 watts of dissipation.

If excessive power dissipation is detrimental, then why not design the circuit for the least amount of dissipation possible? Why not just size the resistor for a very high value of resistance, thus severely limiting current and keeping power dissipation figures very low? Take this circuit, for example, with a 100 kO resistor instead of a 1 kO resistor. Note that both the power supply voltage and the diode's zener voltage are identical to the last example:

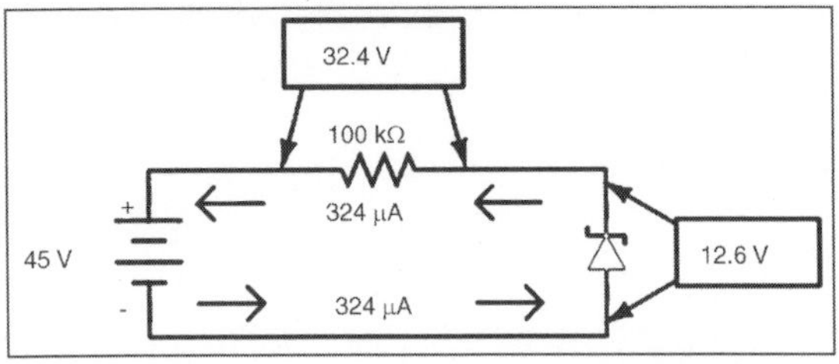

Fig. Zener regulator with 100 kO resistor.

With only 1/100 of the current we had before (324 μA instead of 32.4 mA), both power dissipation figures should be 100 times smaller:

$$P_{resistor} = (324\ \text{mA})(32.4\ \text{V})$$

$$P_{resistor} = 10.498\ \text{mW}$$

$$P_{diode} = (324\ \mu\text{A})(12.6\ \text{A})$$

$$P_{diode} = 4.0824\ \text{mW}$$

Seems ideal, doesn't it? Less power dissipation means lower operating temperatures for both the diode and the resistor, and also less wasted energy in the system, right? A higher resistance value does reduce power dissipation levels in the circuit, but it unfortunately introduces another problem. Remember that the purpose of a regulator circuit is to provide a stable voltage for another circuit. In other words, we're eventually going to power something with 12.6 volts, and this something will have a current draw of its own. Consider our first regulator circuit, this time with a 500Ω load connected in parallel with the zener diode.

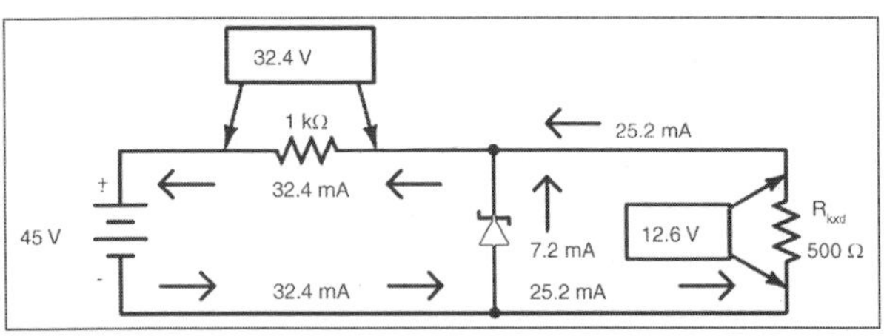

Fig. Zener regulator with 1000 Ω series resistor and 500 Ω load.

If 12.6 volts is maintained across a 500 Ω load, the load will draw 25.2 mA of current. In order for the 1 kΩ series "dropping" resistor to drop

32.4 volts (reducing the power supply's voltage of 45 volts down to 12.6 across the zener), it still must conduct 32.4 mA of current. This leaves 7.2 mA of current through the zener diode. Now consider our "power-saving" regulator circuit with the 100 kΩ dropping resistor, delivering power to the same 500 Ω load. What it is supposed to do is maintain 12.6 volts across the load, just like the last circuit. However, as we will see, it cannot accomplish this task.

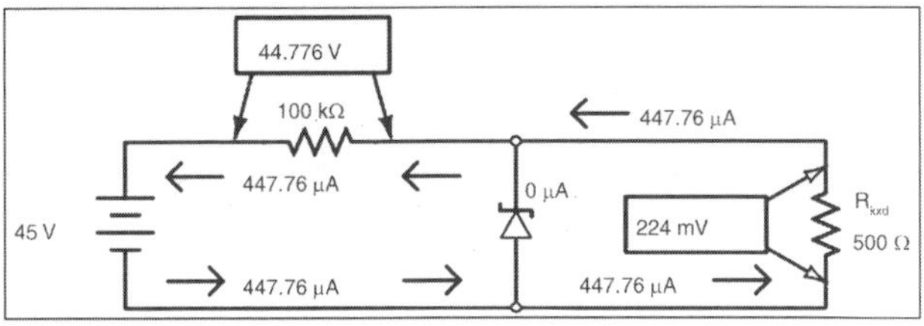

Fig. Zener non-regulator with 100 KΩ series resistor with 500 Ω load.>

With the larger value of dropping resistor in place, there will only be about 224 mV of voltage across the 500 Ω load, far less than the expected value of 12.6 volts! Why is this? If we actually had 12.6 volts across the load, it would draw 25.2 mA of current, as before. This load current would have to go through the series dropping resistor as it did before, but with a new (much larger!) dropping resistor in place, the voltage dropped across that resistor with 25.2 mA of current going through it would be 2,520 volts! Since we obviously don't have that much voltage supplied by the battery, this cannot happen.

The situation is easier to comprehend if we temporarily remove the zener diode from the circuit and analyse the behaviour of the two resistors alone.

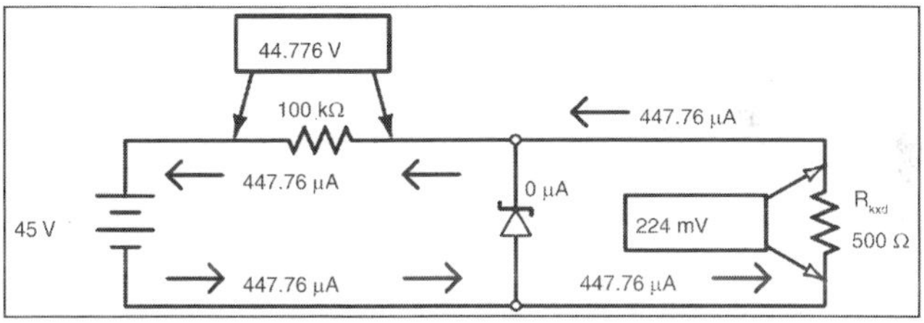

Fig. Non-regulator with Zener removed.

Both the 100 kΩ dropping resistor and the 500 Ω load resistance are in series with each other, giving a total circuit resistance of 100.5 kΩ. With a total voltage of 45 volts and a total resistance of 100.5 kΩ, Ohm's Law (I = E/R) tells us that the current will be 447.76 μA.

Figuring voltage drops across both resistors (E=IR), we arrive at 44.776 volts and 224 mV, respectively. If we were to re-install the zener diode at this point, it would "see" 224 mV across it as well, being in parallel with the load resistance. This is far below the zener breakdown voltage of the diode and so it will not "break down" and conduct current. For that matter, at this low voltage the diode wouldn't conduct even if it were forward-biased! Thus, the diode ceases to regulate voltage. At least 12.6 volts must be dropped across to "activate" it.

The analytical technique of removing a zener diode from a circuit and seeing whether or not enough voltage is present to make it conduct is a sound one. Just because a zener diode happens to be connected in a circuit doesn't guarantee that the full zener voltage will always be dropped across it! Remember that zener diodes work by limiting voltage to some maximum level; they cannot make up for a lack of voltage.

In summary, any zener diode regulating circuit will function so long as the load's resistance is equal to or greater than some minimum value. If the load resistance is too low, it will draw too much current, dropping too much voltage across the series dropping resistor, leaving insufficient voltage across the zener diode to make it conduct. When the zener diode stops conducting current, it can no longer regulate voltage, and the load voltage will fall below the regulation point. Our regulator circuit with the 100 kΩ dropping resistor must be good for some value of load resistance, though. To find this acceptable load resistance value, we can use a table to calculate resistance in the two-resistor series circuit (no diode), inserting the known values of total voltage and dropping resistor resistance, and calculating for an expected load voltage of 12.6 volts:

	$R_{dropping}$	R_{load}	Total	
E		12.6	45	Volts
I				Amps
R	100 k			Ohms

With 45 volts of total voltage and 12.6 volts across the load, we should have 32.4 volts across

	$R_{dropping}$	R_{load}	Total	
E	32.4	12.6	45	Volts
I				Amps
R	100 k			Ohms

With 32.4 volts across the dropping resistor, and 100 kΩ worth of resistance in it, the current through it will be 324 μA:

	$R_{dropping}$	R_{load}	Total	
E	32.4	12.6	45	Volts
I	**324 μ**			Amps
R	100 k			Ohms

↑

Ohm's Law

$$I = \frac{E}{R}$$

Being a series circuit, the current is equal through all components at any given time:

	$R_{dropping}$	R_{load}	Total	
E	32.4	12.6	45	Volts
I	324 μ	324 μ	324 μ	Amps
R	100 k			Ohms

Rule of Series circuit:

$$I_{Total} = I_1 = I_2 = \ldots I_n$$

Calculating load resistance is now a simple matter of Ohm's Law (R = E/I), giving us 38.889 kΩ:

	$R_{dropping}$	R_{load}	Total	
E	32.4	12.6	45	Volts
I	324 μ	324 μ	324 μ	Amps
R	100 k	38.889 k		Ohms

Ohm's Law

$$R = \frac{E}{I}$$

Thus, if the load resistance is exactly 38.889 kΩ, there will be 12.6 volts across it, diode or no diode. Any load resistance smaller than 38.889 kΩ will result in a load voltage less than 12.6 volts, diode or no diode. With the diode in place, the load voltage will be regulated to a maximum of 12.6 volts for any load resistance greater than 38.889 kΩ.

With the original value of 1 kΩ for the dropping resistor, our regulator circuit was able to adequately regulate voltage even for a load resistance as low as 500 Ω. What we see is a tradeoff between power dissipation and acceptable load resistance. The higher-value dropping resistor gave us less power dissipation, at the expense of raising the acceptable minimum load resistance value. If we wish to regulate voltage for low-value load resistances, the circuit must be prepared to handle higher power dissipation.

Zener diodes regulate voltage by acting as complementary loads, drawing more or less current as necessary to ensure a constant voltage drop across the load. This is analogous to regulating the speed of an automobile by braking rather than by varying the throttle position: not only is it wasteful, but the brakes must be built to handle all the engine's power when the driving conditions don't demand it.

Despite this fundamental inefficiency of design, zener diode regulator circuits are widely employed due to their sheer simplicity. In high-power applications where the inefficiencies would be unacceptable, other voltage-regulating techniques are applied. But even then, small zener-based circuits are often used to provide a "reference" voltage to drive a more efficient amplifier circuit controlling the main power.

Zener diodes are manufactured in standard voltage ratings listed in Table below. The table "Common zener diode voltages" lists common voltages for 0.3W and 1.3W parts. The wattage corresponds to die and package size, and is the power that the diode may dissipate without damage.

Common zener diode voltages:

0.5W						
2.7V	3.0V	3.3V	3.6V	3.9V	4.3V	4.7V
5.1V	5.6V	6.2V	6.8V	7.5V	8.2V	9.1V
10V	11V	12V	13V	15V	16V	18V
20V	24V	27V	30V			
1.3W						
4.7V	5.1V	5.6V	6.2V	6.8V	7.5V	8.2V
9.1V	10V	11V	12V	13V	15V	16V
18V	20V	22V	24V	27V	30V	33V
36V	39V	43V	47V	51V	56V	62V
68V	75V	100V	200V			

Zener Diode Clipper: A clipping circuit which clips the peaks of waveform at approximately the zener voltage of the diodes. The circuit has two zeners connected series opposing to symmetrically clip a waveform at nearly the Zener voltage. The resistor limits current drawn by the zeners to a safe value.

<="" a="" style="margin: 0px; padding: 0px; colour: rgb(51, 51, 51); font-family: Verdana, Arial, Tahoma, sans-serif; font-size: 13px; font-style: normal; font-variant: normal; font-weight: normal; letter-spacing: normal; line-height: 18px; orphans: 2; text-align: left; text-indent: 0px; text-transform: none; white-space: normal; widows: 2; word-spacing: 0px; -webkit-text-size-adjust: auto; -webkit-text-stroke-width: 0px; background-colour: rgb(255, 255, 255);">

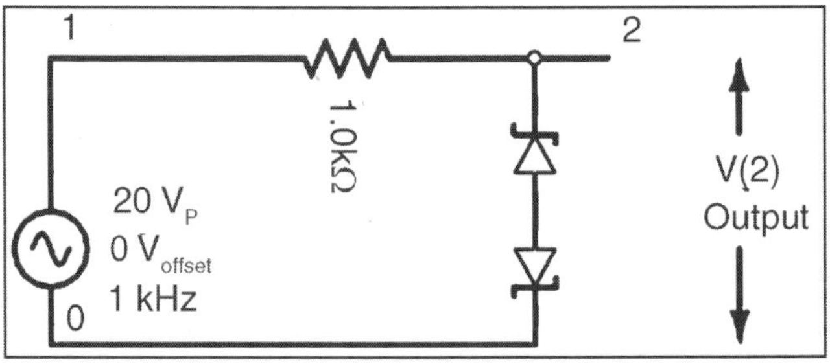

```
*SPICE 03445.eps
D1 4 0 diode
D2 4 2 diode
R1 2 1 1.0k
V1 1 0 SIN(0 20 1k)
Model diode d bv=10
Tran 0.001m 2m
End.
```

ZENER DIODE CLIPPER

<="" a="" style="margin: 0px; padding: 0px;">The zener breakdown voltage for the diodes is set at 10 *V* by the diode model parameter "bv=10" in the spice net list in Figure above. This causes the zeners to clip at about 10 V. The back-to-back diodes clip both peaks. For a positive half-cycle, the top zener is reverse biased, breaking down at the zener voltage of 10 V. The lower zener drops approximately 0.7 *V* since it is forward biased. Thus, a more accurate clipping level is 10+0.7=10.7V. Similar negative half-cycle clipping occurs a -10.7 V. shows the clipping level at a little over ±10 V.

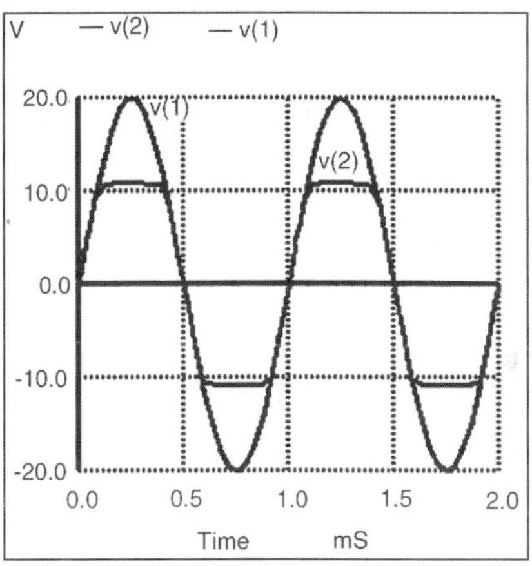

Fig. Zener diode clipper: v(1) input is clipped at waveform v(2).

LINE-FREQUENCY DIODE RECTIFIERS

In most power electronic applications, the power input is in the form of a 50- or 60-Hz sine wave ac voltage provided by the electric utility, that is first converted to a dc voltage. Increasingly, the trend is to use the inexpensive rectifiers with diodes to convert the input ac into dc in an uncontrolled manner, using rectifiers with diodes. In such diode rectifiers, the power flow can only be from the utility ac side to the dc side. A majority of the power electronics applications such as switching dc power supplies, ac motor drives, dc servo drives, and so on, use such uncontrolled rectifiers. In most of these applications, the rectifiers are supplied directly from the utility source without a 60-Hz transformer. The avoidance of this costly and bulky 60-Hz transformer is important in most modem power electronic systems.

The dc output voltage of a rectifier should be as ripple free as possible. Therefore, a large capacitor is connected as a filter on the dc side. As will be shown in this chapter, this capacitor gets charged to a value close to the peak of the ac input voltage. As a consequence, the current through the rectifier is very large near the peak of the 60-Hz ac input voltage and it does not flow continuously; that is, it becomes zero for finite durations during each half-cycle of the line frequency. These rectifiers draw highly distorted current from the utility. Now and even more so in the future, harmonic standards and guidelines will limit the amount of current distortion allowed into the utility, and the simple diode rectifiers may not be allowed.

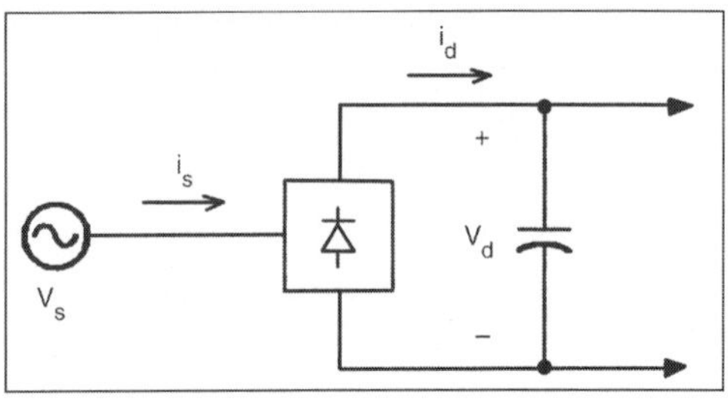

Fig. Block diagram of a rectifier.

Rectifiers with single-phase and three-phase inputs. The diodes are assumed to be ideal in the analysis of rectifiers. In a similar manner, the electromagnetic interference (EMI) filter at the ac input to the rectifiers is ignored, since it does not influence the basic operation of the rectifier.

BASIC RECTIFIER CONCEPTS

Rectification of ac voltages and currents is accomplished by means of diodes. Several simple circuits are considered to illustrate the basic concepts.

Pure Resistive Load

Consider the circuit, with a sinusoidal voltage source v_s. The waveforms, that both the load voltage V_d and the current i have an average (dc) component. Because of the large ripple in V_d and i, this circuit is of little practical significance.

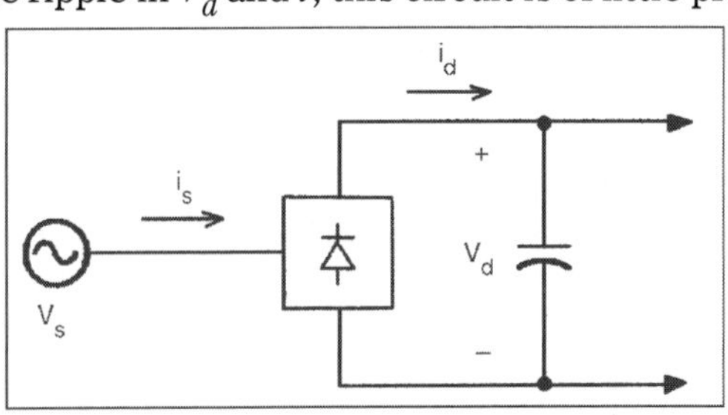

Fig. Basic rectifier with a load resistance.

Inductive load

Let us consider the load to be inductive, with an inductor in series with a resistor. Prior to $t = 0$, the voltage v_s is negative and the current in the circuit is zero.

Subsequent to $t = 0$, the diode becomes forward biased and a current begins to flow. Then, the diode can be replaced by a short, as shown in the equivalent circuit. The current in this circuit is governed by the following differential equation:

$$v_s = Ri + L\frac{di}{dt}$$

Where the voltage across the inductor $v_L = L\,di/dt$. The resulting voltages and current.

Until t_1, $vs > v_R$ (hence $v_L = v_s - v_R$ is positive), the current builds up, and the inductor stored energy increases. Beyond t_1, v_L becomes negative, and the current begins to decrease. After t_2, the input voltage v_s becomes negative but the current is still positive and the diode must conduct because of the inductor stored energy.

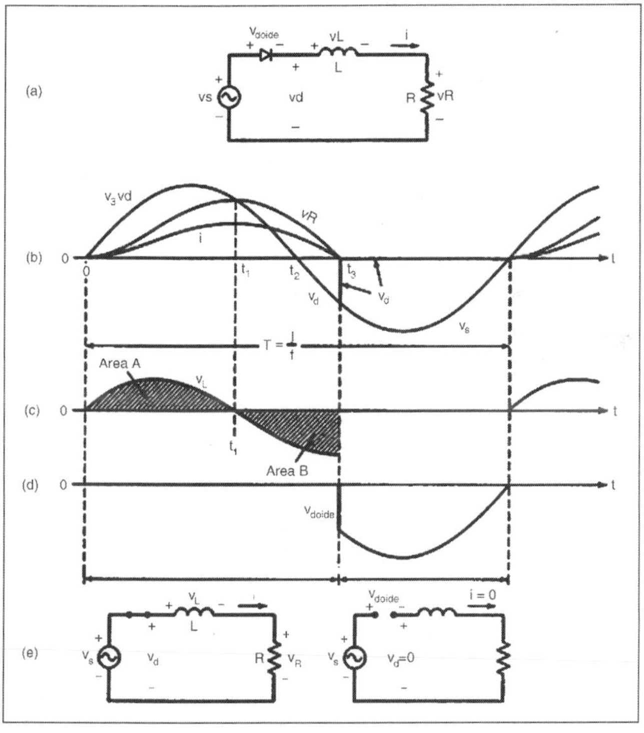

Fig. Basic rectifier with an inductive load.

The instant t_3, when the current goes to zero and the diode stops conducting, can be obtained as follows: The inductor equation $v_L = L\,di/dt$ can be rearranged as,

$$\frac{1}{L}v_L dt = di$$

Integrating both sides of the above equation between zero and t_3 and recognising that $i(0)$ and $i(t_3)$ are both zero give,

$$\frac{1}{L}\int_0^{t_3} v_L dt = \int_{i(0)}^{i(t_3)} di = i(t_3) - i(0) = 0,$$

we can observe that

$$\int_0^{t_3} v_L dt = 0$$

A graphical interpretation of the above equation is as follows: Equation above can be written as,

$$\int_0^{t_1} v_L dt + \int_{t_1}^{t_3} v_L dt = 0$$

Which in terms of the volt-second areas A and B of Fig. is

Area A- Area $B = 0$

Therefore, the current goes to zero at t_3 when area $A = B$ in Fig.

Beyond t_3, the voltages across both R and L are zero and a reverse polarity voltage ($=-v_s$) appears across the diode. These waveforms repeat with the time period $T = 1/f$.

The load voltage V_d becomes negative during the interval from t_2 to t_3. Therefore, in comparison to the case of purely resistive load, the average load voltage is less.

Load With an Internal DC Voltage

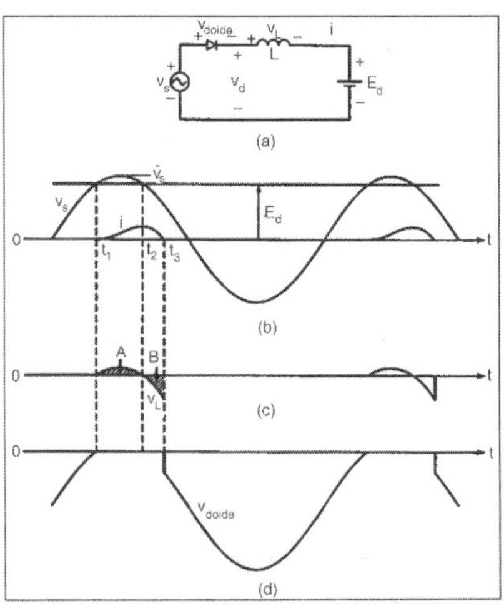

Fig. Basic rectifier with an internal dc voltage.

Next, we will consider the circuit, where the load consists of an inductor L and a dc voltage E_d. The diode begins to conduct at t_1 when v_s, exceeds E_d. The current reaches its peak at t_2 (when v_s is again equal to E_d) and decays to zero at t_3, with t_3 determined by the requirement that the volt-second area A be equal to area B in the plot of v_L. The voltage across the diode

SINGLE-PHASE DIODE BRIDGE RECTIFIERS

A commonly used single-phase diode bridge rectifier. A large filter capacitor is connected on the dc side. The utility supply is modeled as a sinusoidal voltage source v_s in series with its internal impedance, which in practice is primarily inductive. Therefore, it is represented by L_s. To improve the line-current waveform, an inductor may be added in series on the ac side, which in effect will increase the value of L_s. The objective of this chapter is to thoroughly analyse the operation of this circuit. Although the circuit appears simple, the procedure to obtain the associated voltage and current waveforms in a closed form is quite tedious. Therefore, we will simulate this circuit using PSpice and MATLAB. However, we will next analyse many simpler and hypothetical circuits in order to gain insight into the operation of the circuit.

Idealised Circuit with $L_S = 0$

As a first approximation to the circuit, we will assume L_s to be zero and replace the dc side of the rectifier by a resistance R or a constant dc current source I_d, respectively. It should be noted in the circuit, that although it is very unlikely that a pure resistive load will be supplied through a diode rectifier, this circuit models power-factor-corrected rectifiers.

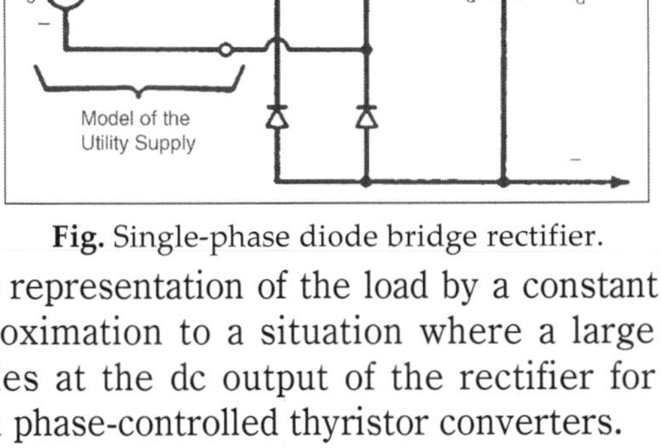

Fig. Single-phase diode bridge rectifier.

Similarly, the representation of the load by a constant dc current in the circuit is an approximation to a situation where a large inductor may be connected in series at the dc output of the rectifier for filtering. This is commonly done in phase-controlled thyristor converters.

The circuits are redrawn, which shows that this circuit consists of two groups of diodes: the top group with diodes 1 and 3 and the bottom group with diodes 2 and 4. With $L_s = 0$, it is easy to see the operation of each group of diodes. The current *id* flows continuously through one diode of the top group and one diode of the bottom group. In the top group, the cathodes of the two diodes are at a common potential. Therefore, the diode with its anode at the highest potential will conduct i_d. That is, when v_s, is positive, diode 1 will conduct i_d and v_s, will appear as a reverse-bias voltage across diode 3. When v_s goes negative, the current *id* shifts (commutates) instantaneously to diode 3 since $L_s = 0$. A reverse-bias voltage appears across diode 1.

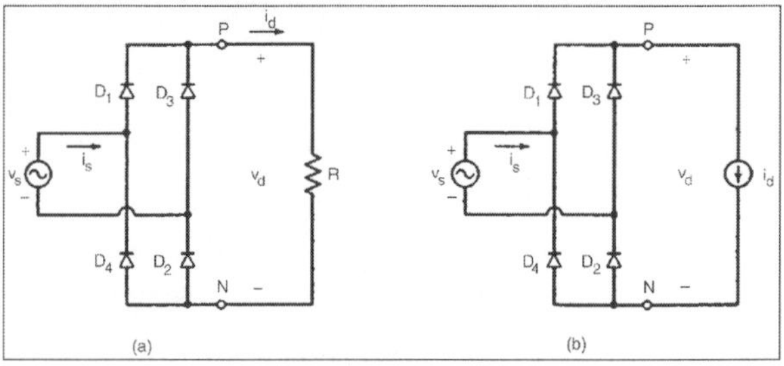

Fig. Idealised diode bridge rectifiers with $L_s = 0$

In the bottom group, the anodes of the two diodes are at a common potential. Therefore, the diode with its cathode at the lowest potential will conduct i_d. That is, when v_s is positive, diode 2 will carry i_d and v_s will appear as a reverse-bias voltage across diode 4. When v_s goes negative, the current i_d instantaneously commutates to diode 4 and a reverse-bias voltage appears across diode 2. The voltage and current waveforms in the circuits. There are several items worth noting. In both circuits, when v_s is positive, diodes

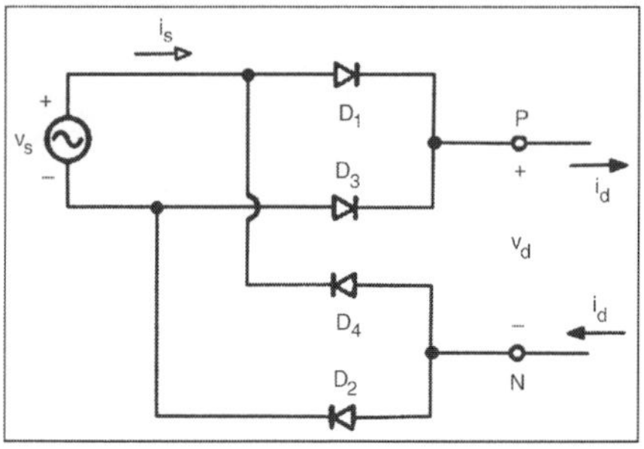

Fig. Redrawn rectifiers.

1 and 2 conduct and $v_d = v_s$, and $i_s = i_d$. When v_s goes negative, diodes 3 and 4 conduct and, therefore, $v_d = -v_s$, and $i_s = -i_d$. Therefore, at any time, the dc-side output voltage of the diode rectifier can be expressed as

$$v_d(t) = |v_s|$$

Similarly, the ac-side current can be expressed as

$$i_s = \begin{cases} i_d & if\ v_s > 0 \\ -i_d & if\ v_s < 0 \end{cases}$$

And the transition between the two values is instantaneous due to the assumption of zero L_s. The average value V_{do}, (where the subscript o stands for the idealised case with $L_s = 0$) of the dc output voltage in both circuits can be obtained by assigning an arbitrary time origin $t = 0$ and then integrating $v_s = \sqrt{2}V_s \sin \omega t$ over one-half time period (where $\omega = 2\pi f$ and $\omega T/2 = \pi$):

$$V_{do} = \frac{1}{(T/2)} \int_0^{T/2} \sqrt{2}V_s \sin \omega t \, dt = \frac{1}{\omega T/2}\left(\sqrt{2}V_s \cos \omega t\right)\Big|_{T/2}^{0} = \frac{2}{\pi}\sqrt{2}V_s$$

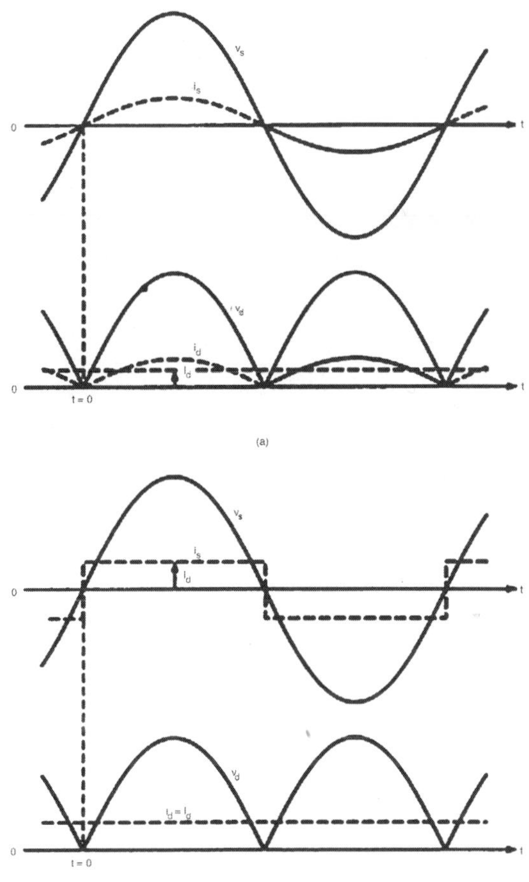

Fig. Waveforms in the rectifiers.

Therefore,

$$V_{do} = \frac{2}{\pi}\sqrt{2}V_s = 0.9V_s$$

where V_s, is the rms value of the input voltage. With $i_d(t) = I_d$, v_s and i_s waveforms along with the fundamental-frequency component i_{s1}. Applying the basic definition of the rms value to the i_s waveform in this idealised case yields,

$$I_s = I_d$$

By Fourier analysis of i_s, the fundamental and the harmonic components have the following rms values in this idealised case:

$$I_{s1} = \frac{2}{\pi}\sqrt{2}I_d = 0.9I_d$$

$$I_{sh} = \begin{cases} 0 & \textit{for even values of } h \textit{ and} \\ I_{s1}/h & \textit{for odd values of } h \end{cases}$$

The harmonic components in i_s. The total harmonic distortion can be calculated by Equation above to be,

THD = 48.43 per cent

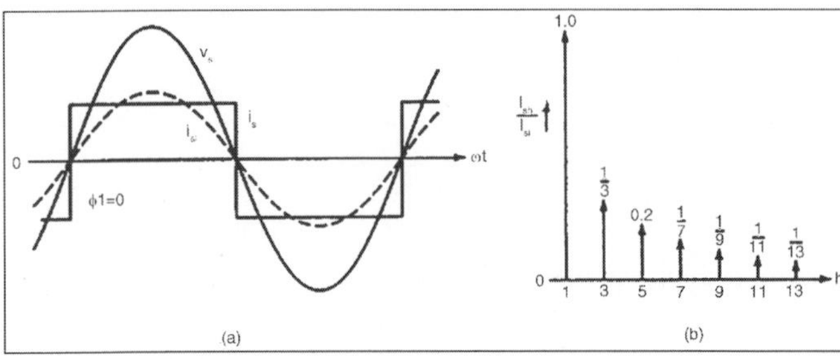

Fig. Line current i_s in the idealised case.

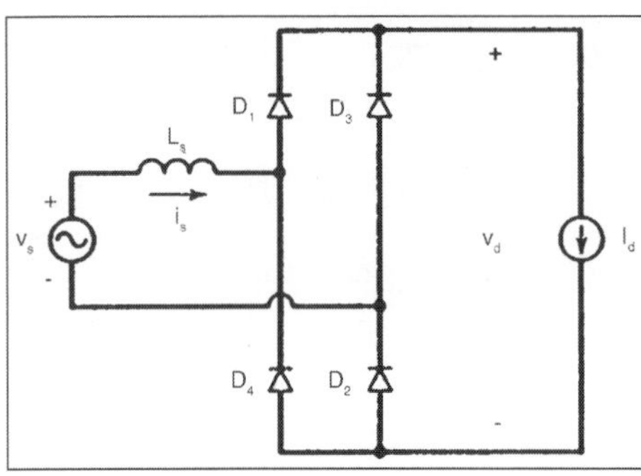

Fig. Single-phase rectifier with L_s.

By visual inspection of i_s waveform, it is apparent that i_{s1} is in phase with the v_s waveform. Therefore from Fig.,

$$\text{DPF} = 1.0$$

and

$$\text{PF} = \text{DPF}\ I_{s1}/I_s = 0.9$$

Effect of L_s on Current Commutation

Next, we will look at the effect of a finite ac-side inductance L_s on the circuit operation. We will assume that the dc side can be represented by a constant dc current I_d. Due to a finite L_s, the transition of the ac-side current i_s from a value of + I_d to $-I_d$ (or vice versa) will not be instantaneous. The finite time interval required for such a transition is called the current commutation time, or the commutation interval u, and this process where the current conduction shifts from one diode (or a set of diodes) to the other is called the current commutation process.

In order to understand this process fully, let us first consider a simple hypothetical circuit with two diodes supplied by a sinusoidal voltage source $v_s = \sqrt{2}V_s \sin \omega t$. The output is represented by a constant dc current source I_d. For comparison purposes, v_s, v_d, and i_s, waveforms with $Ls = 0$.

Prior to time $\omega t = 0$, the voltage v_s is negative and the current I_d is circulating through D_2 with $v_d = 0$ and $i_s = 0$. When v_s becomes positive at $\omega t = 0$, a forward-bias voltage appears across D_1 and it begins to conduct. With a finite L_s, the buildup of i_s can be obtained from the circuit redrawn. Since D_2 is conducting, it provides a short-circuit (with $v_d = 0$, assuming an ideal diode) path through which i_s can build up.

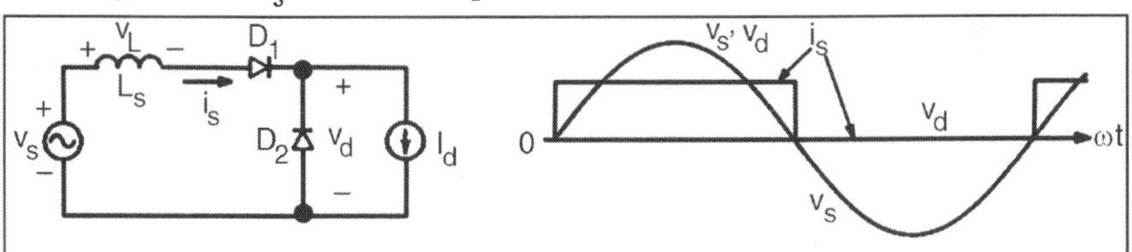

Fig. Basic circuit to illustrate current commutation. Waveforms assume $Ls = 0$

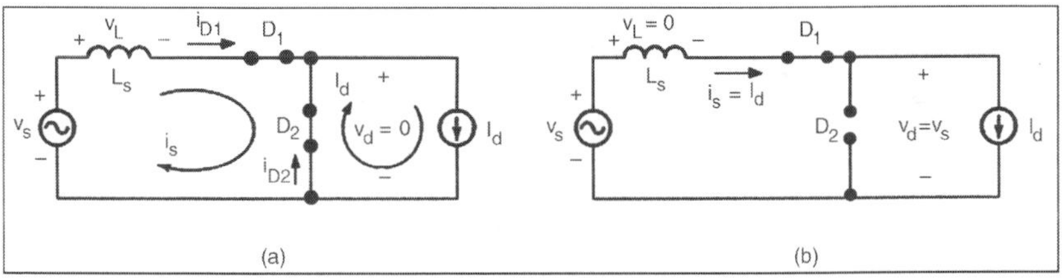

Fig. (a) Circuit during the commutation. (b) Circuit after the current commutation is completed

The two mesh currents shown are I_d and i_s. In terms of these mesh currents, the diode current $i_{D2} = I_d - i_s$. Therefore, as i_s builds up to a value I_d during the commutation interval $\omega t = u$, i_{D2} is positive and D_2 conducts in the circuit. The current i_s cannot exceed I_d since it will result in a negative value of i_{D2} that is not possible. As a consequence, the diode D_2 stops conducting at $\omega t = u$, resulting in the circuit. The waveforms are plott as a function of ωt.

It is clear from the above introduction that the current i_s through the inductor starts with a value of zero at the beginning of the commutation interval and ends up with a value of I_d at the end. Therefore, to obtain the length of the commutation interval u, we should consider the inductor equation. During the commutation interval, the input ac voltage appears as a current commutation voltage across the inductor:

$$v_L = \sqrt{2}V_s \sin \omega t = L_s \frac{di_s}{dt} \quad 0 < \omega t < u$$

The right side of the above equation can be written as $\omega L_s\, di_s/d(\omega t)$. Therefore,

$$\sqrt{2}V_s \sin \omega t\, d(\omega t) = \omega L_s di_s$$

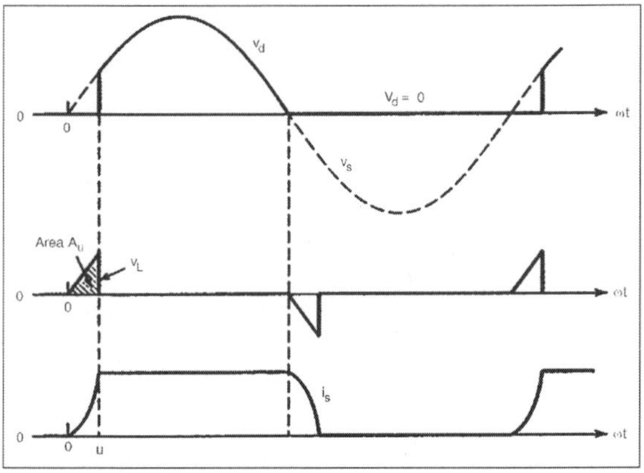

Fig. Waveforms in the basic circuit.

Integrating both sides of Equation above and recognising that i_s goes from zero to I_d during the commutation interval from zero to u, we get,

$$\int_0^u \sqrt{2}V_s \sin \omega t\, d(\omega t) = \omega L_s \int_0^{I_d} di_s = \omega L_s I_d$$

The left side is the integral of the inductor voltage v_L during the commutation interval. The above voltage integral is the same as the volt-radian area A_u:

$$A_u = \int_0^u \sqrt{2}V_s \sin \omega t\, d(\omega t) = \sqrt{2}V_s(1 - \cos u)$$

Combining Eqs. above yields,

$$A_u = \sqrt{2}V_s(1 - \cos u) = \omega L_s I_d$$

The important observation from Equation above is that the integral of the commutation voltage over the commutation interval can always be calculated by the product of ω, L_s, and the change in the current through L_s, during commutation. From Equation,

$$\cos u = 1 - \frac{\omega L_s I_d}{\sqrt{2}V_s}$$

Equation above confirms that if $L_s = 0$, $\cos u = 1$ and the current commutation will be instantaneous with $u = 0$. For a given frequency ω, the commutation interval u increases with L_s, and I_d and decreases with increasing voltage V_s.

The finite commutation interval reduces the average value of the output voltage. $L_s = 0$, the average value V_{do} of v_d is,

$$V_{do} = \frac{1}{2\pi}\int_0^{\pi} \sqrt{2}V_s \sin\omega t\, d(\omega t) = \frac{2\sqrt{2}}{2\pi}V_s = 0.45V_s$$

With a finite L_s and hence a non-zero u, $v_d = 0$ during the interval u. Therefore,

$$V_d = \frac{1}{2\pi}\int_u^{\pi} \sqrt{2}V_s \sin\omega t\, d(\omega t)$$

Which can be written as,

$$V_d = \frac{1}{2\pi}\int_u^{\pi} \sqrt{2}V_s \sin\omega t\, d(\omega t) - \frac{1}{2\pi}\int_0^{u} \sqrt{2}V_s \sin\omega t\, d(\omega t)$$

Substituing Eqs. above yields,

$$V_d = 0.45V_s - \frac{area\ A_u}{2\pi} = 0.45V_s - \frac{\omega L_s}{2\pi}I_d$$

Where the reduction in the average output voltage by ΔV_d from V_{do} is,

$$\Delta V_d = \frac{area\ A_u}{2\pi} = \frac{\omega L_s}{2\pi}I_d$$

We will now extend this analysis to the circuit, redrawn in Fig. Once again, we need to consider the current commutation process. Prior to $\omega t = 0$, diodes 3 and 4 are conducting I_d and $i_s = -I_d$. The circuit is carefully redrawn to show the current commutation process during $0 < \omega t < u$. Subsequent to $t = 0$, v_s becomes positive and diodes 1 and 2 become forward biased because of the short-circuit path provided by the conducting diodes 3 and 4. The three mesh currents, where the two commutation currents i_u are equal, based on the assumption of identical diodes. All four diodes conduct during the commutation interval, and therefore, $v_d = 0$. In terms of these mesh currents, we can express diode currents and the line current i_s during the commutation interval as,

$$i_{D1} = i_{D2} = i_u$$

$$i_{D3} = i_{D4} = I_d - i_u$$

and

$$i_s = -I_d + 2i_u$$

Where i_u builds up from zero at the beginning to I_d at the end of the commutation interval. Therefore, at w$t = u$, $i_{D1} = i_{D2} = I_d$ and $i_s = I_d$. During this commutation of current from diodes 3 and 4 to diodes 1 and 2, the current through inductor L_s changes from $-I_d$ to I_d.

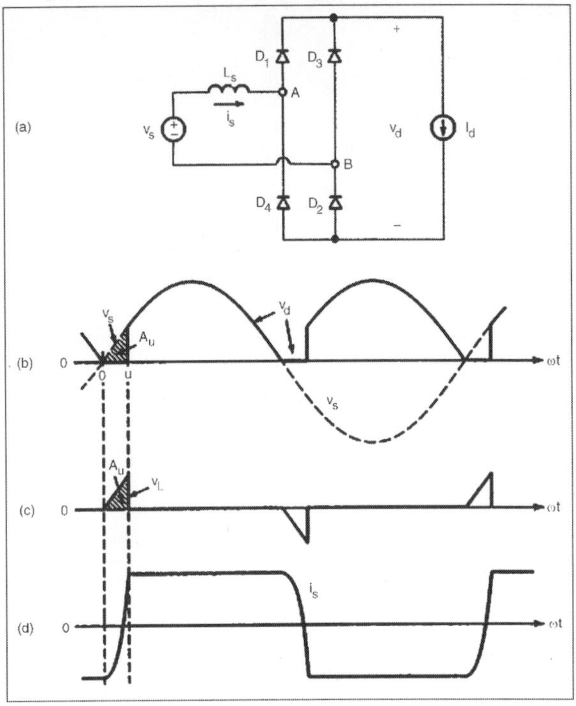

Fig. (a) Single-phase diode rectifier with L_s. (b) Waveforms.

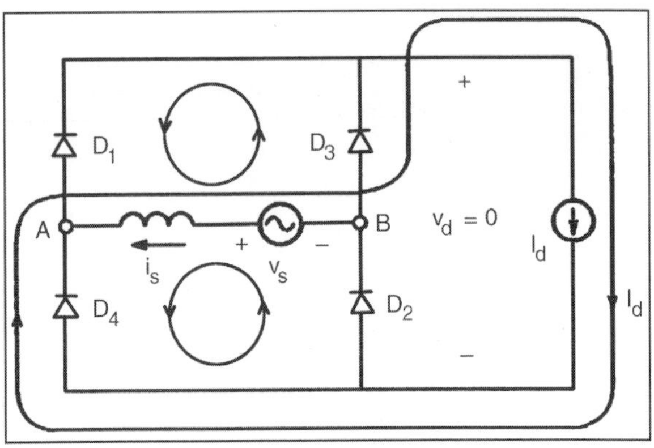

Fig. Redrawn circuit

Following the analysis previously carried out on the hypothetical circuit, volt-radian area Au in the waveforms can be written from Eqs. above as,

$$volt-radian\ area\ A_u = \int_0^u \sqrt{2}V_s \sin \omega t\, d(\omega t) = \omega L_s \int_{-I_d}^{I_d} di_s = 2\omega L_s I_d$$

Where the lower limit of integration now is $i_s(0) = -I_d$. Therefore,

$$A_u = \sqrt{2}V_s(1-\cos u) = 2\omega L_s I_d$$

and

$$\cos u = 1 - \frac{2\omega L_s}{\sqrt{2}V_s} I_d$$

A similar commutation takes place one-half cycle later when i_s, goes from I_d to $-I_d$. In this circuit, the average value of v_d in the idealised case (with $L_s = 0$) was calculated in Equation above as $V_{do} = 0.9V_s$. Therefore in the presence of L_s, the average value V_d can be calculated, following the procedure outlined previously by Eqs. above. Alternatively, we can calculate V_d by inspecting, where compared to the idealised case, the area A_u is "lost" every half-cycle from the integral of voltage V_d. Therefore,

$$V_d = V_{do} - \frac{area\ A_u}{\pi} = 0.9V_s - \frac{2\omega L_s I_d}{\pi}$$

Constant DC-side Voltage $v_d(t) = V_d$

Next, we will consider the circuit, where the assumption is that the dc-side voltage is constant. It is an approximation to the circuit with a large value of *C*. Another assumption here is that the circuit conditions are such that the current *Id* is zero during the zero crossing of v_s, as shown by the waveforms. Under these conditions, the equivalent circuit is drawn. Consider the waveforms. When v_s exceeds V_d at θ_b, diodes 1 and 2 begin to conduct. The current reaches its peak at θ_p, beyond which v_L becomes negative. The current becomes zero at θ_f when the volt-second areas *A* and *B* become equal and negative of each other. The current remains zero until $\pi + \theta_b$. With a given value of V_d, the average value I_d of the dc current can be calculated by the following procedure:

1. The angle θ_b can be calculated from the equation,

$$V_d = \sqrt{2}V_s \sin\theta_b$$

2. The inductor voltage starts at zero at θ_b, and becomes zero at θ_p prior to becoming negative. From symmetry,

$$\theta_p = \pi - \theta_b$$

3. When the current is flowing, the inductor voltage V_L is given by,

$$v_L = L_s \frac{di_d}{dt} = \sqrt{s}V_s \sin(\omega t) - V_d$$

And its integral with respect to w*t* can be written as,

$$\omega L_s \int_{\theta_b}^{\theta} di_d = \int_{\theta_b}^{\theta} (\sqrt{2}V_s \sin\omega t - V_d) d(\omega t)$$

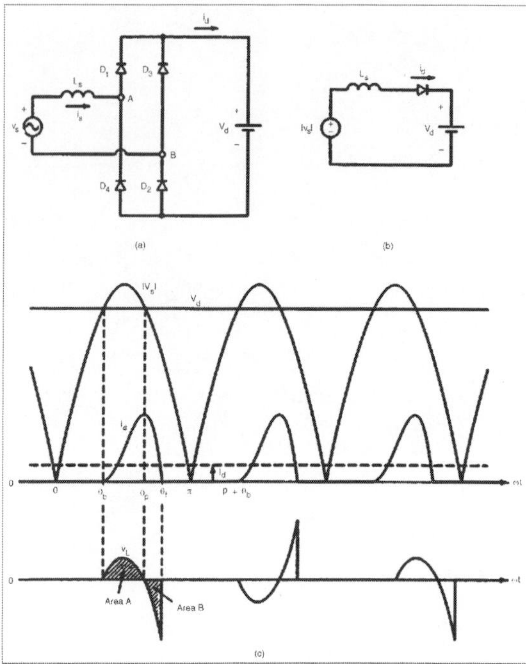

Fig. (a) Rectifier with a constant dc-side voltage.
(b) Equivalent circuit. (c) Waveforms.

Where q > q_b. Recognising that i_d at q_b is zero, Equation above results in,

$$i_d(\theta) = \frac{1}{\omega L_s}\int_{\theta_b}^{\theta}\left(\sqrt{2}V_s \sin\omega t - V_d\right)d\omega t$$

4. The angle θ_f at which i_d goes to zero can be obtained from Equation above as,

$$0 = \int_{\theta_b}^{\theta_f}\left(\sqrt{2}V_s \sin\omega t - V_d\right)d(\omega t)$$

It corresponds to area $A = B$.

5. The average value I_d Of the dc current can be obtained by integrating $i_d(\theta)$ from θ_b to θ_f and then dividing by π:

$$I_d = \frac{\int_{\theta_b}^{\theta_f} i_d(\theta)\,d\theta}{\pi}$$

It is intuitively obvious that for given circuit parameters, I_d will depend on the value of V_d and vice versa. To present the relationship between the two in a general manner, we will normalise V_d by V_{do}, and I_d by $l_{\text{short circuit}}$, where,

$$I_{\text{short circuit}} = \frac{V_s}{\omega L_s}$$

Is the rms current that will flow if the ac voltage source v_s was short circuited through L_s. Following the above procedure results in the plot, where the current reaches zero as *Vd* approaches the peak value of the ac input voltage.

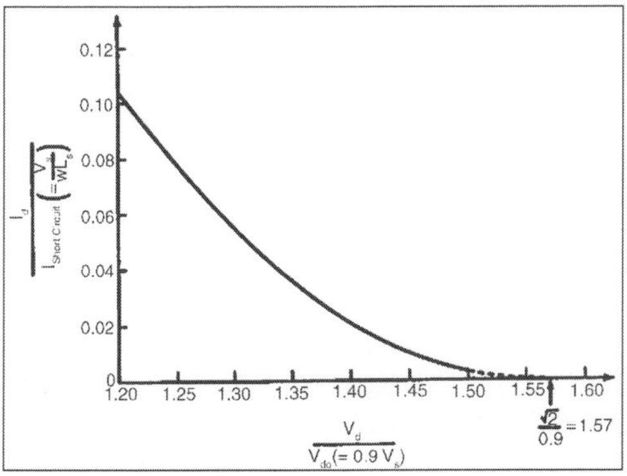

Fig. Normalised I_d versus V_d in the rectifier.

CHARACTERISTIC OF RECTIFIER

The approximation of a constant dc voltage $v_d(t) = V_d$ may be reasonable if the capacitance in the practical circuit is large. That is, if the load were replaced by an equivalent resistance R_{load}, the time constant $C_d R_{\text{load}}$ is much larger than the line-frequency cycle time, resulting in a very small ripple in v_d. This approximation allows us to present the characteristics of the rectifier in a general fashion. The various quantities are plotted as a function of the dc current I_d, normalised with respect to the short-circuit current $I_{\text{short circuit}}$ in order to combine the effects of L_s and frequency ω in the same plots.

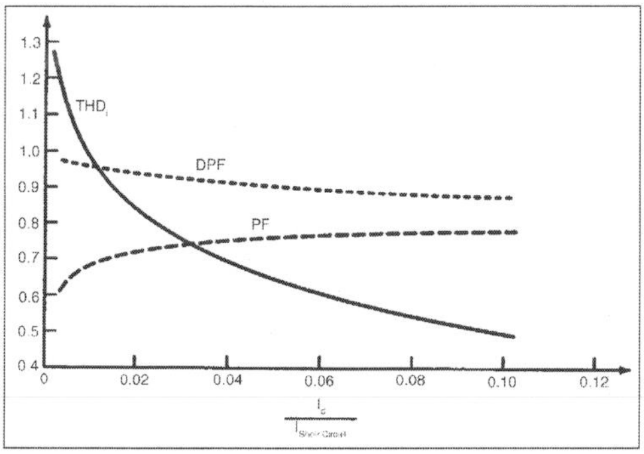

Fig. Total harmonic distortion, DPF and PF in the rectifier.

For a given value of l_d, increasing L_s results in a smaller $I_{\text{short circuit}}$ and hence a larger $I_d/l_{\text{short circuit}}$. Therefore, that increasing L_s results in improved i_s waveform with a lower THD, a better power factor, and a lower (improved) crest factor.

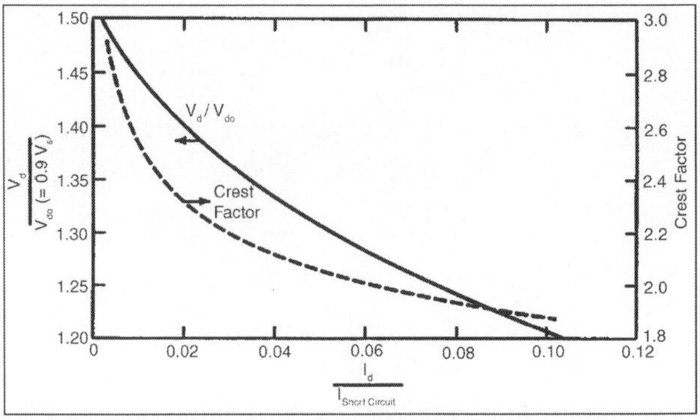

Fig. Normalised V_d and the crest factor in the rectifier.

PRACTICAL DIODE BRIDGE RECTIFIERS

Having considered the simplified circuits we are now ready to consider the practical circuit. The load is represented by an equivalent resistance R_{load}. In this circuit, there will be some ripple in the capacitor voltage, and therefore, it must be analysed differently than the circuit.

A circuit can be easily analysed by a circuit simulation programme such as PSpice. However, for educational purposes we will first analytically calculate the circuit waveforms.

Analytical Calculations Under a Highly Discontinuous Current

The circuit operating conditions are assumed to result in a highly discontinuous i_d, similar to the waveform, where i_d goes to zero prior to the zero crossing of v_s every half-cycle. Then the equivalent circuit can be used to calculate the voltages and currents. If the above condition is not met, the current commutation must be included, which makes analytical calculations difficult.

In order to describe the system, the state variables chosen are the inductor current i_d and the capacitor voltage v_d. During each half-cycle of line frequency, there are two distinct intervals, similar to those shown earlier (where $t_b = \theta_b/\omega$ and $t_f = \theta f/\omega$):

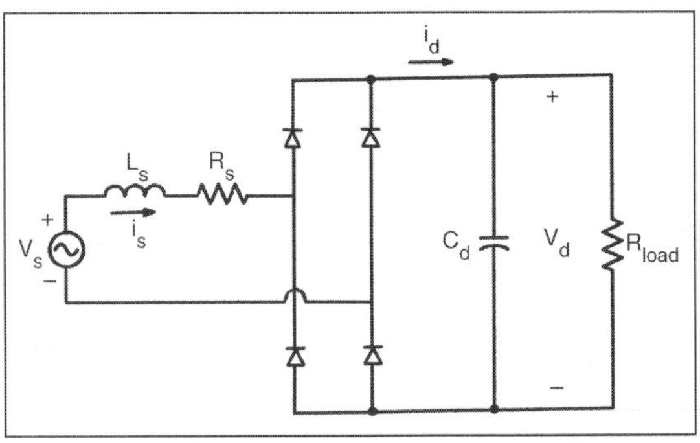

Fig. Practical diode-bridge rectifier with a filter capacitor.

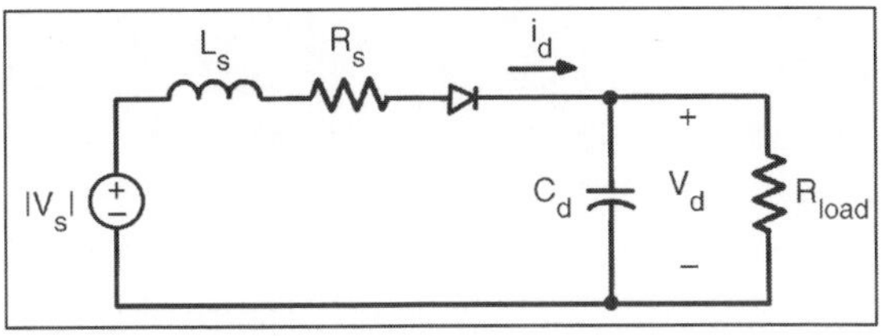

Fig. Equivalent circuit.

(a) $t_b < t < t_f$. When the current is flowing during $t_b < t < t_f$, where t_b is the beginning of conduction and t_f is the final conduction time, the following equations describe the circuit every half-cycle of line frequency:

$$|v_s| = R_s i_d + L_s \frac{di_d}{dt} + v_d \quad (using\ KVL)$$

and

$$i_d = C_d \frac{dv_d}{dt} + \frac{v_d}{R_{load}} \quad (using\ KCL)$$

where KVL and KCL are the Kirchhoff voltage and current laws. Rearranging the above equations in the state variable form during $t_b < t < t_f$ yields

$$\begin{bmatrix} \frac{di_d}{dt} \\ \frac{dv}{dt} \end{bmatrix} = \begin{bmatrix} -\frac{R_s}{L_s} & -\frac{1}{L_s} \\ \frac{1}{C_d} & -\frac{1}{C_d R_{load}} \end{bmatrix} \cdot \begin{bmatrix} i_d \\ v_d \end{bmatrix} + \begin{bmatrix} 1/L_s \\ 0 \end{bmatrix} \cdot |v_s|$$

The state variable vector x consists of i_d and v_d. The state transition matrix is,

$$A = \begin{bmatrix} -\frac{R_s}{L_s} & -\frac{1}{L_s} \\ \frac{1}{C_d} & -\frac{1}{C_d R_{load}} \end{bmatrix}$$

and

$$b = \begin{bmatrix} 1/L_s \\ 0 \end{bmatrix}$$

Using Eqs. for the trapezoidal rule of integration yields,

$$x(t) = Mx(t - \Delta t) + N\left[|v_s(t)| - |v_s(t - \Delta t)|\right]$$

Where,

$$x(t) = Mx(t - \Delta t) + N\left[|v_s(t)| - |v_s(t - \Delta t)|\right]$$

4(b) $t_f < t < t_b + \frac{1}{2}T$. During the interval if $t_f < t < t_b + \frac{1}{2}T$, when the diode bridge is not conducting, $i_d = 0$

and

$$\frac{dv_d}{dt} = -\frac{1}{C_d R_{load}} v_d$$

The solution to Equation above can be expressed as,

$$v_d(t) = v_d(t_f) e^{-(t-t_f)/(C_d R_{load})}$$

In the solution of Eqs. above, we need the time instant tb at which the current conduction starts. Since it is not known prior to the solution, we will use an estimated value. An exact value of t_b will result in the beginning of current conduction exactly one-half cycle later. Therefore, we will use this condition to check for the accuracy of our choice of t_b and slowly increment tb from its initially chosen value until we reach the exact value within a small tolerance.

Line-Voltage Distortion

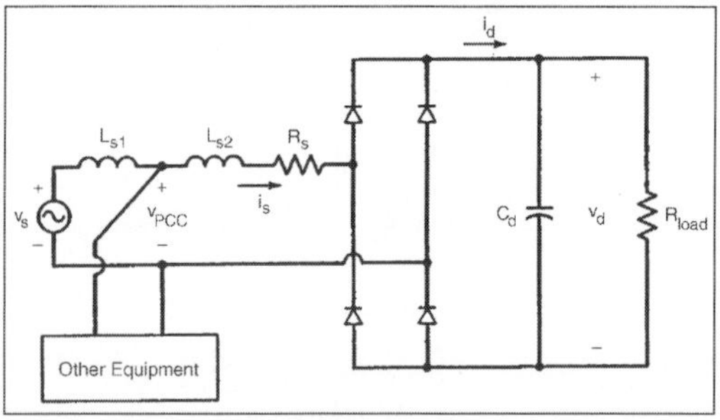

Fig. Line-voltage notching and distortion.

Distorted currents drawn by loads such as the diode bridge rectifiers can result in distortion in the utility voltage waveform. For example, consider the circuit. Here L_{s1} represents the internal impedance of the utility source and L_{s2} may be intentionally added as a part of the power electronics equipment. A resistance R_s is included that also can be used to represent the diode resistances.

The voltage across other equipment at the point of common coupling (PCC) is,

$$v_{PCC} = v_s - L_{s1}\frac{di_s}{dt}$$

Where v_s is assumed to be sinusoidal.

Expressing is in Equation above in terms of its fundamental and harmonic components yields,

$$v_{PCC} = \left(v_s - L_{s1}\frac{di_{s1}}{dt}\right) - L_{s1}\sum_{h\neq 1}\frac{di_{sh}}{dt}$$

Where,

$$(v_{PCC})_1 = v_s - L_{s1}\frac{di_{s1}}{dt}$$

And the voltage distortion component due to the harmonics is,

$$(v_{PCC})_{di_s} = -L_{s1}\sum_{h\neq 1}\frac{di_{sh}}{dt}$$

THREE-PHASE, FULL-BRIDGE RECTIFIERS

In industrial applications where three-phase ac voltages are available, it is preferable to use three-phase rectifier circuits, compared to single-phase rectifiers, because of their lower ripple content in the waveforms and a higher power-handling capability. The three-phase, six-pulse, full-bridge diode rectifier is a commonly used circuit arrangement. A filter capacitor is connected at the dc side of the rectifier.

Similar to the analysis of single-phase, full-bridge rectifiers, we will begin with simplified circuits prior to the discussion of the circuit.

Idealised Circuit With L_S=0

In the circuit, the ac-side inductance L_s is assumed to be zero and the dc side is replaced by a constant dc current I_d. We will see later that replacing the dc current I_d by a load resistance R_{load} makes little difference in the circuit operation.

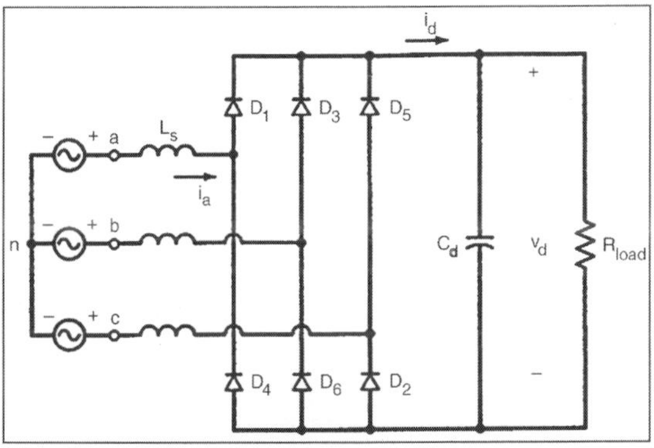

Fig. Three-phase, full-bridge rectifier.

With L_s = 0, the current I_d flows through one diode from the top group and one from the bottom group.

Similar to the discussion in the case of single-phase rectifiers, in the top group, the diode with its anode at the highest potential with conduct and the other two become reversed biased. In the bottom group, the diode with its cathode at the lowest potential will conduct and the other two become reverse biased.

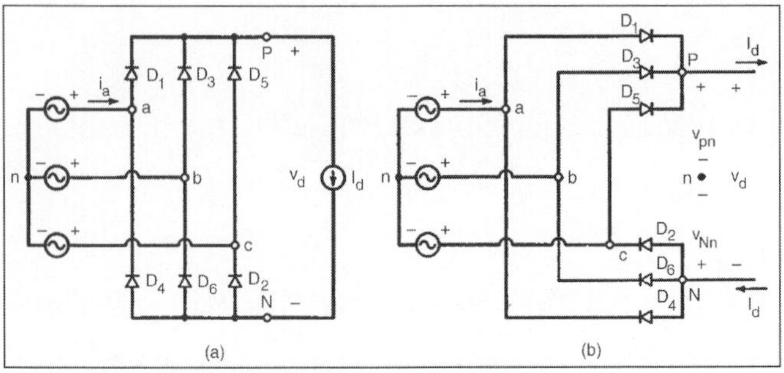

Fig. Three-phase rectifier with a constant dc current.

The voltage waveforms in the circuits, where v_{Pn} is the voltage at the point P with respect to the ac voltage neutral point n. Similarly, v_{Nn} is the voltage at the negative dc terminal N. Since I_d flows continuously, at any time, v_{Pn} and v_{Nn} can be obtained in terms of one of the ac input voltages v_{an}, v_{bn}, and v_{cn}. Applying KVL in the circuit on an instantaneous basis, the dc-side voltage is,

$$v_d = v_{Pn} - v_{Nn}$$

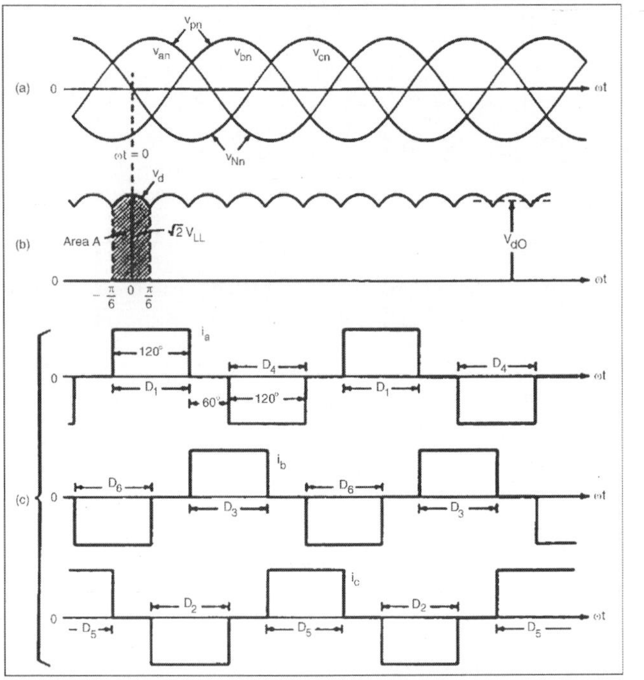

Fig. Waveforms in the circuit.

The instantaneous waveform of v_d consists of six segments per cycle of line frequency. Hence, this rectifier is often termed a six-pulse rectifier. Each segment belongs to one of the six line-to-line voltage combinations. Each diode conducts for 120°. Considering the phase a current waveform,

$$i_a = \begin{cases} I_d & \text{when diode 1 is conducting} \\ -I_d & \text{when diode 4 is conducting} \\ 0 & \text{when neither diode 1 or 4 is conducting} \end{cases}$$

The commutation of current from one diode to the next is instantaneous, based on the assumption of $L_s = 0$.

The diodes are numbered in such a way that they conduct in the sequence 1, 2, 3... Next, we will compute the average value of the output dc voltage and rms values of the line currents, where the subscript o is added due to the assumption of $L_s = 0$.

To obtain the average value of the output de voltage, it is sufficient to consider only one of the six segments and obtain its average over a 60° or $\pi/3$-rad interval. Arbitrarily, the time origin $t = 0$ is chosen when thc line-to-line voltage v_{ab} is at its maximum. Therefore,

$$v_d = v_{ab} = \sqrt{2}V_{LL}\cos\omega t \qquad -\frac{1}{6}\pi < \omega t < \frac{1}{6}\pi$$

where V_{LL} is the rms value of line-to-line voltages. By integrating v_{ab}, the volt-second area A is given by,

$$A = \int_{-\pi/6}^{\pi/6} \sqrt{2}V_{LL}\cos\omega t\, d(\omega t) = \sqrt{2}V_{LL}$$

and therefore dividing A by the $\pi/3$ interval yields,

$$V_{do} = \frac{1}{\pi/3}\int_{-\pi/6}^{\pi/6} \sqrt{2}V_{LL}\cos\omega t\, d(\omega t) = \frac{3}{\pi}\sqrt{2}V_{LL} = 1.35V_{LL}$$

One of the phase voltages and the corresponding phase current (labeled v_s and i_s) are redrawn. Using the definition of rms current in the phase current waveform, the rms value of the line current i_s in this idealised case is,

$$I_s = \sqrt{\frac{2}{3}}I_d = 0.816I_d$$

By means of Fourier analysis of i_s in this idealised case, the fundamental-frequency component i_{s1} has an rms value,

$$I_{s1} = \frac{1}{\pi}\sqrt{6}I_d = 0.78I_d$$

The harmonic components I_{sh} can be expressed in terms of the fundamental-frequency component as,

$$I_{sh} = I_{s1}/h$$

where $h = 5, 7, 11, 13....$ The even and triplen harmonics are zero. Since i_{s1} is in phase with its utility phase voltage,

$$\text{DPF} = 1.0$$

and therefore,

$$PF = \frac{3}{\pi} = 0.955$$

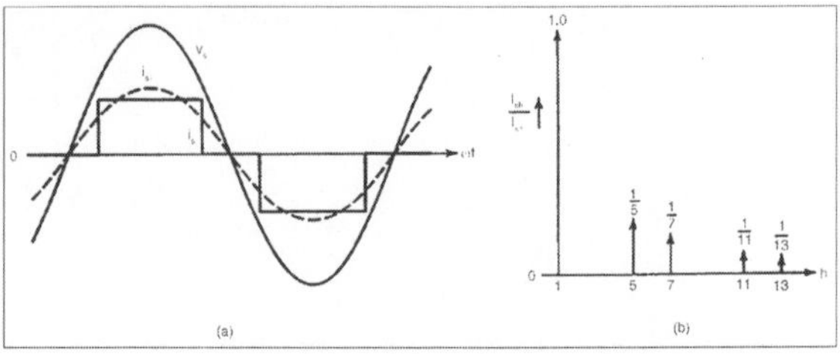

Fig. Line current in a three-phase rectifier in the idealised case with $L_s = 0$ and a constant dc current

The voltage waveforms will be identical if the load on the dc side is represented by a resistance R_{load} instead of a current source I_d. The phase currents will also flow during identical intervals. The only difference will be that the current waveforms will not have a flat top.

Effect of L_s on Current Commutation

Next, we will include L_s on the ac side and represent the dc side by a current source $i_d = I_d$. Now the current commutations will not be instantaneous. We will look at only one of the current commutations because all others are identical in a balanced circuit. Consider the commutation of current from diode 5 to diode 1, beginning at t or $\omega t = 0$ (the time origin is chosen arbitrarily). Prior to this, the current i_d is flowing through diodes 5 and 6. The subcircuit pertinent to this current commutation.

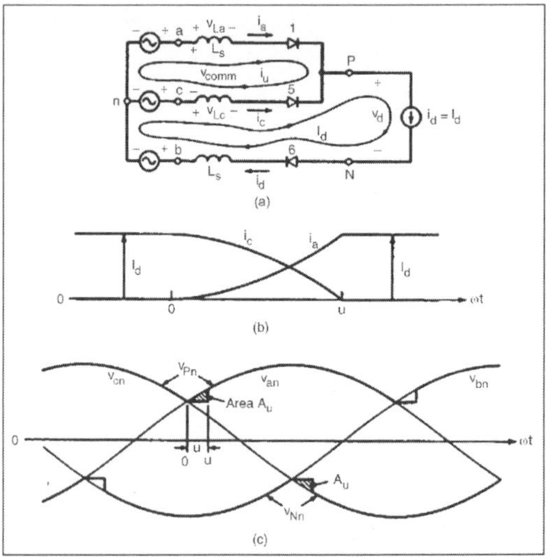

Fig. Three-phase rectifier with a finite L_s and a constant dc current, Current commutation process.

The current commutation only involves phases a and c, and the commutation voltage responsible is $v_{comm} = v_{an} - v_{cn}$. The two mesh currents i_u and I_d are labeled. The commutation current i_u flows due to a short-circuit path provided by the conducting diode 5. In terms of the mesh currents, the phase currents are,

$$i_a = i_u$$

and

$$i_c = I_d - i_u$$

These are plotted, where i_u builds up from zero to I_d at the end of the commutation interval $\omega t_u = u$. In the circuit,

$$v_{La} = L_s \frac{di_a}{dt} = L_s di_u / dt$$

and

$$v_{Lc} = L_s \frac{di_c}{dt} = -L_s \frac{di_u}{dt}$$

Noting that $i_c = I_d - i_u$ and therefore $di_c/dt = d(I_d - i_u)/dt = -di_u/dt$. Applying KVL in the upper loop in the circuit and using the above equations yield,

$$v_{comm} = v_{an} - v_{cn} = v_{La} - v_{Lc} = 2L_s \frac{di_u}{dt}$$

Therefore from the above equation,

$$L_s \frac{di_u}{dt} = \frac{v_{an} - v_{cn}}{2}$$

The commutation interval u can be obtained by multiplying both sides of Equation 5-78 by ω and integrating:

$$\omega L_s \int_0^{I_d} di_u = \int_0^u \frac{v_{an} - v_{cn}}{2} d(\omega t)$$

where the time origin is assumed to be at the beginning of the current commutation. With this choice of time origin, we can express the line-to-line voltage $(v_{an} - v_{cn})$ as,

$$v_{an} - v_{cn} = \sqrt{2} V_{LL} \sin \omega t$$

Using Equations yields,

$$\omega L_s \int_0^{I_d} di_u = \omega L_s I_d = \frac{\sqrt{2} V_{LL}(1 - \cos u)}{2}$$

or,

$$\cos u = 1 - \frac{2\omega L_s I_d}{\sqrt{2} V_{LL}}$$

If the current commutation was instantaneous due to zero L_s, then the voltage v_{Pn} will be equal to v_{an} beginning with $\omega t = 0$. However, with a finite L_s, during $0 < \omega t < \omega t_u$

$$v_{Pn} = v_{an} - L_s \frac{di_u}{dt} = \frac{v_{an} + v_{cn}}{2}$$

Where the voltage across $L_s[= L_s (di_u/dt)]$ is the drop in the voltage v_{Pn} during the commutation interval. The integral of this voltage drop is the area A_u, which according to Equation 5-81 is,

$$A_u = \omega L_s I_d$$

This area is "lost" every 60° (π/3 rad) interval. Therefore, the average dc voltage output is reduced from its V_{do} value, and the voltage drop due to commutation is,

$$\Delta V_d = \frac{\omega L_s I_d}{\pi / 3} = \frac{3}{\pi} \omega L_s I_d$$

Therefore, the average dc voltage in the presence of a finite commutation interval is,

$$V_d = V_{do} - \Delta V_d = 1.35 V_{LL} - \frac{3}{\pi} \omega L_s I_d$$

where V_{do} is the average voltage with an instantaneous commutation due to $L_s = 0$, given by Equation above.

Distortion in the Line-Current Waveforms

It is useful to know the power factor, total harmonic distortion, and the dc output voltage in the practical circuit. However, to present these results in a generalised manner requires the assumption of a constant dc output voltage.

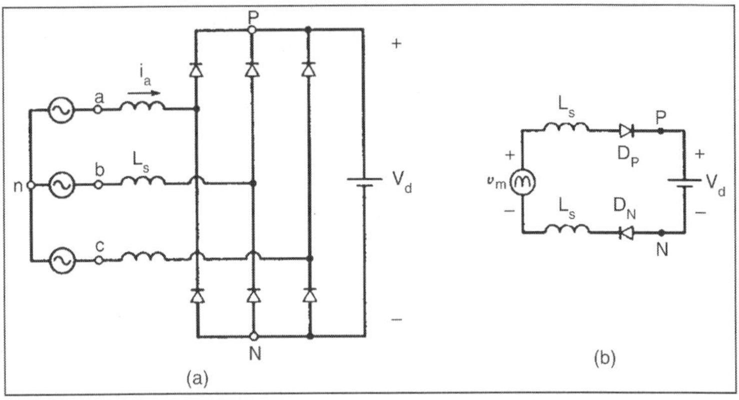

The dc-side current I_d is normalised by the per-phase short-circuit curtent. This per-phase short-circuit current can be obtained in terms of the line-to-line input ac voltages:

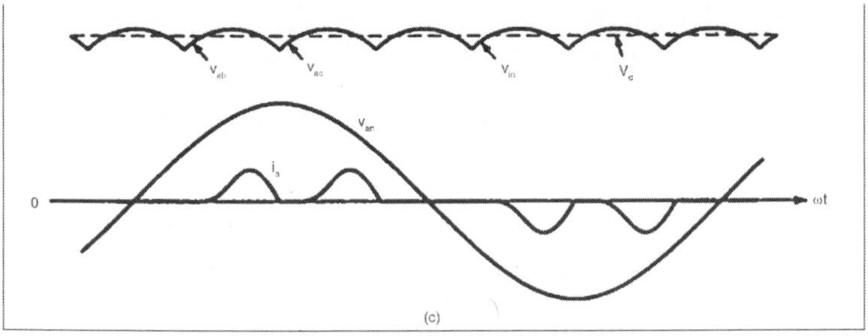

Fig. (a) Three-phase rectifier with a finite L_s and a constant dc voltage. (b) Equivalent circuit. (c) Waveforms.

$$I_{\text{short circuit}} = \frac{V_{LL} / \sqrt{3}}{\omega_1 L_s}$$

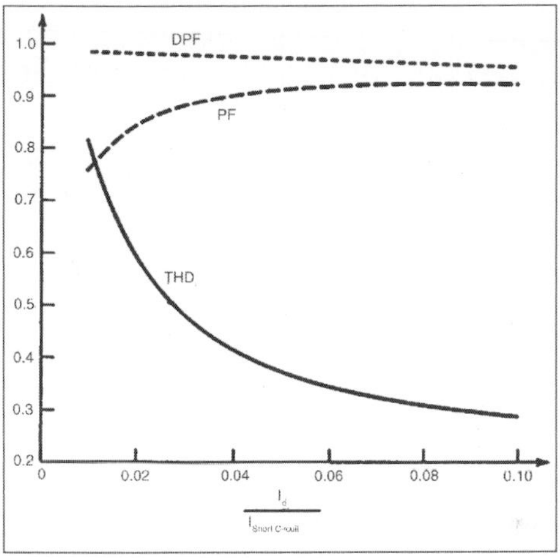

Fig. Total harmonic distortion, DPF, and PF in the rectifier.

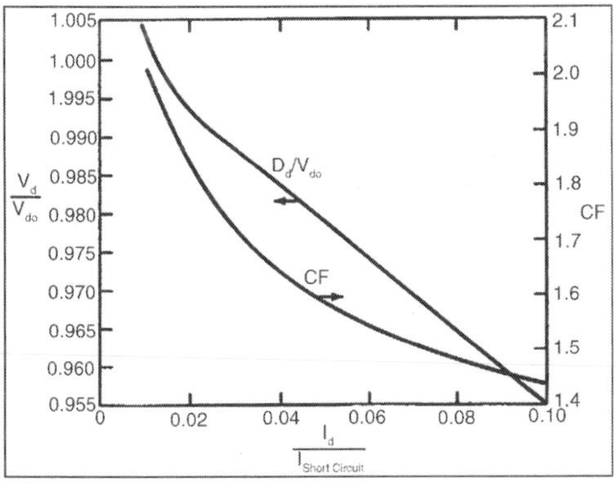

Fig. Normalised Vd and crest factor in the rectifier.

The plots of PF, DPF, THD, crest factor (*CF*), and V_d/V_{do} as a function of I_d normalised by $I_{short\ circuit}$.

COMPARISON OF SINGLE-PHASE AND THREE-PHASE RECTIFIERS

The line current in a single-phase rectifier contains significantly more distortion compared to a three-phase rectifier. This results in a much poorer power factor in a single-phase rectifier compared to a three-phase rectifier. This is confirmed by comparing for single-phase rectifiers with the results for three-phase rectifiers. The displacement power factor (cos ϕ_1) is high in both rectifiers.

The ripple in the dc cucrent is smaller in a three-phase rectifier in comparison to a single-phase rectifier. The ripple current, which flows through the filter capacitor, dictates the capacitance and the current-handling capability required of the filter capacitor. Therefore, in some applications, the filter capacitance required may be much smaller in a three-phase rectifier compared to a single-phase rectifier.

In a three-phase rectifier, the maximum regulation in the dc voltage V_d from no-load to a full-load condition will generally be less than 5 per cent. This regulation is often much larger in single-phase rectifiers.

Based on the foregoing discussion and the fact that the use of single-phase rectifiers in three-phase, four-wire systems introduces large currents in the neutral (even in a balanced system), it is always preferable to use a three-phase rectifier over a single-phase rectifier.

POWER DIODES IN THE FULL WAVE RECTIFIER

Power Diodes tutorial we discussed ways of reducing the ripple or voltage variations on a direct DC voltage by connecting capacitors across the load resistance. While this method may be suitable for low power applications it is unsuitable to applications which need a "steady and smooth" DC supply voltage. One method to improve on this is to use every half-cycle of the input voltage instead of every other half-cycle. The circuit which allows us to do this is called a Full Wave Rectifier.

Like the half wave circuit, a full wave rectifier circuit produces an output voltage or current which is purely DC or has some specified DC component. Full wave rectifiers have some fundamental advantages over their half wave rectifier counterparts. The average (DC) output voltage is higher than for half wave, the output of the full wave rectifier has much less ripple than that of the half wave rectifier producing a smoother output waveform.

In a Full Wave Rectifier circuit two diodes are now used, one for each half of the cycle. A multiple winding transformer is used whose secondary winding is split equally into two halves with a common centre tapped connection, (*C*). This configuration results in each diode conducting in turn when its anode terminal is positive with respect to the transformer centre point *C* producing an output during both half-cycles, twice that for the half wave rectifier so it is 100 per cent efficient.

FULL WAVE RECTIFIER CIRCUIT

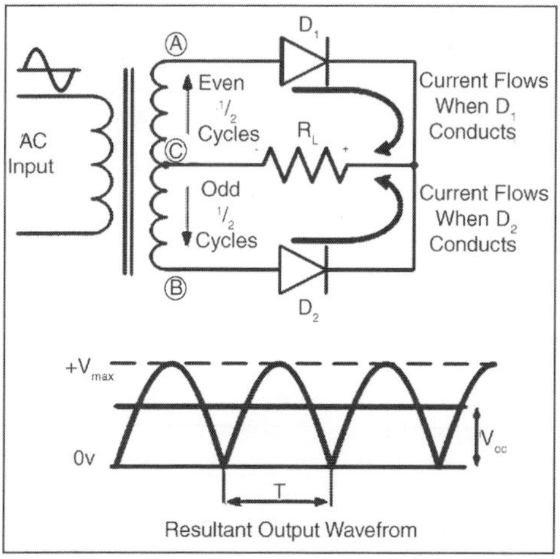

The full wave rectifier circuit consists of two *power diodes* connected to a single load resistance (R_L) with each diode taking it in turn to supply current to the load. When point A of the transformer is positive with respect to point C, diode D_1 conducts in the forward direction as indicated by the arrows. When point B is positive (in the negative half of the cycle) with respect to point C, diode D_2 conducts in the forward direction and the current flowing through resistor R is in the same direction for both half-cycles. As the output voltage across the resistor R is the phasor sum of the two waveforms combined, this type of full wave rectifier circuit is also known as a "bi-phase" circuit.

As the spaces between each half-wave developed by each diode is now being filled in by the other diode the average DC output voltage across the load resistor is now double that of the single half-wave rectifier circuit and is about 0.637 V_{max} of the peak voltage, assuming no losses.

$$V_{d.c.} = \frac{2V_{max}}{\pi} = 0.637V_{max} = 09V_{RMS}$$

Where: V_{MAX} is the maximum peak value in one half of the secondary winding and V_{RMS} is the rms value.

The peak voltage of the output waveform is the same as before for the half-wave rectifier provided each half of the transformer windings have the same rms voltage value. To obtain a different DC voltage output different transformer ratios can be used. The main disadvantage of this type of full wave rectifier circuit is that a larger transformer for a given power output is required with two separate but identical secondary windings making this type of full wave rectifying circuit costly compared to the "Full Wave Bridge Rectifier" circuit equivalent.

THE FULL WAVE BRIDGE RECTIFIER

Another type of circuit that produces the same output waveform as the full wave rectifier circuit above, is that of the Full Wave Bridge Rectifier. This type of single phase rectifier uses four individual rectifying diodes connected in a closed loop "bridge" configuration to produce the desired output. The main advantage of this bridge circuit is that it does not require a special centre tapped transformer, thereby reducing its size and cost. The single secondary winding is connected to one side of the diode bridge network and the load to the other side as shown below.

The Diode Bridge Rectifier

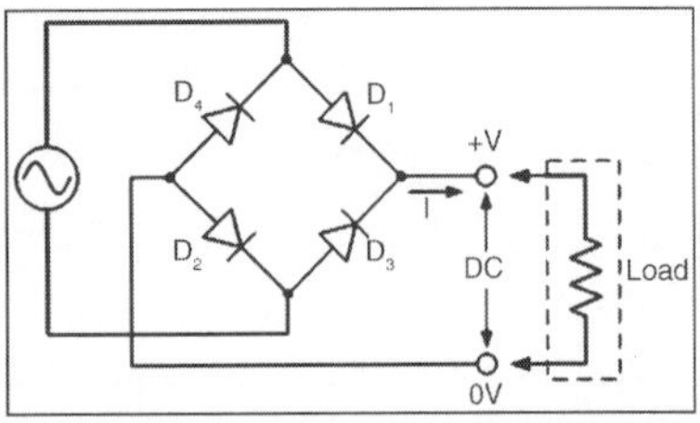

The four diodes labelled D_1 to D_4 are arranged in "series pairs" with only two diodes conducting current during each half cycle. During the positive half cycle of the supply, diodes D_1 and D_2 conduct in series while diodes D_3 and D_4 are reverse biased and the current flows.

The Positive Half-cycle

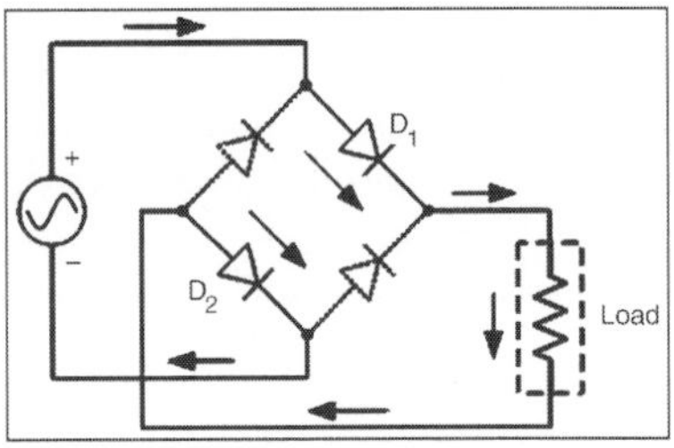

During the negative half cycle of the supply, diodes $D3$ and $D4$ conduct in series, but diodes $D1$ and $D2$ switch "OFF" as they are now reverse biased. The current flowing through the load is the same direction as before.

The Negative Half-cycle

As the current flowing through the load is unidirectional, so the voltage developed across the load is also unidirectional the same as for the previous two diode full-wave rectifier, therefore the average DC voltage across the load is $0.637V_{max}$.

However in reality, during each half cycle the current flows through two diodes instead of just one so the amplitude of the output voltage is two voltage drops ($2 \times 0.7 = 1.4V$) less than the input V_{MAX} amplitude. The ripple frequency is now twice the supply frequency (*e.g.,* 100Hz for a 50Hz supply)

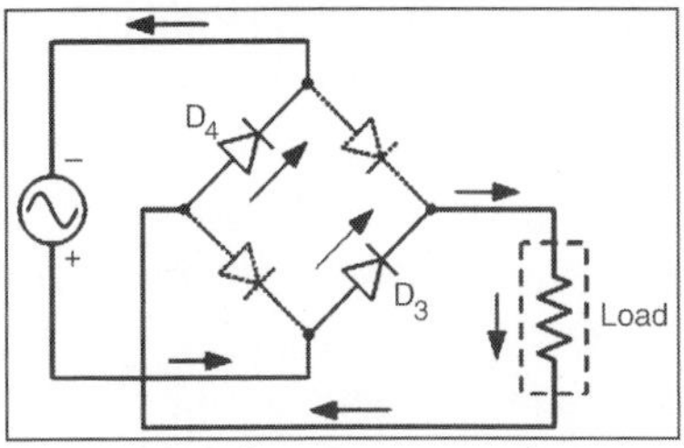

Although we can use four individual power diodes to make a full wave bridge rectifier, pre-made bridge rectifier components are available "off-the-shelf" in a range of different voltage and current sizes that can be soldered directly into a PCB circuit board or be connected by spade connectors.

The image to the right shows a typical single phase bridge rectifier with one corner cut off. This cut-off corner indicates that the terminal nearest to the corner is the positive or +ve output terminal or lead with the opposite (diagonal) lead being the negative or -ve output lead. The other two connecting leads are for the input alternating voltage from a transformer secondary winding.

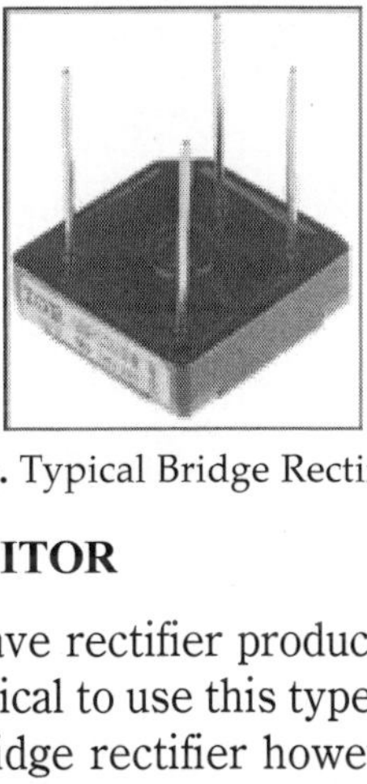

Fig. Typical Bridge Rectifier

THE SMOOTHING CAPACITOR

The single phase half-wave rectifier produces an output wave every half cycle and that it was not practical to use this type of circuit to produce a steady DC supply. The full-wave bridge rectifier however, gives us a greater mean DC value (0.637 V_{max}) with less superimposed ripple while the output waveform is twice that of the frequency of the input supply frequency. We can therefore increase its average DC output level even higher by connecting a suitable smoothing capacitor across the output of the bridge circuit as shown below.

Full-wave Rectifier with Smoothing Capacitor

The smoothing capacitor converts the full-wave rippled output of the rectifier into a smooth DC output voltage. Generally for DC power supply circuits the smoothing capacitor is an Aluminium Electrolytic type that has a capacitance value of 100uF or more with repeated DC voltage pulses from the rectifier charging up the capacitor to peak voltage. However, their are two important parameters to consider when choosing a suitable smoothing capacitor and these are its Working Voltage, which must be higher than the no-load output value of the rectifier and its Capacitance Value, which determines the amount of ripple that will appear superimposed on top of the DC voltage. Too low a capacitance value and the capacitor has little effect on the output waveform. But if the smoothing capacitor is sufficiently large enough (parallel capacitors can be used) and the load current is not too large, the output voltage will be almost as smooth as pure DC. As a general rule of thumb, we are looking to have a ripple voltage of less than 100mV peak to peak.

The maximum ripple voltage present for a Full Wave Rectifier circuit is not only determined by the value of the smoothing capacitor but by the frequency and load current, and is calculated as:

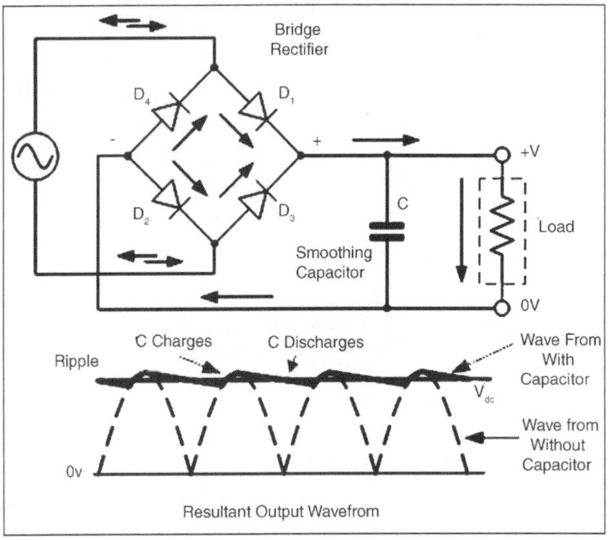

Bridge Rectifier Ripple Voltage

$$V_{RIPPLE} = \frac{I_{LOAD}}{fC} \text{ volts}$$

Where: I is the DC load current in amps, f is the frequency of the ripple or twice the input frequency in Hertz, and C is the capacitance in Farads.

The main advantages of a full-wave bridge rectifier is that it has a smaller AC ripple value for a given load and a smaller reservoir or smoothing capacitor than an equivalent half-wave rectifier. Therefore, the fundamental frequency of the ripple voltage is twice that of the AC supply frequency (100Hz) where for the half-wave rectifier it is exactly equal to the supply frequency (50Hz).

The amount of ripple voltage that is superimposed on top of the DC supply voltage by the diodes can be virtually eliminated by adding a much improved p-filter (pi-filter) to the output terminals of the bridge rectifier. This type of low-pass filter consists of two smoothing capacitors, usually of the same value and a choke or inductance across them to introduce a high impedance path to the alternating ripple component.

Another more practical and cheaper alternative is to use an off the shelf 3-terminal voltage regulator IC, such as a LM78xx (where "xx" stands for the output voltage rating) for a positive output voltage or its inverse equivalent the LM79xx for a negative output voltage which can reduce the ripple by more than 70dB (Datasheet) while delivering a constant output current of over 1 amp. In the next tutorial about diodes, we will look at the Zener Diode which takes advantage of its reverse breakdown voltage characteristic to produce a constant and fixed output voltage across itself.

5

Network Circuits Theory

NETWORK

DRM IS PASSIVE AND ACTIVE

Passive DRM protects its content from onlookers who do not have a DRM-enabled client. Encryption is generally used for Passive DRM, so that the content is meaningless garbage unless you have the right bits in your client. I consider this 'passive' protection, because the data is inaccessible by default and only becomes accessible if you have the right kind of client, with the right key.

Active DRM, then, would be a scheme where protection is only provided if the client in use is one that is correctly coded to block access where it has not been specifically granted. This is a scheme in which the data is readily accessible to most normal viewers/players, but has a special code that tells a DRM-enabled viewer/player to hide the content from people who haven't been approved.

The whole problem is his two categories are a false distinction. You can't arbitrarily draw a line through a system and say "this is passive, this is active." For your CSS example, if you consider a given player's decryption code along with an arbitrary encrypted DVD, you have a system with both active and passive elements. If you leave out either of those elements, you have a disc that won't play or a player with no disc, the only perfectly secure system (assuming your cryptography is good.) When judging the efficiency of new compression schemes, the size of the decoder is added to the size of the compressed data to get a fair assessment of its efficiency. Otherwise you could win contests with a one-byte file and a 10 GB decoder programme that simply contains all the actual data.

Whichever way you design a system, complexity is being pushed from one party to another but never eliminated. For DVD, where most of the complexity is in the player, there is a huge variety of player implementations that each have their own bugs. The author of every disc needs to test against many combinations of players because of that problem.

Likewise, if you push the complexity onto the disc by including executable code there, the player gets simpler but the disc could be buggy. However, in that case, the content author will get a bad reputation for the buggy disc.

This doesn't just apply to DRM. While he might consider a MPEG4-AVC video file as "passive" in his terminology, it is really a complex series of instructions to the decoder. Look at the number of different but valid ways to encode video and you'll see it's closer to a programme than to "passive" data.

Now in his definition for "Active DRM", he is not actually describing the general class of software protection techniques. He is describing a system that is poorly-designed, often due to an attempt to retrofit DRM onto an existing system without it. Ofcourse it makes sense that if you have two ways to access the content, one with DRM and the other without, the additional complexity makes no sense to the end-user or mass copiers. It may make economic sense to the content author, but they have to weigh the potential risks to their business also (annoying users vs. stopping some casual copying.)

Even assuming his terminology makes sense, the Windows Media Centre system he references is actually a combination of "active" and "passive". The cable video stream is encrypted ("passive"), and the Windows DRM component is "active". In particular, it has a "black box" DLL that checks the host environment and hashes various items to derive a key, hence the problem.

All I can distill from what Alun says is "an unprotected system is made more complex by adding DRM." I agree, but this doesn't say anything larger about "active" versus "passive" DRM.

Full disclosure: I was previously one of the designers of the Blu-ray protection layer (BD+), a unique approach to disc protection that involves both cryptography and software protection. You can consider me biased, but my analysis should be able to stand on its own.

So, if you encrypt a file, and hand me the encrypted bits on a disk, that's what I would call "passive DRM", because its protection does not require me to be running the DRM client code. Only its unlocking requires me to run the DRM client code (or a reasonable facsimile thereof) and get a key.

That latter kind of DRM can only function in an environment where users can be prevented from _not_ running the DRM client code. This is why sometimes successful attempts by various content producers to pass legislation that prevents users from choosing what software to run – or not run – on their own machines.

Sure, you can say that there's a continuum between these two extremes – and that the fact that "Passive DRM" requires a key to be given to the client certainly makes it possible to hack – but I think the distinction is worth making.

So, now that I've further explained myself, do you agree with me that there is a valid distinction between the two types of protection? Are we simply quibbling about names?

- *Encrypted file + DRM client ("ProtectionX")*: DRM client attempts to verify environment before decrypting file. Assumes file is encrypted with standard symmetric crypto so that it can't be retrieved without at least one instance of DRM client.
- *Unprotected CD + autorun assassin software ("Assassin")*: CD is unprotected 16 bit samples readable by standard consumer grade equipment. In one specific environment and configuration (Windows with autorun), the assassin programme runs and attempts to interfere with any attempt to copy data.

  What I'm trying to tell you is that your example of a file with a flag that says "do not copy" or a disc with autorun assassin software that is otherwise unprotected is a special case. It's an attempt to retrofit some kind of client-side control onto a format that doesn't already have it, not a type of pre-defined DRM.

No one, especially content producers, asks for an unprotected format and assumes they'll bolt on something later. All you're seeing is the fact that some formats made it out of the gate without protection (CD) or were weak and eventually broken (DVD-CSS). In those cases, some people make the effort to add optional "assassin" software, even if it only slows down some small per centage of users.

You and I both agree that this is ineffective, easily bypassed, and can interfere with legitimate use. But some companies have analysed the cost/ benefit tradeoff and based on that information continue to deploy this approach rather than accept the alternative (do nothing).

Side note: The reason why this subject is particularly important to me is that some people are already using the concepts "active" and "passive" in terms of DRM. The claim is that crypto-based systems (*i.e.*, AACS) are "passive" and software protection (*i.e.*, BD+) are "active". This is non-sense since you have to consider the software that decrypts the disc part of the system, and thus there are no true "passive" systems.

Type 2 is no DRM at all, in essence. Whether it's a "broadcast flag" on television, or a CD with an executable root-kit, there are legitimate systems that won't even see that the content should be protected, and the temptation to deal with that is to somehow outlaw such systems. It's effectively already happened – while multi-region DVD players are common and inexpensive in the UK, in the US, I find it difficult to find a player that is advertised as multi-region. While they're clearly not illegal, the marketplace here has somehow been coerced into making them unavailable.

Maybe my proposed two types should be "DRM" and "not DRM":) Well, it's some kind of restriction, although it's certainly not in the same class as something designed to be DRM from the beginning. It's similar to people saying XOR with a constant is "encrypting the data" — nowhere near accurate. In that case, I call it "masking the data" since it is slightly transformed. So what's the term for "unprotected, but with a couple speedbumps bolted on the side"?

DUAL-FREQUENCY-BAND PATCH ANTENNA

A dual-frequency-band patch antenna capable of resonating at two different operating frequencies. The dual-frequency-band patch antenna includes two conductive plates that are placed side by side, with one conductive plate sized to resonate at a first frequency and the other conductive plate sized to resonate at a second frequency. The conductive plates are electrically joined by an electrical connection and share a single feed from a transceiver.

The conductive plates are separated from a ground plane by a dielectric medium, which according to the present invention is very thin in terms of the wavelength at which the antenna operates ($i(*i.e.*), 'low-profile'). When the antenna is excited at the first frequency, the first conductive plate resonates like a conventional patch antenna. The second conductive plate (204), however, acts like a parasitic patch, thereby increasing the overall bandwidth and efficiency. Conversely, when the antenna is excited at the second frequency, the second conductive plate (204) resonates and the first conductive plates acts like a parasitic patch.

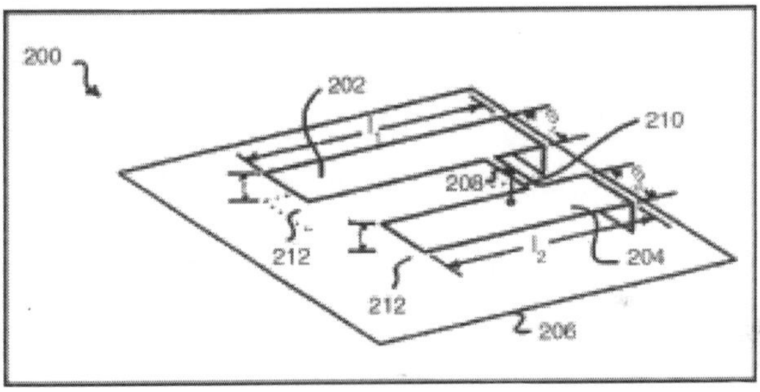

PASSIVE ELEMENTS

Most of the Circuit elements have at least two leads (electrical terminals). They are characterized by voltage across the terminals and current flowing through the device, *see Figure* this is V-I characterization of device.

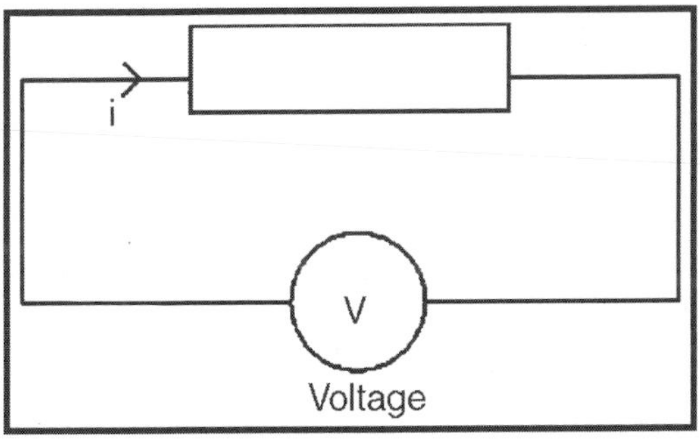

Resistance

Across this element, if we applied a voltage source and observe the current, then we will observer that $v \alpha i$, that is, $v = Ri$, where R is the resistance measured in Ohms Figure.

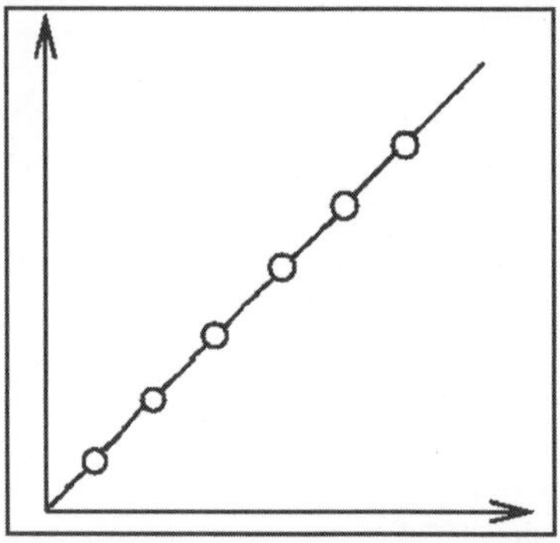

Higher the value of R, larger the voltage required to achieve the same current. Voltage proportional to current - is Ohm's law (It was deduced hueristically by experiments for metals, by George Simon Ohm). For lamp, this law is not true, as with increase in current, temperature of bulb increases, causing the increase in resistance. Hence Ohms law is not strictly true for lamp. But for most of the practical purposes and for this course the Ohm's law holds true for the resistive circuit elements.

The resistive elements (resistances) can be fixed or variable. Commonly use resistor types are carbon film and wirewound. The example of variable resistor is potentiometre.

For a material with length l and cross-sectional area A, the resistance will be,

$$R = \frac{P^l}{A},$$

where P is specific resistivity (property of material). Inverse of resistance is conductance.

$$G = \frac{1}{R},$$

measured in mhos or Seimens.

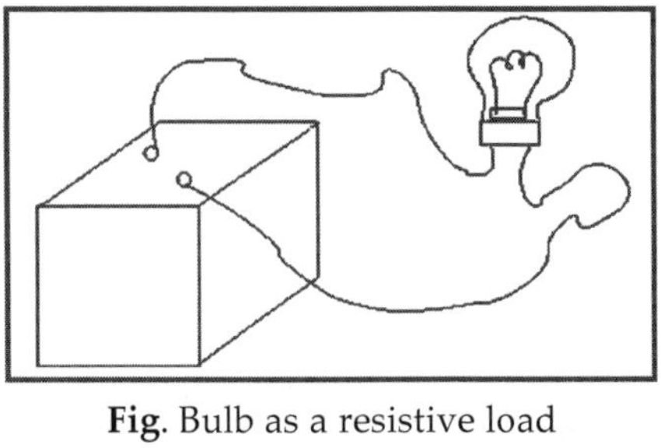

Fig. Bulb as a resistive load

In Fig. property of interest is resistance between A and B. It is dependent on details of wire, connector, filament, material, and shape. It can be abstracted as simple resistance Figure.

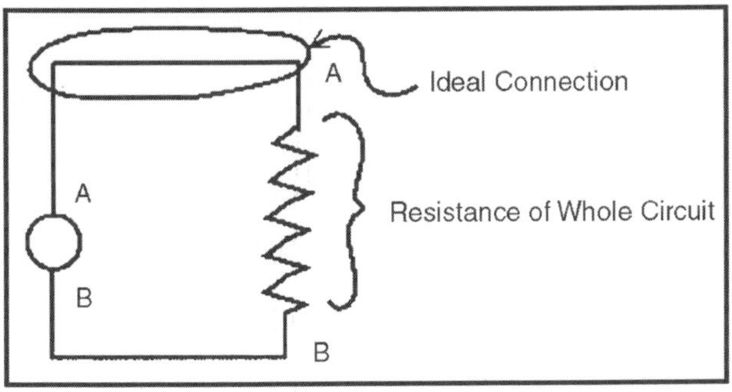

Fig. Bulb abstracted as a simple resistance

Inductance

It another important basic circuit element. Current flowing in a wire causes generation of magnetic field intensity *(H)*. *H* is independent of material medium surrounding the current carrying wire. The *H* leads to magnetic flux density.

$$B = \mu_r \mu_0 H .$$

Here $\mu 0$ is absolute permeability of vaccum. μr is relative permeability of material where B is measured.

The flux flowing around wire links with the conducting wire. And if the flux linkage changes it lead to generation of EMF (electromotive force) which try to oppose the change in flux. This means it tries to nullify the change in current.

The current i causes production of magnetic flux,

$$E = \frac{d\phi}{dt}$$

is the EMF of a single turn. Here, $BA = f$.

Thus, the total EMF,

$$v = N\frac{d\phi}{dt},$$

where N is the number of turns.

Defining the inductance,

$$-v = L\frac{di}{dt}, L$$

is inductance and measured in Henry. See the output current of sinusoidal v applied across an inductor in Figure.

In lumped model, inductance is considered only due to element. The inductance due to wires connecting it to other elements is neglected.

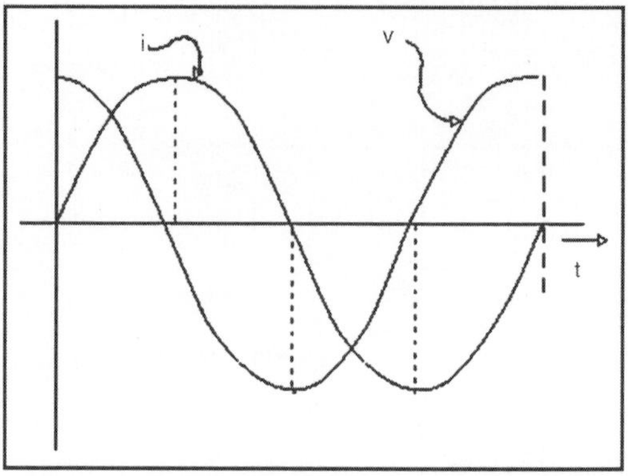

Fig. V and I as function of t for an Inductor

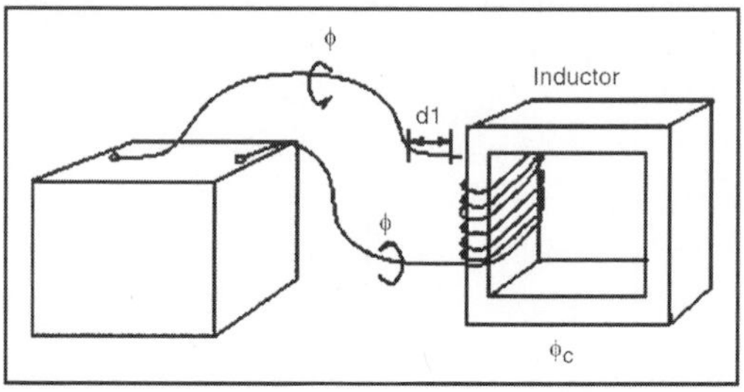

Fig. Physical Implementation of an Inductor

Analysis of Circuit

To find voltage or currents in an element of interest. One can also find voltage and current in all the elements of circuit. Lumped simplified model of resitance, inductance and capacitance.

$$Capacitance: i = c\frac{dv}{dt}, c \text{ capacitance.}$$

The magnitude of charge on either plate is given by $q = cv$.

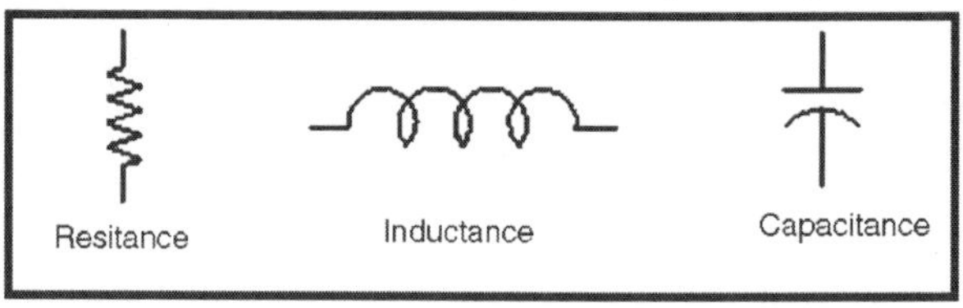

Fig. Symbols for Resitance, Inductance and Capacitance

Lumped Model

Shape, material, wire, connectors - effect of each is assumed to be due to single entity shown by the symbols in the diagram. In actual resistance, inductance and capacitance are distributed all across the circuits. For most practical purpose, lumped model- satisfactory.

Series and Parallel Connections

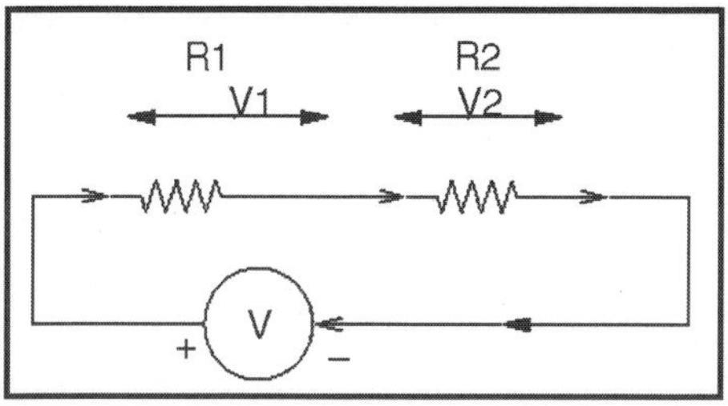

Fig. A Series Connection of Resis tors

Kirchoff's Volatage Law

In a circuit, if your start from a point A and tranverses the circuit in any fashion and reaches back to point A, the total sum of potential changes should be zero. This has to be true since, same point cannot have two different potentials.

Kirchoff's Current Law

At any point in the network, total amount of current entering and leaving the point has to be equal to rate of accumulation of charge at the point. Since, in the circuits ordinarily the points where one circuit element is connected to other circuit element (these points are called nodes) do not store charge sum of incomming current has to be equal to sum of outgoing currents.

Thus, for series model, we get relations:

$$v_1 + v_2 = v$$

$$iR_1 + iR_2 = v$$

$$i(R_1 + iR_2) = v$$

Thus equivalent resistance of series connection is:

$$R = R_1 + R_2$$

For parallel connection, we get the relations:

$$v_1 = v_2 = v$$

$$i_1R_1 = i_2R_2 = v$$

$$i = i_1 + i_2 = \frac{v}{R_1} + \frac{v}{R_2}$$

$$i\frac{R_1R_2}{R_1 + R_2} = v$$

Thus, equivalent resistance here is:

$$R = \frac{R_1R_2}{R_1 + R_2}.$$

Similarly, we calculate equivalent inductance and capacitance for series and parallel cases.

Inductances in Series

$$v = v_1 + v_2 = L_1 \frac{di}{dt} + L_2 \frac{di}{dt}$$

$$v = (L_1 + L_2)\frac{di}{dt}$$

Thus,

$$L = L_1 + L_2$$

Inductances in Parallel

$$v_1 = v_2 = v$$

$$v_1 = L_1 \frac{di_1}{dt}$$

$$v_2 = L_2 \frac{di_2}{dt}$$

$$\frac{v_1}{L_1} + \frac{v_2}{L_2} = v(\frac{1}{L_1} + \frac{1}{L_2}) = \frac{d(i_1 + i_2)}{dt} = \frac{di}{dt}$$

Thus,

$$L = \frac{L_1 L_2}{L_1 + L_2}$$

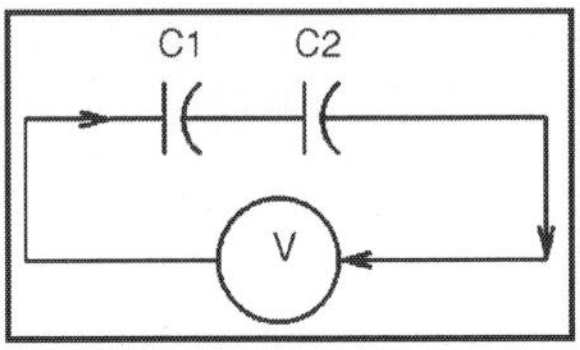

Fig. Series Connection of Capacitances

Capacitances in Series

$$i = C_1 \frac{dv_1}{dt} = C_2 \frac{dv_2}{dt}$$

$$i = (\frac{1}{C_1} + \frac{1}{C_2}) = \frac{d(v_1 + v_2)}{dt}$$

$$i = \frac{C_1 C_2}{C_1 + C_2} \frac{dv}{dt}$$

Thus,

$$C = \frac{C_1 C_2}{C_1 + C_2}$$

Capacitances in parallel:

$$i_1 = C_1 \frac{dv}{dt}$$

$$i = i_1 + i_2 = (C_1 + C_2)\frac{dv}{dt}$$

Thus,

$$C = C_1 + C_2$$

Inductance of a Solenoid (Coil)

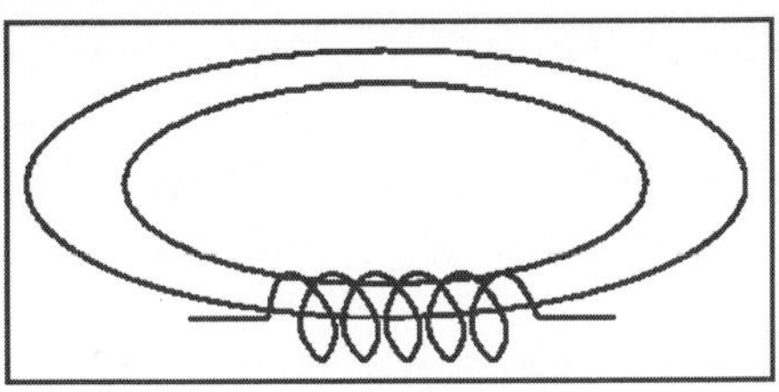

Fig. A Solenoid Diagram Showing Magnetic circuit path Lengths

Define l = magnetic circuit path length

A = magnetic circuit crossectional area.

Inductance: Assuming that H is same in the closed path of length l.

$$\oint \overline{H} \cdot \overline{dl} = Ni$$

$$H.l = Ni$$

$$H = \frac{Ni}{L}$$

$$B = \mu H$$

$$B = \frac{\mu Ni}{l}$$

Thus,

$$\phi = \frac{\mu NiA}{l}$$

Now,

$$v = L\frac{di}{dt} = N\frac{d\phi}{dt}$$

Thus

$$, v = \frac{\mu N^2 A}{l} \frac{di}{dt}$$

Which gives

$$L = \frac{\mu N^2 A}{l}$$

Here $H.l = Ni$ is magnetomotive force (equivalent of eletromotive force - EMF in magnetic domain), and flux ϕ is equivalent of current in magnetic domain. The Magnetic reluctance,

$$= \frac{1}{\mu A}$$

is then equivalent of resistance in magentic domain.

Linearity: When elemental change in cause ΔC, always leads to same elemental change in effect, *i.e.*, ΔE, then the system is said to be linear. In general, for a system let input $x(t)$ lead to output $y(t)$, and a small perturbation in $\delta x(t)$ causes a small perturbation $\delta y(t)$ in output. If perturbation $\delta x(t)$ alwasy leads to same perturbation of $\delta y(t)$ in the output irrespective of any $x(t)$, then the system is linear.

The implication of the above is that if input $x_1(t)$ causes output $y_1(t)$, and $x_2(t)$ causes $y_1(t)$, then $Ax_1(t)+Bx_2(t)$will cause an output of $Ay_1(t)+By_2(t)$ This is principal of superposition.

NETWORK THEOREMS

Anyone who's studied geometry should be familiar with the concept of a theorem: a relatively simple rule used to solve a problem, derived from a more intensive analysis using fundamental rules of mathematics.

At least hypothetically, any problem in math can be solved just by using the simple rules of arithmetic (in fact, this is how modern digital computers carry out the most complex mathematical calculations: by repeating many cycles of additions and subtractions!), but human beings aren't as consistent or as fast as a digital computer. We need "shortcut" methods in order to avoid procedural errors.

In electric network analysis, the fundamental rules are Ohm's Law and Kirchhoff's Laws. While these humble laws may be applied to analyse just about any circuit configuration (even if we have to resort to complex algebra to handle multiple unknowns), there are some "shortcut" methods of analysis to make the math easier for the average human.

As with any theorem of geometry or algebra, these network theorems are derived from fundamental rules. In this chapter, I'm not going to delve into the formal proofs of any of these theorems.

If you doubt their validity, you can always empirically test them by setting up example circuits and calculating values using the "old" (simultaneous equation) methods versus the "new" theorems, to see if the answers coincide.

OBJECTIVES

The objective of the DC Network Theorems Circuit Board and Course is to familiarize students with the Kirchhoff's Current and Voltage Laws, Kirchhoff's Laws Combined, Kirchhoff's Solution with Two Sources, Superposition, Thevenin Circuits, Thevenizing a Bridge Circuit, Thevenin/ Norton Conversion, and Delta to Y or Y to Delta. The course provides comprehensive, hands-on instruction in the terminology, principles, and applications of the circuitry used in DC networks.

NORTON'S THEOREM

Norton's Theorem states that it is possible to simplify any linear circuit, no matter how complex, to an equivalent circuit with just a single current source and parallel resistance connected to a load. Just as with Thevenin's Theorem, the qualification of "linear" is identical to that found in the Superposition Theorem: all underlying equations must be linear (no exponents or roots).

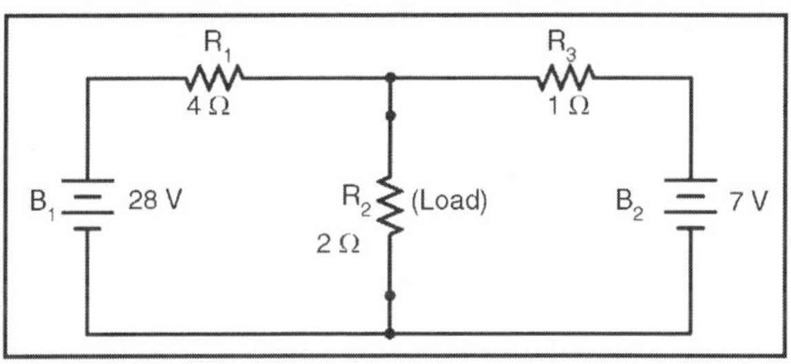

...after Norton conversion...

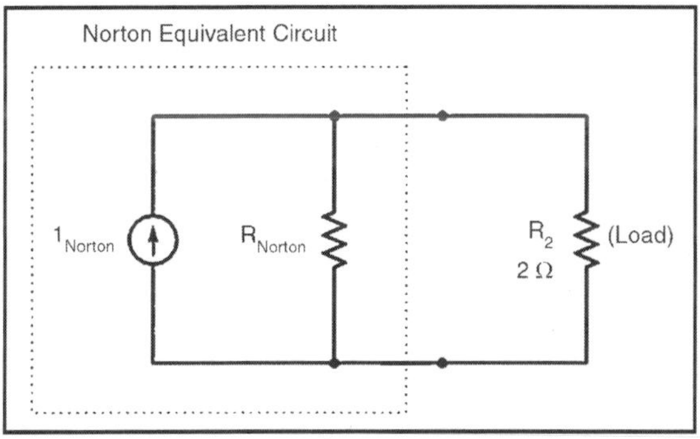

Remember that a *current source* is a component whose job is to provide a constant amount of current, outputting as much or as little voltage necessary to maintain that constant current.

As with Thevenin's Theorem, everything in the original circuit except the load resistance has been reduced to an equivalent circuit that is simpler to analyse. Also similar to Thevenin's Theorem are the steps used in Norton's Theorem to calculate the Norton source current (I_{Norton}) and Norton resistance (R_{Norton}). As before, the first step is to identify the load resistance and remove it from the original circuit:

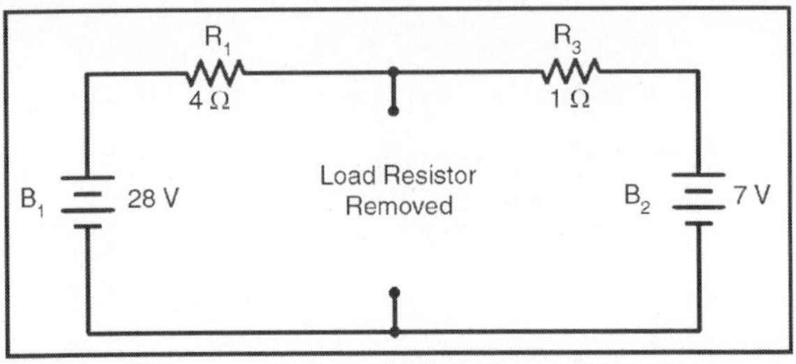

Then, to find the Norton current (for the current source in the Norton equivalent circuit), place a direct wire (short) connection between the load points and determine the resultant current. Note that this step is exactly opposite the respective step in Thevenin's Theorem, where we replaced the load resistor with a break (open circuit):

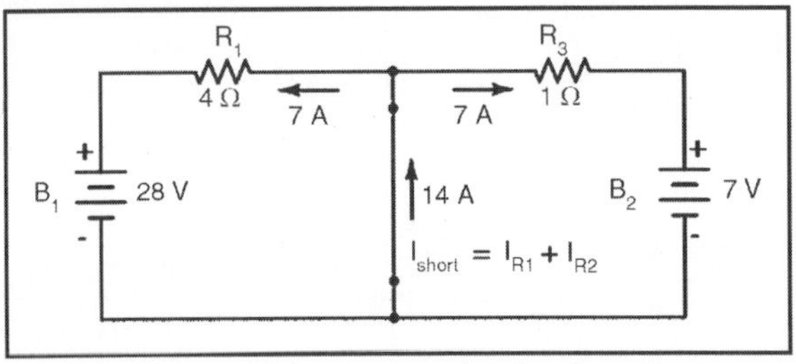

With zero voltage dropped between the load resistor connection points, the current through R_1 is strictly a function of B_1's voltage and R_1's resistance: 7 amps (I=E/R).

Likewise, the current through R_3 is now strictly a function of B_2's voltage and R_3's resistance: 7 amps (I=E/R). The total current through the short between the load connection points is the sum of these two currents: 7 amps + 7 amps = 14 amps. This figure of 14 amps becomes the Norton source current (I_{Norton}) in our equivalent circuit:

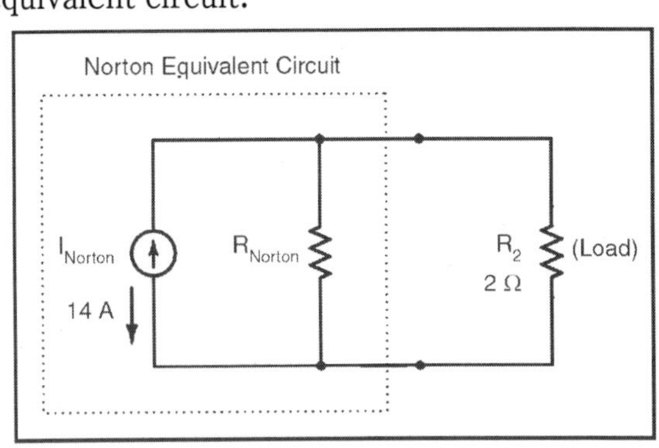

Remember, the arrow notation for a current source points in the direction opposite that of electron flow. Again, apologies for the confusion. For better or for worse, this is standard electronic symbol notation. To calculate the Norton resistance (R_{Norton}), we do the exact same thing as we did for calculating Thevenin resistance ($R_{Thevenin}$): take the original circuit (with the load resistor still removed), remove the power sources (in the same style as we did with the Superposition Theorem: voltage sources replaced with wires and current sources replaced with breaks), and figure total resistance from one load connection point to the other:

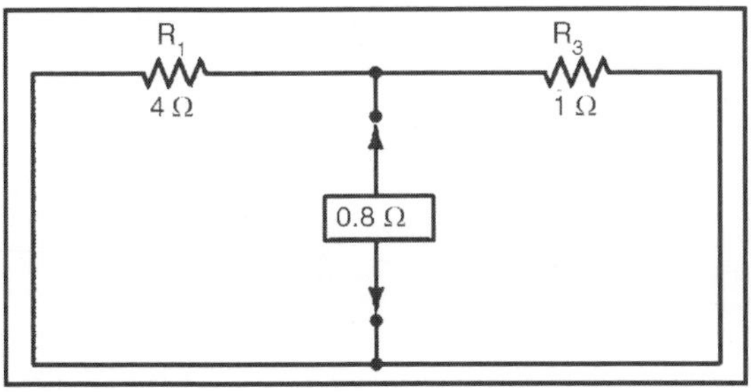

Now our Norton equivalent circuit looks like this:

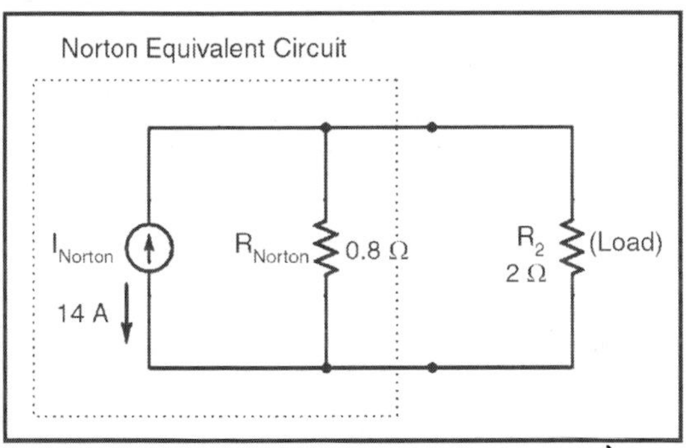

If we re-connect our original load resistance of 2 Ù, we can analyse the Norton circuit as a simple parallel arrangement:

	R_{Norton}	R_{Load}	Total	
E	8	8	8	Volts
I	10	4	14	Amps
R	0.8	2	571.43m	Ohms

As with the Thevenin equivalent circuit, the only useful information from this analysis is the voltage and current values for R_2; the rest of the information is irrelevant to the original circuit. However, the same advantages seen with Thevenin's Theorem apply to Norton's as well: if we wish to analyse load resistor voltage and current over several different values of load resistance, we can use the Norton equivalent circuit again and again, applying nothing more complex than simple parallel circuit analysis to determine what's happening with each trial load.

DESCRIPTION ABOUT NETWORK ANALYSIS

Generally speaking, *network analysis* is any structured technique used to mathematically analyse a circuit (a "network" of interconnected components). Quite often the technician or engineer will encounter circuits containing multiple sources of power or component configurations which defy simplification by series/parallel analysis techniques. In those cases, he or she will be forced to use other means.

This chapter presents a few techniques useful in analyzing such complex circuits. To illustrate how even a simple circuit can defy analysis by breakdown into series and parallel portions, take start with this series-parallel circuit:

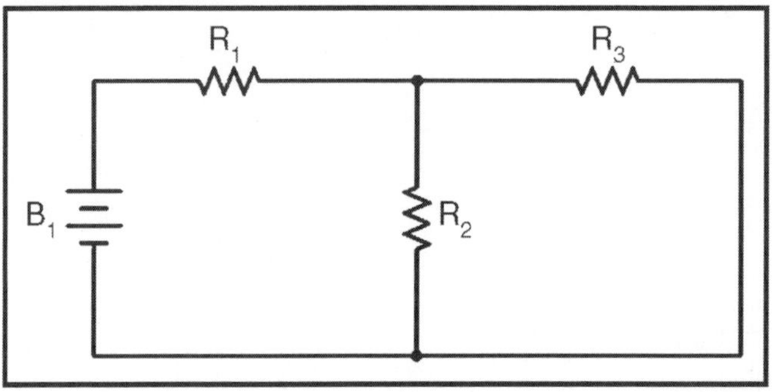

To analyse the above circuit, one would first find the equivalent of R_2 and R_3 in parallel, then add R_1 in series to arrive at a total resistance. Then, taking the voltage of battery B_1 with that total circuit resistance, the total current could be calculated through the use of Ohm's Law (I=E/R), then that current figure used to calculate voltage drops in the circuit. All in all, a fairly simple procedure. However, the addition of just one more battery could change all of tha0t:

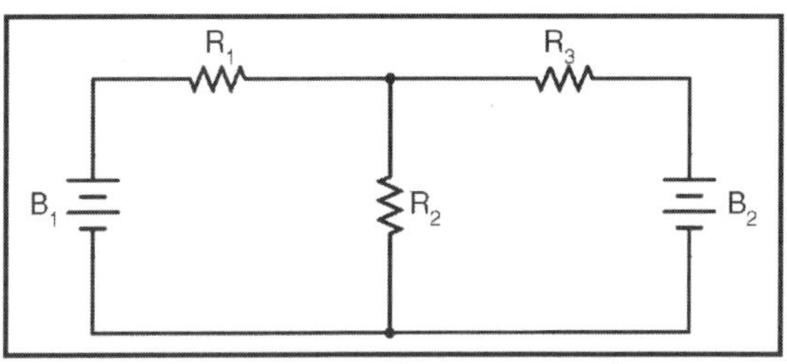

Resistors R_2 and R_3 are no longer in parallel with each other, because B_2 has been inserted into R_3's branch of the circuit. Upon closer inspection, it appears there are *no* two resistors in this circuit directly in series or parallel with each other. This is the crux of our problem: in series-parallel analysis, we started off by identifying sets of resistors that *were* directly in series or parallel with each other, reducing them to single equivalent resistances.

It should be clear that this seemingly simple circuit, with only three resistors, is impossible to reduce as a combination of simple series and simple parallel sections: it is something different altogether. However, this is not the only type of circuit defying series/parallel analysis:

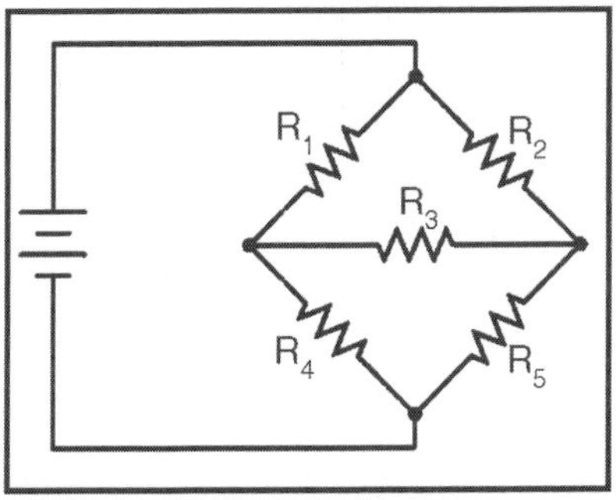

Here we have a bridge circuit, and for the sake of example we will suppose that it is *not* balanced (ratio R_1/R_4 not equal to ratio R_2/R_5).

If it were balanced, there would be zero current through R_3, and it could be approached as a series/parallel combination circuit (R_1—R_4//R_2—R_5). However, any current through R_3makes a series/parallel analysis impossible. R_1 is not in series with R_4 because there's another path for electrons to flow through R_3. Neither is R_2 in series with R_5 for the same reason. Likewise, R_1 is not in parallel with R_2 because R_3 is separating their bottom leads. Neither is R_4 in parallel with R_5.

Although it might not be apparent at this point, the heart of the problem is the existence of multiple unknown quantities. At least in a series/parallel combination circuit, there was a way to find total resistance and total voltage, leaving total current as a single unknown value to calculate (and then that current was used to satisfy previously unknown variables in the reduction process until the entire circuit could be analysed). With these problems, more than one parametre (variable) is unknown at the most basic level of circuit simplification.

With the two-battery circuit, there is no way to arrive at a value for "total resistance," because there are *two* sources of power to provide voltage and current (we would need *two* "total" resistances in order to proceed with any Ohm's Law calculations). With the unbalanced bridge circuit, there is such a thing as total resistance across the one battery (paving the way for a calculation of total current), but that total current immediately splits up into unknown proportions at each end of the bridge, so no further Ohm's Law calculations for voltage ($E=IR$) can be carried out.

So what can we do when we're faced with multiple unknowns in a circuit? The answer is initially found in a mathematical process known as *simultaneous equations* or *systems of equations*, whereby multiple unknown variables are solved by relating them to each other in multiple equations. In a scenario with only one unknown (such as every Ohm's Law equation we've dealt with thus far), there only needs to be a single equation to solve for the single unknown:

- E = 1R (E is unknown; 1 and R are known).
- $I=\frac{E}{R}$ (I is unknown; E and R are known)
- $R=\frac{E}{1}$ (R is unknown; E and 1 are known)

However, when we're solving for multiple unknown values, we need to have the same number of equations as we have unknowns in order to reach a solution. There are several methods of solving simultaneous equations, all rather intimidating and all too complex for explanation in this chapter. However, many scientific and programmable calculators are able to solve for simultaneous unknowns, so it is recommended to use such a calculator when first learning how to analyse these circuits. Later on we'll see that some clever people have found tricks to avoid having to use simultaneous equations on these types of circuits.

ACTIVE ELEMENTS

Active components are those that have gain or directionality. They include Semiconductors (Diode, Transistors, Operational Amplifier and other Solid State Devices) and Thermionic Valves (Vacuum Tubes).

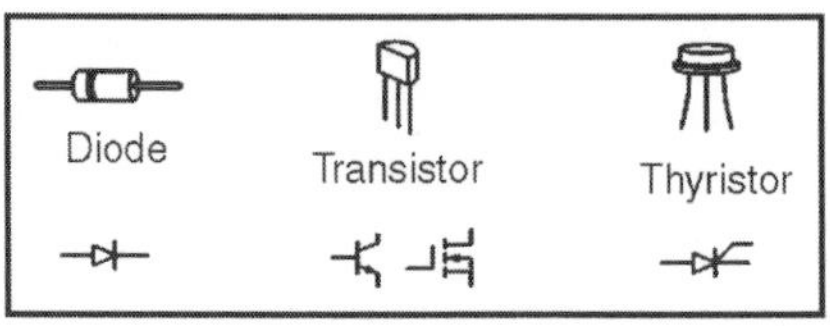

VOLTAGE AND CURRENT SOURCES

Real voltage sources can be represented as ideal voltage sources in series with a resistance r, the ideal voltage source having zero resistance. Real current sources can be represented as ideal current sources in parallel with a resistance r, the ideal current source having infinite resistance.

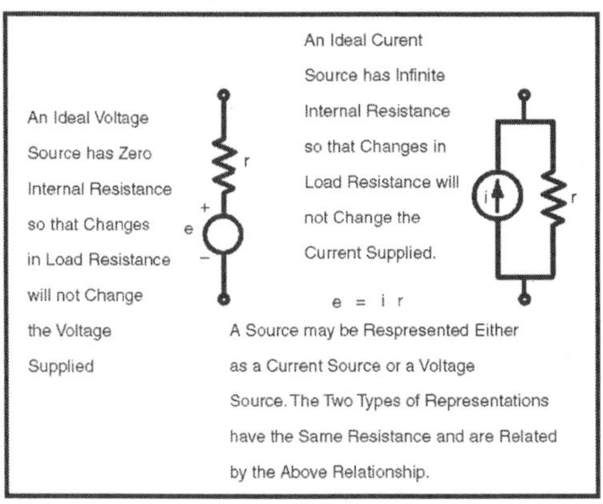

CURRENT VOLTAGE SOURCES

The concept of an ideal voltage source is pretty simple, and it was really embedded in the previous discussion.

- An ideal voltage source is a voltage source that maintains the same voltage across the source's terminals no matter what current is drawn from the terminals of the source or what current flows into the terminals.
- That's it in a nutshell. If the source is a DC Source, we can plot a voltage current plot for an ideal voltage source. The plot is shown below. However, we need to define terms. Here is a circuit symbol for an ideal voltage source.

In this symbol, we assume the following:

- The voltage across the terminals is denoted as V_t.
- The load current flowing from the source to a load (presumably a load is attached when the source is in a circuit) is denoted as I_L.
- With those definitions, here is the source symbol. It's just a circle with polarity indicated.

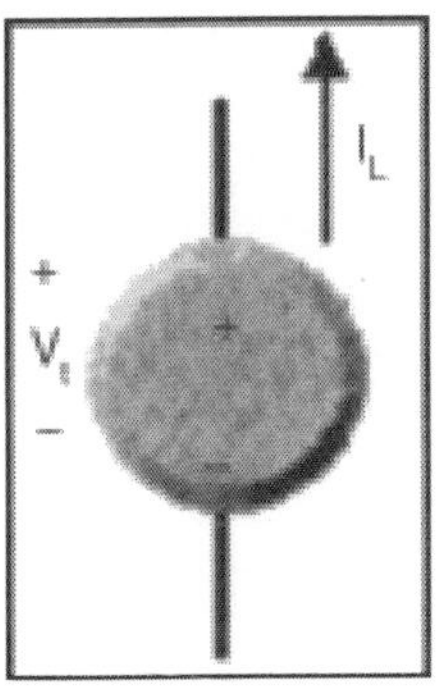

And, here is the plot of terminal voltage against load current.

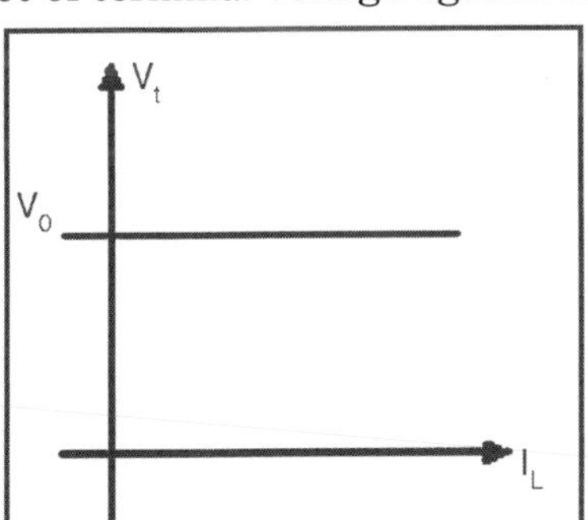

Given the Discussion above, we can say:

- V_t = constant, no matter what the load current is.

  That's pretty much the description of the ideal voltage source. It's not too complex, but it is an important concept. In the next section we'll look at how you can put this concept to use. For the rest of this section we'll look at ideal current sources starting next.

IDEAL CURRENT SOURCE IN SIMPLE MODE

An ideal current source is a simple model for many current sources. It is reminiscent of the ideal voltage source but with voltage and current interchanged. Here is the story.

- There is a special circuit symbol for an ideal current source.

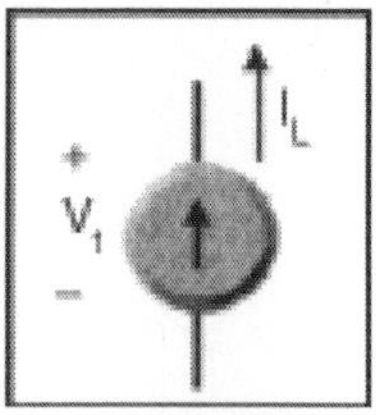

- I_L = constant, no matter what the terminal voltage is.
- The plot of load current against terminal voltage is similar to the plot for an ideal voltage source, but voltage and current are interchanged. Here is the plot.

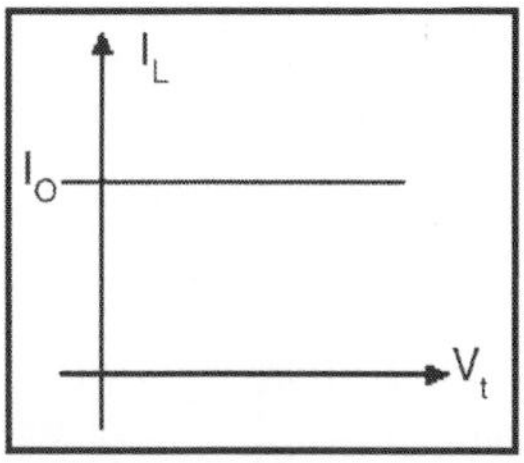

Notice that an ideal current source is somewhat similar to an ideal voltage source. However, when you use an ideal source - usually when doing circuit analysis- there is a significant difference in the analysis. However, that's getting ahead of the story. We first have to worry about how you would "use" an ideal source, when we know that there is no such thing as an ideal source, *i.e.*, a source that is "perfect" in some way.

USING IDEAL SOURCES

The idea of using ideal sources is something that you may rebel at. After all, there is no such thing as an ideal source anywhere in the world. You can't pull an ideal source off the shelf in the lab, so why are we even talking about them? The answer to that question is that you use ideal sources when you have a non-ideal (a real source) source in a circuit. There are two important things to note.

- There are some sources that are very good sources and that can be modelled as ideal sources. (And when that happens, be grateful.) Some situations like that include the following.

- A power supply in the lab. Many times you connect a power supply to some electronic circuit, for example, and when you connect the circuit you find that the output voltage from the power supply doesn't change measurably. (After all, power supply designers try to make that happen!) In that case, the power supply might be considered to be an ideal source at least as long as you are working on that particular circuit.

- There are many sources that do not perform ideally. However, it has proven to be possible to construct models of real sources, and those models often contain ideal source in combination with other ideal elements (like resistors, etc.).
- The vinin and Nortonequivalent circuits are examples of models of real sources that can account for loading effects (*i.e.*, drawing enough current from the source to change the output voltage) and they are widely used in circuit analysis. You will even find that manufacturers give you parametre values for Thevinin and Norton equivalents on the front panel of many instruments like function generators.

You often have situations in which the sources that you use can be approximated with ideal sources.

Shown below is a bridge circuit powered by a battery. Often a battery maintains a pretty constant voltage across the terminals, so you may be able to replace the battery with an ideal voltage source when you analyse the circuit.

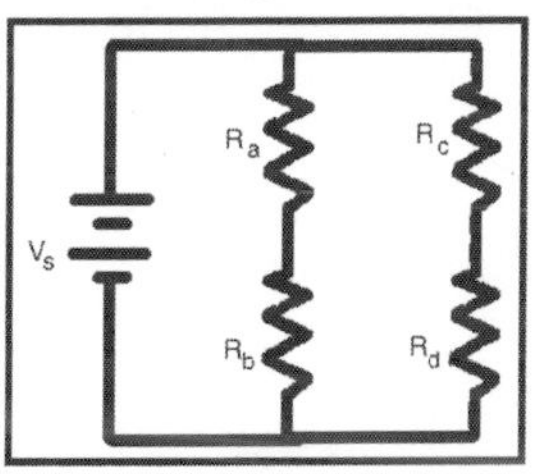

Here's the circuit with an ideal voltage source substituted for the battery. At this point, you may know how to do the analysis so you're ready to go. If you don't know how to do the analysis, you'll get there in these lessons. to go to the first lesson on circuit analysis.

However, you may want to spend more time looking at sources, particular sources that are not ideal.

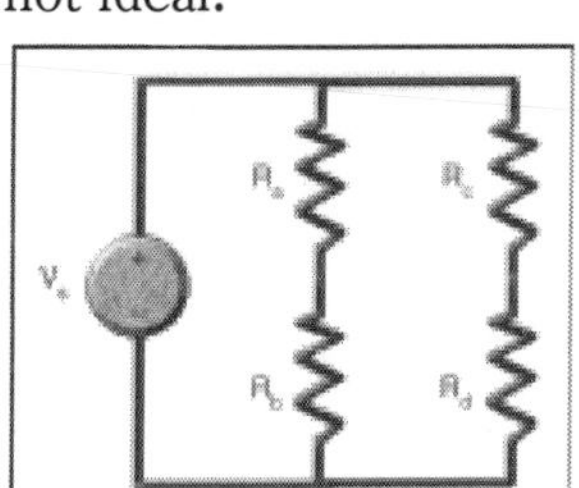

PC Controlled Voltage or Current Source

This project was begun as a means to charge and cycle NiCad batteries but has become a versatile tool for experiments requiring either a controlled voltage up to +30V and/or a current of +/- 2.5 Amps. If all you want to do is charge and recycle NiCad batteries, then, with hindsight, I'd advise you to buy one of the commerical units available. The circuit I will describe will do the job but it is experimental and has many short comings. It is more the beginning of a useful tool than one that has been fully developed. This circuit and the related software have taken me a good 2 years to reach even a marginal degree of reliability despite it's apparently simple actions.

My complete assembly uses the Discovery Series K-2805 Parrallel Port Interface Kit as distributed by Dick Smith Electronics as the source of control voltages and, if used as a charger, the Battery Cell measurements. Any other PC interface with at least 4 analogue inputs and 2 analogue outputs, both with 0 to +5V range, could be put to use.

Below is the circuit for one of the two identical voltage/current sources included in my construction.

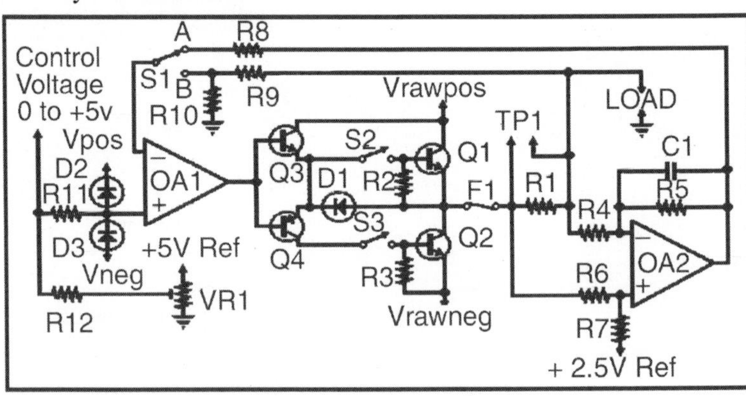

Vrawpos/neg is supplied in my construction by a seperate power supply module which comprises a centre tapped transformer, a 4amp rectifier and two 10000uF capacitors supplying unregulated 18/36V peak DC.

Typically I use this circuit with Prawpos at +18V and Prawneg and Ground connected to the centre tap. More typically Prawpos would be connected to +18V, Prawneg to -18V and Ground to the centre tap. For higher voltages, Ground can be connected to the -18V supply but great care must be taken this is not connecting a -18V voltage to the PC interface ground which may be earthed through the PC.

The load in my case is a NiCad battery of between 1.2 and 24V, but it could really be anything; active loads like batteries are just more of a challenge to control.

Note: To discharge a 1.2V battery will required a negative power source, but for larger voltages, Prawneg may be connected to ground as described above.

The components are:

Label	Component
OA1,OA2	2 parts of 4 OpAmp I.C. LM324
Q1,Q2	2N2055 (TO3 package NPN) or 2N3055
Q3	BC547 (any small NPN)
Q4	BC557 (any small PNP)
D1,D2,D3	signal diode rated over 30V
C1	1nF Green
F1	3A Fast Blow
R1	0.1R 5W
R2,R3	470R 1/2W
R4,R6	10K 1/4W
R5,R7	100K 1/4W
R8	47K 1/4W
R9	270K 1/4W
R10	47K 1/4W
R11	1K 1/4W
R12	100K 1/4W
VR1	500K

Q1 is sources voltage to the load and Q2 sinks voltage from the load. Naturally they get hot and so need good heat sinking and some thermal isolation from other components and sensitive fingers. I was lucky to find a preloved aluminium box with an internal frame on which I mounted the transistors. Air holes either end have further helped to keep the box reasonably cool.

Source or Sink actions can be disabled by opening S2 or S3 respectively. R2 and R3 then clamp the transistor base to prevent transistor operation. Even then, some minor leakage will be apparent through R2 and R3 themselves; a better solution may be to have S2 and S3 in circuit to the load but their rating would need to be increased accordingly.

Transistors Q3 and Q4 buffer the controlling action of opamp OA1 which, throught the magic of feedback, eliminates any dead zone due to the Vbe voltage drops in Q3, Q4 or Q1 (*i.e.*, without feedback, any control voltage would first have to overcome these voltage drops before current could flow to or from the load).

The link between the base of Q3 and the emitter of Q4 is to enable lower discharge voltages to ground when Vrawneg is to ground (enabling discharge to +0.7V with S2 closed, +1.1V with S2 open). This will only work with D1 there to allow the emitter of Q4 to rise above ground. D1 is also there to prevent Base to Emitter breakdown of Q4 (which occurs when the output of OA1 exceeds +6V); it doesn't really prevent it, but it does limit the current from OA1s output through the base to the emitter. The same it true for Q3 if the output of OA1 goes below -6V. Trying to understand this apparently simple circuit has caused me a few headaches. As mentioned already, to reliably discharge a 1.2V cell a split supply is really necessary but it is a bit scary letting a PC connect a -15V to a +1.2V cell. A voltage between 0 and +5V is fed to OA1 either from a PC interface or, when the interface is unplugged, from the manually operated VR1 (it assumes that the impedance of the Analogue Voltage from the PC interface is much lower than R12). I have used this circuit to charge batteries using only the manual contol but VR1 does not provide a stable control for low currents and frequent adjustment is necessary (not to mention a good alarm clock to remind you to turn it off). Still the manual control is particularly useful for quick diagnostics.

With SW1 in position A, Current Source Mode, negative feedback to OA1 comes from a current measurement taken across R1 via OA2. Since the OA2 is referenced to +2.5V, 0V on the control voltage should result in -2.5Amps (sinking) through R1 (assuming a suitable load is connected).

If the control voltage rises to +5.0V, then +2.5Amps (sourcing) through R1 should result. If the load imedance is too high, the output voltage at the load will rise or fall towards the supply rail voltages (more on this in a moment).

With SW1 in position B, Voltage Source Mode, negative feedback to OA1 comes from the output voltage, divided by a factor of 6 by R9 and R10. A control voltage of 0V will result in 0V at the output, +5V will (should) result in +30V (if Vpos is > +30V) at the output. I am thinking R10 could be connected to Vneg to allow negative outputs but it seemed more useful to have the output voltage controlled relative to ground.

For the purposes of Battery charging, SW1 remains in position A. With the opamp used (LM324) and the the supply conditioning I provide for the opamp (see below), the maximum supply voltage in current source mode is Vrawpos - 4.5V (13.5V for 18V supply). As such, running this system off a car battery will only allow you to charge a 5 cell NiCad. Using an OpAmp which can source and sink closer to supply voltage and possibly connecting the OpAmp supply directly to Vrawpos (quite okay for a 13.5V car battery supply) may allow you to charge 7 cell packs off a car battery but I haven't tried it. I think we'll leave fast charging off car batteries to a more dedicated unit. I originally thought it would be useful to measure both output voltage and output current and since both feedback signals should be in the 0 to 5V range, these can be measured with the PC Interface that I use which has 10 analogue inputs. Current measurement turns out to be of no particular use in battery charging as any errors in the control of the source or sink current arise from the +2.5V reference voltage which in turn are reflected in the measurement with a few extra errors thrown in.

The current measurement may be of more use where this device is being used as a voltage source or in analysing someone elses charger have done this to determine the parametres for an unlabelled NiCad pack that came with a toy car and plug pack charger. Current can also be measured by a sensitive voltmetre across R1 using test poit TP1. Below is the circuit I used to supply the I.C.s and voltage references:

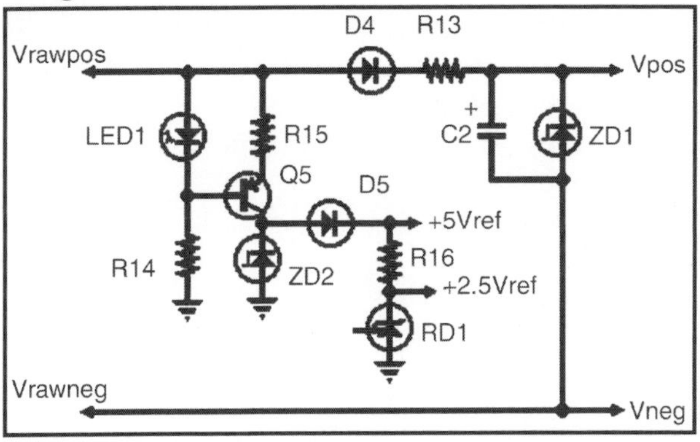

The components are:

Label	Component
D1	Rectifier Diode rated over 30V and 100 mA
ZD1	Zener Diode 30V 1W
ZD2	Zener Diode 5.1V 400mW
Q5	BC557 PNP
RD1	LM336-2.5V Voltage Reference
D5	Small Rectifier Diode (~30mA duty)
LED1	Whatever
R13	330R 1W
R14	2.2K 1/4W
R15	110R 1/4W
R16	1K 1/4W
C2	10uF 50VW

The LM324 is happy with voltages up to 30V, so this supply is clamped at 30V with ZD1 and C2 removes a good deal of ripple from the unregulated supply. The value of R13 was found by trial and error to match the I.C. supply current. D4 is there to prevent the inevidable mistake.

The voltage reference circuit draws more curreThe LM324 is happy with voltages nt than I would have liked so I have used a current source from the unregulated supply to feed 5V Zener ZD2 which in turn feeds the 2.5V reference circuit. Originally I used a simple resistor to feed ZD2 but with the supply varying from 10 to 36V, the current to ZD2 also varied greatly and all sorts of problems followed. Q5 controls the current to approx. 20mA by using the voltage drop through LED1 as a reference voltage.

The diode is there to drop the "5V" level to about 4.7V. This is connected in my unit to the 5V supply in the PC interface (through a 1K resistor) so that both operate at the exact same level.

This was a good idea when the 2.5V reference was created by a simple divider, but now that I use a 2.5V reference diode, this link and diode D5 are nolonger necessary and I have only shown it as this is what is in my construction. While it shows only one 2.5V ref. circuit (R16 and RD1), I have in my construction 2 circuits, one for each voltage current source circuit. This was necessary before I fitted Q5 to provide a constant current to ZD2. I suspect two 2.5V reference circuits are nolonger necessary but I haven't tried it yet.

One of the problems in using this Current Source for battery charging is that, until the software has taken control, this unit will initially attempt to sink 2.5Amps from the battery (control voltage at 0V). As such, either the battery must be left disconnected or S3 left open until the software has got itself together. If the computer crashes for some reason the results would be unpleasant. The simplest solution I have come up with so far is to disable the +2.5V reference circuit with the following alternative supply circuit.

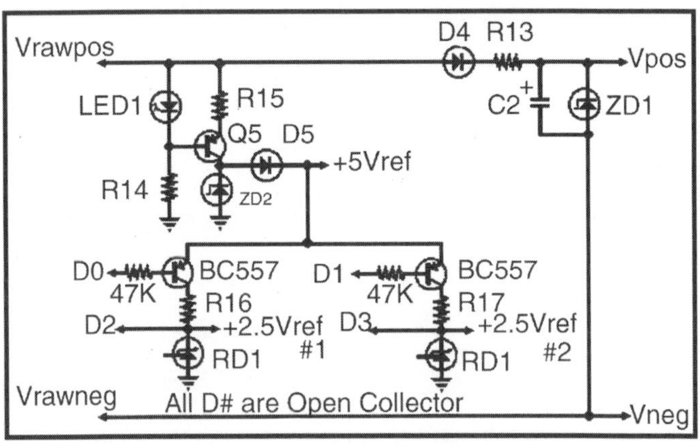

While D0 to D3 are open collector outputs from the PC Interface which must be enabled by the software before they sink any current. As such, while D0 and D1 are not enabled following boot up, the 2.5V reference will float between 0V (as will occur with zero control volts) and about 0.8V. On my unit at boot up, this results in about 100mA discharge of the connected battery. Not exactly passive, but not disasterous either.

As it happens, this modification provides additional versatility. By connecting D2 and D3 to the 2.5V reference voltages, these can be clamped to rise no greater than 0.9V. When this is so, 5V control voltage will now correspond to 5Amps source. Given the right power supply (mine could not handle this load for long) and the right load, this could be useful. There would be no affect in the voltage source mode. This is the end of description of the circuit regarding voltage and current source control. The software (charger) I've written for an IBM PC compatible (based on Intel Microprocessor 8086 series and above) computer and the Discovery Series PC Interface, has a manual test mode and a battery charger mode. Future modes intended for the voltage/current source will be added as they occur to me (or you). I will be considering release of the source code on a GNU licence with the hope that useful modification would be released back to me. In the mean while, if you would like the source code, my current e-mail address is "rpact" with an ISP called "dingoblue.net.au".

6

Combinational Analysis and Circuits Design

COMBINATIONAL ANALYSIS PROCEDURE

The bulk of the Combinational Analysis module is accessed through a single window of that name allowing you to view truth tables and Boolean expressions. This window can be opened in two ways.

WINDOW MENU

Select Combinational Analysis, and the current Combinational Analysis window will appear. If you haven't viewed the window before, the opened window will represent no circuit at all.

Only one Combinational Analysis window exists within Logisim, no matter how many projects are open. There is no way to have two different analysis windows open at once.

PROJECT MENU

From a window for editing circuits, you can also request that Logisim analyse the current circuit by selecting the Analyse Circuit option from the Project menu. Before Logisim opens the window, it will compute Boolean expressions and a truth table corresponding to the circuit and place them there for you to view.

For the analysis to be successful, each input must be attached to an input pin, and each output must be attached to an output pin. Logisim will only analyse circuits with at most eight of each type, and all should be single-bit pins. Otherwise, you will see an error message and the window will not open.

In constructing Boolean expressions corresponding to a circuit, Logisim will first attempt to construct a Boolean expressions corresponding exactly to the gates in the circuit. But if the circuit uses some non-gate components (such as a multiplexer), or if the circuit is more than 100 levels deep (unlikely), then it will pop up a dialog box telling you that deriving Boolean expressions was impossible, and Logical will instead derive the expressions based on the truth table, which will be derived by quietly trying each combination of inputs and reading the resulting outputs.

After analyzing a circuit, there is no continuing relationship between the circuit and the Combinational Analysis window. That is, changes to the circuit will not be reflected in the window, nor will changes to the Boolean expressions and/or truth table in the window be reflected in the circuit. Of course, you are always free to analyse a circuit again; and, as we will see later, you can replace the circuit with a circuit corresponding to what appears in the Combinational Analysis window.

LIMITATIONS

Logical will not attempt to detect sequential circuits: If you tell it to analyse a sequential circuit, it will still create a truth table and corresponding Boolean expressions, although these will not accurately summarize the circuit behaviour. (In fact, detecting sequential circuits is provably impossible, as it would amount to solving the Halting Problem. Of course, you might hope that Logical would make at least some attempt - perhaps look for flip-flops or cycles in the wires - but it does not.) As a result, the Combinational Analysis system should not be used indiscriminately: Only use it when you are indeed sure that the circuit you are analyzing is indeed combinational! Logical will make a change to the original circuit that is perhaps unexpected: The Combinational Analysis system requires that each input and output have a unique name that conforming to the rules for Java identifiers. (Roughly, each character must either a letter or a digit, and the first character must be a letter.

No spaces allowed!) It attempts to use the pins' existing labels, and to use a list of defaults if no label exists. If an existing label doesn't follow the Java-identifier rule, then Logical will attempt to extract a valid name from the label if at all possible. Incidentally, the ordering of the inputs in the truth table will match their top-down ordering in the original circuit, with ties being broken in left-right order.

ARITHMETIC CIRCUIT DESIGN

Suppose we wanted to build a device that could add two binary bits together. Such a device is known as a half-adder, and its gate circuit looks like this:

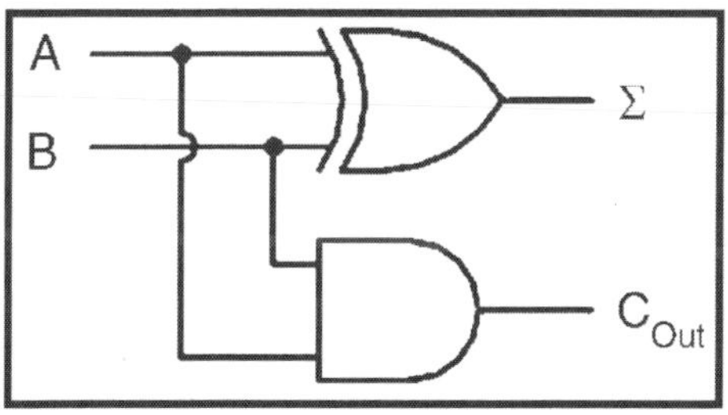

The Σ symbol represents the "sum" output of the half-adder, the sum's least significant bit (LSB). C_{out} represents the "carry" output of the half-adder, the sum's most significant bit (MSB). If we were to implement this same function in ladder (relay) logic, it would look like this:

Either circuit is capable of adding two binary digits together. The mathematical "rules" of how to add bits together are intrinsic to the hard-wired logic of the circuits. If we wanted to perform a different arithmetic operation with binary bits, such as multiplication, we would have to construct another circuit. The above circuit designs will only perform one function: add two binary bits together. To make them do something else would take re-wiring, and perhaps different compo entry.

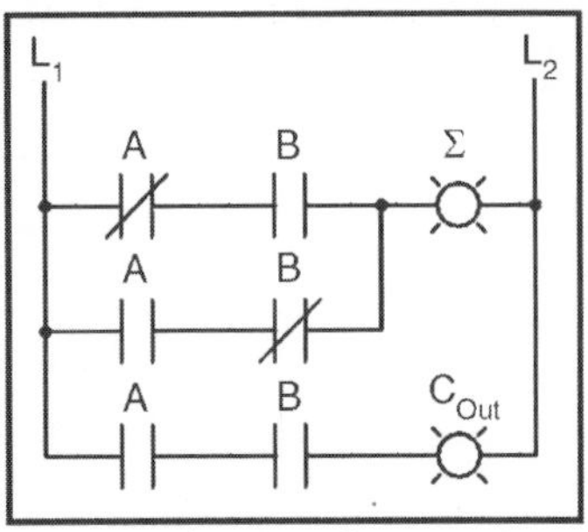

In this sense, digital arithmetic circuits aren't much different from analog arithmetic (operational amplifier) circuits: they do exactly what they're wired to do, no more and no less. We are not, however, restricted to designing digital computer circuits in this manner. It is possible to embed the mathematical "rules" for any arithmetic operation in the form of digital data rather than in hard-wired connections between gates. The result is unparalleled flexibility in operation, giving rise to a whole new kind of digital device: the programmable computer.

LOOK-UP TABLES

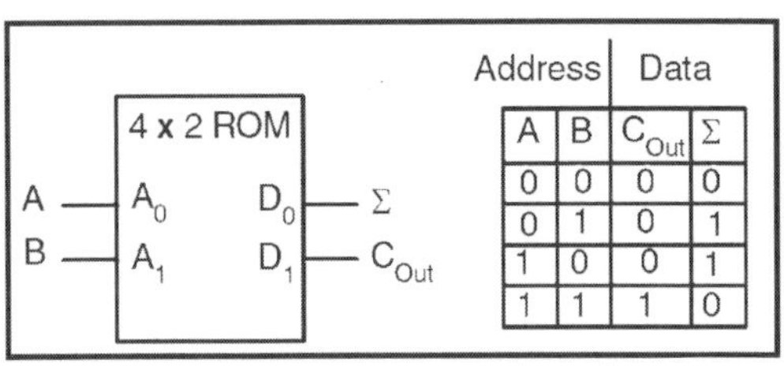

A	B	C_{Out}	Σ
0	0	0	0
0	1	0	1
1	0	0	1
1	1	1	0

Having learned about digital memory devices in the last chapter, we know that it is possible to store binary data within solid-state devices. Those storage "cells" within solid-state memory devices are easily addressed by driving the "address" lines of the device with the proper binary value(s). Suppose we had a ROM memory circuit written, or programmed, with certain data, such that the address lines of the ROM served as inputs and the data lines of the ROM served as outputs, generating the characteristic response of a particular logic function. Theoretically, we could programme this ROM chip to emulate whatever logic function we wanted without having to alter any wire connections or gates.

Consider the following example of a 4 × 2 bit ROM memory (a very small memory!) programmed with the functionality of a half adder:

If this ROM has been written with the above data (representing a half-adder's truth table), driving the A and B address inputs will cause the respective memory cells in the ROM chip to be enabled, thus outputting the corresponding data as the Ó (Sum) and C_{out} bits. Unlike the half-adder circuit built of gates or relays, this device can be set up to perform any logic function at all with two inputs and two outputs, not just the half-adder function. To change the logic function, all we would need to do is write a different table of data to another ROM chip. We could even use an EPROM chip which could be re-written at will, giving the ultimate flexibility in function.

It is vitally important to recognize the significance of this principle as applied to digital circuitry. Whereas the half-adder built from gates or relays processes the input bits to arrive at a specific output, the ROM simply remembers what the outputs should be for any given combination of inputs. This is not much different from the "times tables" memorized in grade school: rather than having to calculate the product of 5 times 6 (5 + 5 + 5 + 5 + 5 + 5 = 30), school-children are taught to remember that 5 × 6 = 30, and then expected to recall this product from memory as needed. Likewise, rather than the logic function depending on the functional arrangement of hard-wired gates or relays (hardware), it depends solely on the data written into the memory (software).

Such a simple application, with definite outputs for every input, is called a look-up table, because the memory device simply "looks up" what the output(s) should to be for any given combination of inputs states. This application of a memory device to perform logical functions is significant for several reasons:

- Software is much easier to change than hardware.
- Software can be archived on various kinds of memory media (disk, tape), thus providing an easy way to document and manipulate the function in a "virtual" form; hardware can only be "archived" abstractly in the form of some kind of graphical drawing.
- Software can be copied from one memory device (such as the EPROM chip) to another, allowing the ability for one device to "learn" its function from another device.
- Software such as the logic function example can be designed to perform functions that would be extremely difficult to emulate with discrete logic gates (or relays!).

The usefulness of a look-up table becomes more and more evident with increasing complexity of function. Suppose we wanted to build a 4-bit adder circuit using a ROM. We'd require a ROM with 8 address lines (two 4-bit numbers to be added together), plus 4 data lines (for the signed output):

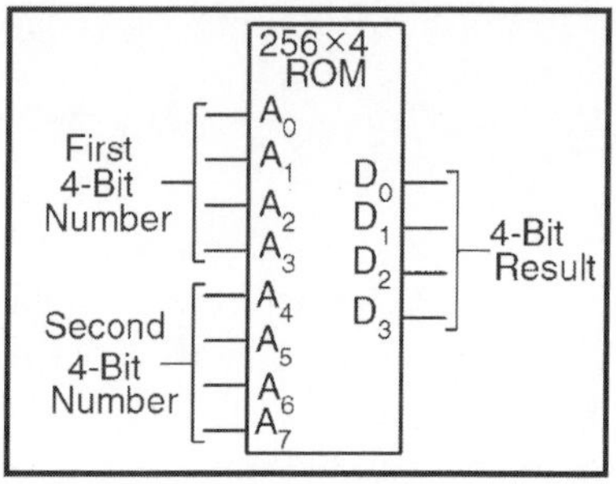

With 256 addressable memory locations in this ROM chip, we would have a fair amount of programming to do, telling it what binary output to generate for each and every combination of binary inputs. We would also run the risk of making a mistake in our programming and have it output an incorrect sum, if we weren't careful. However, the flexibility of being able to configure this function (or any function) through software alone generally outweighs that costs.

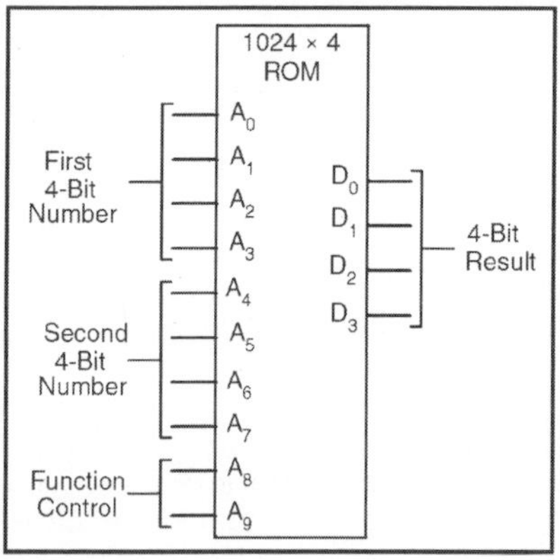

Consider some of the advanced functions we could implement with the above "adder." We know that when we add two sets of numbers in 2's complement signed notation, we risk having the answer overflow. For instance, if we try to add 0111 (decimal 7) to 0110 (decimal 6) with only a 4-bit number field, the answer we'll get is 1001 (decimal -7) instead of the correct value, 13 (7 + 6), which cannot be expressed using 4 signed bits.

If we wanted to, we could avoid the strange answers given in overflow conditions by programming this look-up table circuit to output something else in conditions where we know overflow will occur (that is, in any case where the real sum would exceed +7 or -8). One alternative might be to programme the ROM to output the quantity 0111 (the maximum positive value that can be represented with 4 signed bits), or any other value that we determined to be more appropriate for the application than the typical overflowed "error" value that a regular adder circuit would output.

It's all up to the programmer to decide what he or she wants this circuit to do, because we are no longer limited by the constraints of logic gate functions. The possibilities don't stop at customised logic functions, either. By adding more address lines to the 256 × 4 ROM chip, we can expand the look-up table to include multiple functions:

With two more address lines, the ROM chip will have 4 times as many addresses as before (1024 instead of 256). This ROM could be programmed so that when A8 and A9 were both low, the output data represented the sum of the two 4-bit binary numbers input on address lines A0 through A7, just as we had with the previous 256 × 4 ROM circuit. For the addresses A8=1 and A9=0, it could be programmed to output the difference (subtraction) between the first 4-bit binary number (A0 through A3) and the second binary number (A4 through A7).

For the addresses A8=0 and A9=1, we could programme the ROM to output the difference (subtraction) of the two numbers in reverse order (second - first rather than first - second), and finally, for the addresses A8=1 and A9=1, the ROM could be programmed to compare the two inputs and output an indication of equality or inequality. What we will have then is a device that can perform four different arithmetical operations on 4-bit binary numbers, all by "looking up" the answers programmed into it. If we had used a ROM chip with more than two additional address lines, we could programme it with a wider variety of functions to perform on the two 4-bit inputs. There are a number of operations peculiar to binary data (such as parity check or Exclusive-ORing of bits) that we might find useful to have programmed in such a look-up table.

Devices such as this, which can perform a variety of arithmetical tasks as dictated by a binary input code, are known as Arithmetic Logic Units (ALUs), and they comprise one of the essential components of computer technology. Although modern ALUs are more often constructed from very complex combinational logic (gate) circuits for reasons of speed, it should be comforting to know that the exact same functionality may be duplicated with a "dumb" ROM chip programmed with the appropriate look-up table(s). In fact, this exact approach was used by IBM engineers in 1959 with the development of the IBM 1401 and 1620 computers, which used look-up tables to perform addition, rather than binary adder circuitry. The machine was fondly known as the "CADET," which stood for "Can't Add, Doesn't Even Try."

A very common application for look-up table ROMs is in control systems where a custom mathematical function needs to be represented. Such an application is found in computer-controlled fuel injection systems for automobile engines, where the proper air/fuel mixture ratio for efficient and clean operation changes with several environmental and operational variables. Tests performed on engines in research laboratories determine what these ideal ratios are for varying conditions of engine load, ambient air temperature, and barometric air pressure. The variables are measured with sensor transducers, their analog outputs converted to digital signals with A/D circuitry, and those parallel digital signals used as address inputs to a high-capacity ROM chip programmed to output the optimum digital value for air/fuel ratio for any of these given conditions.

Sometimes, ROMs are used to provide one-dimensional look-up table functions, for "correcting" digitized signal values so that they more accurately represent their real-world significance.

An example of such a device is a thermocouple transmitter, which measures the millivoltage signal generated by a junction of dissimilar metals and outputs a signal which is supposed to directly correspond to that junction temperature.

Unfortunately, thermocouple junctions do not have perfectly linear temperature/voltage responses, and so the raw voltage signal is not perfectly proportional to temperature.

By digitizing the voltage signal (A/D conversion) and sending that digital value to the address of a ROM programmed with the necessary correction values, the ROM's programming could eliminate some of the nonlinearity of the thermocouple's temperature-to-millivoltage relationship, so that the final output of the device would be more accurate. The popular instrumentation term for such a look-up table is a digital characterizer.

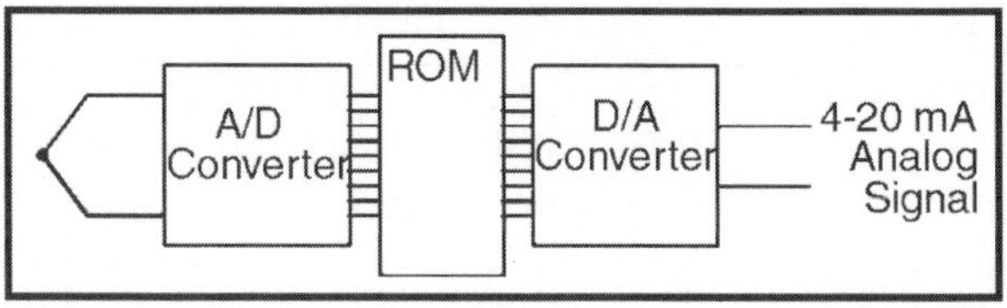

Another application for look-up tables is in special code translation. A 128 x 8 ROM, for instance, could be used to translate 7-bit ASCII code to 8-bit EBCDIC code:

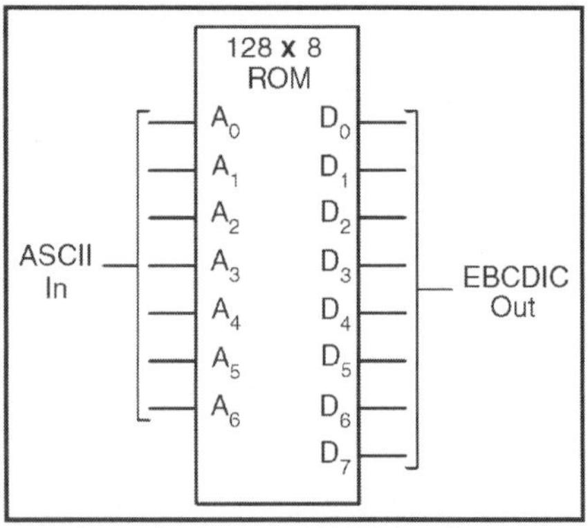

Again, all that is required is for the ROM chip to be properly programmed with the necessary data so that each valid ASCII input will produce a corresponding EBCDIC output code.

DESIGN PROCEDURE

Logic circuits for digital systems may be combinational or sequential. A combinational circuit consists of logic gates whose outputs at any time are determined by combining the values of the applied inputs using logic operations. A combinational circuit performs an operation that can be specified logically by a set of Boolean expression. In addition to using logic gates, sequential circuits employ elements that store bit values.

Sequential circuit outputs are a function of inputs and the bit value in storage elements. These values, in turn, are a function of previously applied inputs and stored values.

As a consequence, the outputs of a sequential circuit depend not only on the presently applied values of the inputs, but also on pas inputs, and the behaviour of the circuit must be specified by a sequence in time of inputs and internal stored bit values.

A combinational circuit consists of input variables, output variables, logic gates and interconnections.

The interconnected logic gates accept signals from the inputs and generate signals at the output. The n input variables come from the environment of the circuit, and the m output variables are available for use by the environment. Each input and output variable exists physically as a binary signal that represents logic 1 or logic 0.

For n input variables, there are 2 ^ n possible binary input combinations. For each binary combination of the input variables, there is one possible binary value on each output.

Thus, a combinational circuit can be specified by a truth table that lists the output values for each combination of the input variables. A combinational circuit can also be described by m Boolean function, one for each output variable. Each such function is expressed as function of the n input variables.

COMBINATIONAL CIRCUIT DESIGN

The design of combinational circuit starts from a specification of the problem and culminates in a logic diagram or set of Boolean equations from which the logic diagram can be obtained.

The procedure involves the following steps:

- From the specifications of the circuit, determine the required number of inputs and outputs, and assign a letter symbol to each.
- Derive the truth table that defines the required relation ship between inputs and outputs.
- Obtain the simplified Boolean functions of each outputs as function of the input variables.
- Draw the logic diagram.
- Verify the correctness of the design.

MAGNITUDE COMPARATOR

74L85 4-BIT MAGNITUDE COMPARATOR

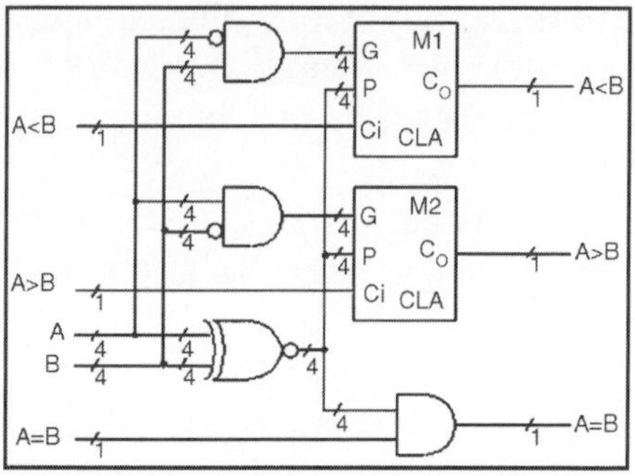

STATISTICS

11 inputs; 3 outputs; 33 gates;

FUNCTION

The 74L85 magnitude comparator can be functionally modeled as above. This is a simplification of implementing a magnitude comparator by a carry function with an inverted input bus as shown.

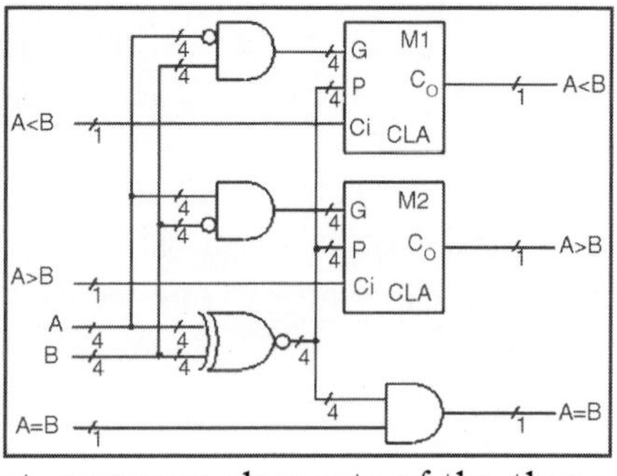

Using this concept, common elements of the three comparator functions A < B, A > B, and A = B are combined to construct the model shown above, which maps directly onto the gate-level realization of the 74L85.

ARITHMETIC LOGIC UNITS

Devices which can perform a variety of arithmetical tasks as dictated by a binary input code, are known as*Arithmetic Logic Units* (ALUs), and they comprise one of the essential components of computer technology. Although modern ALUs are more often constructed from very complex combinational logic (gate) circuits for reasons of speed, it should be comforting to know that the exact same functionality may be duplicated with a "dumb" ROM chip programmed with the appropriate look-up table(s).

In fact, this exact approach was used by IBM engineers in 1959 with the development of the IBM 1401 and 1620 computers, which used look-up tables to perform addition, rather than binary adder circuitry. The machine was fondly known as the "CADET," which stood for "Can't Add, Doesn't Even Try." A very common application for look-up table ROMs is in control systems where a custom mathematical function needs to be represented. Such an application is found in computer-controlled fuel injection systems for automobile engines, where the proper air/fuel mixture ratio for efficient and clean operation changes with several environmental and operational variables.

Tests performed on engines in research laboratories determine what these ideal ratios are for varying conditions of engine load, ambient air temperature, and barometric air pressure. The variables are measured with sensor transducers, their analog outputs converted to digital signals with A/D circuitry, and those parallel digital signals used as address inputs to a high-capacity ROM chip programmed to output the optimum digital value for air/fuel ratio for any of these given conditions.

Sometimes, ROMs are used to provide one-dimensional look-up table functions, for "correcting" digitized signal values so that they more accurately represent their real-world significance. An example of such a device is a *thermocouple transmitter*, which measures the millivoltage signal generated by a junction of dissimilar metals and outputs a signal which is supposed to *directly* correspond to that junction temperature.

Unfortunately, thermocouple junctions do not have perfectly linear temperature/voltage responses, and so the raw voltage signal is not perfectly proportional to temperature. By digitizing the voltage signal (A/D conversion) and sending that digital value to the address of a ROM programmed with the necessary correction values, the ROM's programming could eliminate some of the nonlinearity of the thermocouple's temperature-to-millivoltage relationship, so that the final output of the device would be more accurate. The popular instrumentation term for such a look-up table is a digital *characterizer*.

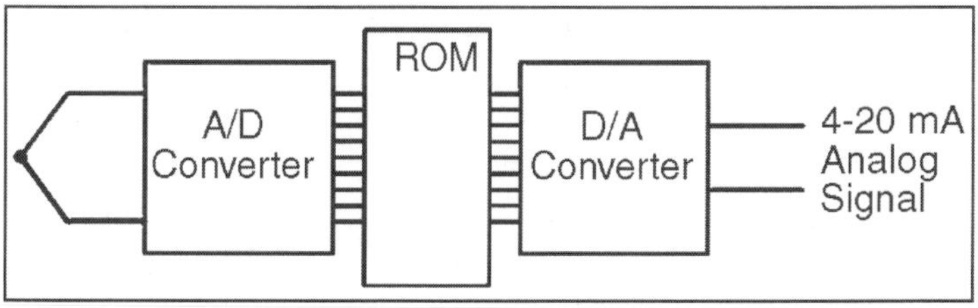

Another application for look-up tables is in special code translation. A 128 × 8 ROM, for instance, could be used to translate 7-bit ASCII code to 8-bit EBCDIC code:

Again, all that is required is for the ROM chip to be properly programmed with the necessary data so that each valid ASCII input will produce a corresponding EBCDIC output code.

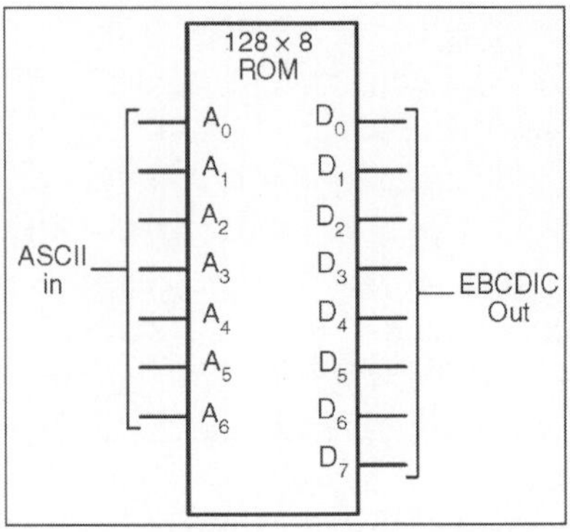

CIRCUITS DECODERS GENERATION

In both the multiplexer and the demultiplexer, part of the circuits decode the address inputs, *i.e.* it translates a binary number of n digits to 2n outputs, one of which (the one that corresponds to the value of the binary number) is 1 and the others of which are 0.

It is sometimes advantageous to separate this function from the rest of the circuit, since it is useful in many other applications. Thus, we obtain a new combinatorial circuit that we call the decoder. It has the following truth table (for n = 3):

a2	a1	a0	d7	d6	d5	d4	d3	d2	d1	d0
0	0	0	0	0	0	0	0	0	0	1
0	0	1	0	0	0	0	0	0	1	0
0	1	0	0	0	0	0	0	1	0	0
0	1	1	0	0	0	0	1	0	0	0
1	0	0	0	0	0	1	0	0	0	0
1	0	1	0	0	1	0	0	0	0	0
1	1	0	0	1	0	0	0	0	0	0
1	1	1	1	0	0	0	0	0	0	0

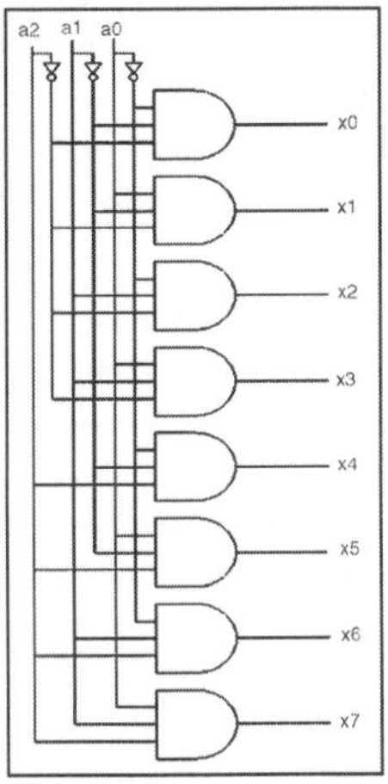

Fig. Circuit diagram for the decoder.

DECODER GENERATION

X86's decoder is generated using an ISA description like all the other ISAs, although how it does that is a bit different. Most of the instructions for most of the other ISAs are defined by passing chunks of code that perform the instruction into an instruction format. The format is basically a template which puts wraps that bit of code in the structure needed to support it and you have your instruction object.

Because almost all of the instructions in x86 are microcoded and many can be encoded in multiple ways and hence appear in the decoder more than once, and because the same non-trivial decoding rules apply to many different instructions, X86 uses the decoder as a layer of indirection and defines the majority of its instructions elsewhere.

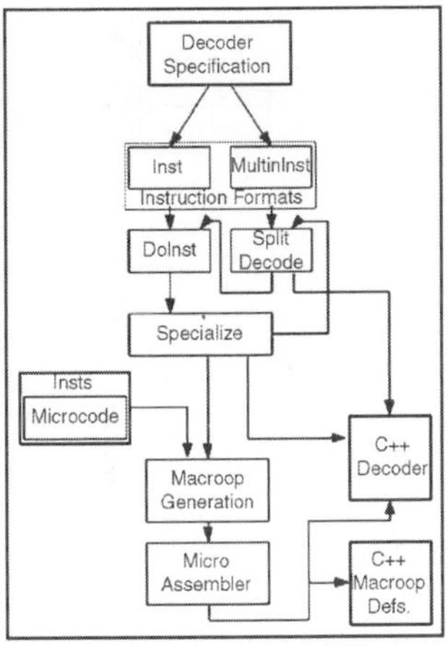

X86 almost exclusively uses only two different instruction formats, Inst and MultiInst. MultiInst is just a compact way of describing multiple related Insts. An Inst essentially selects an instruction like XOR and provides a specification for its operands. Inside the instruction format, the instruction name and operand specification are passed to a python function called"specializeInst" which figures out what to do with it. If the operand specification describes more than one version of the instruction, for instance one that uses memory and one that uses registers, the instruction's information is passed into another funciton,"doSplitDecode", which separates out those versions and passes each individually back through the same system.

This goes on until the instructions have been fully split out and code has been generated to figure out what version to use. As a nice bonus, the MultiInst format doesn't add much complexity to this model since it can simply jump right into doSplitDecode and continue as normal.

There is one additional format for string instructions that works similar to MultiInst, except instead of specializing the instruction based on its operands, it specializes it based on its prefixes. There are also a few instructions that don't use this system, but because there are so few and because they work like the instructions in other ISAs I won't describe them here.

At this point, the code for selecting the right version of an instruction is put into the C++ decoder function. Almost all of this function is built this way, with the minor exception of small bits of logic that glue everything together and make large scale distinctions the number of opcode bytes.

3 TO 8 DECODER

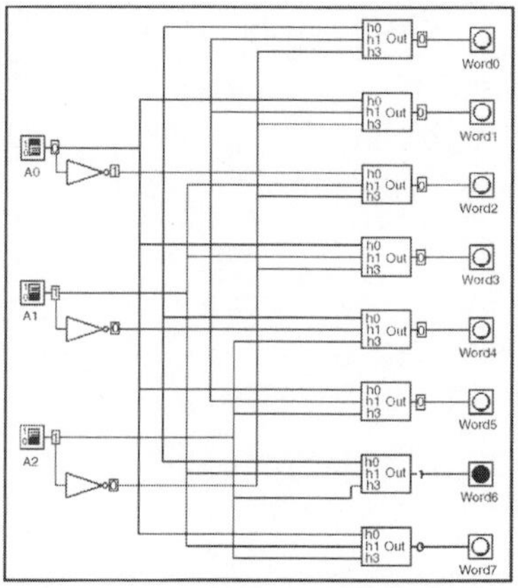

The 3 to 8 decoder unit takes 3 address lines as input and outputs 8 address enable lines. Which of the 8 output is enabled is dependant upon the configuration of the 3 address line. If A0, A1, and A2 are all 0, then address 0 is enabled. If A0=0, A1=1, A2=0, then address 3 is enables. With 3 address lines, the number of words that can be addressed is 8 (2^3=8).

ENCODERS

An encoder is a circuit that changes a set of signals into a code. Lets begin making a 2-to-1 line encoder truth table by reversing the 1-to-2 decoder truth table.

D_1	D_0	A
0	1	0
1	0	1

This truth table is a little short. A complete truth table would be,

D_1	D_0	A
0	0	
0	1	0
1	0	1
1	1	

One question we need to answer is what to do with those other inputs? Do we have them generate an additional error output? In many circuits this problem is solved by adding sequential logic in order to know not just what input is active but also which order the inputs became active.

A more useful application of combinational encoder design is a binary to 7-segment encoder. The seven segments are given according,

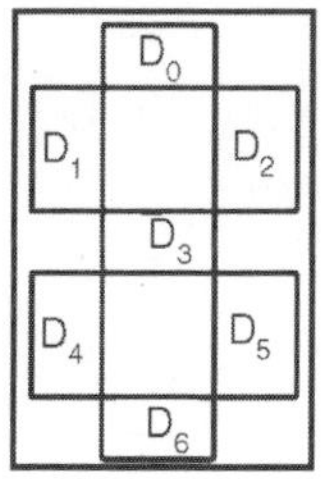

Our truth table is:

I_3	I_2	I_1	I_0	D_6	D_5	D_4	D_3	D_2	D_1	D_0
0	0	0	0	1	1	1	0	1	1	1
0	0	0	1	0	0	1	0	0	1	0
0	0	1	0	1	0	1	1	1	0	1
0	0	1	1	1	0	1	1	0	1	1
0	1	0	0	0	1	1	1	0	1	0
0	1	0	1	1	1	0	1	0	1	1
0	1	1	0	1	1	0	1	1	1	1
0	1	1	1	1	0	1	0	0	1	0
1	0	0	0	1	1	1	1	1	1	1
1	0	0	1	1	1	1	1	0	1	1

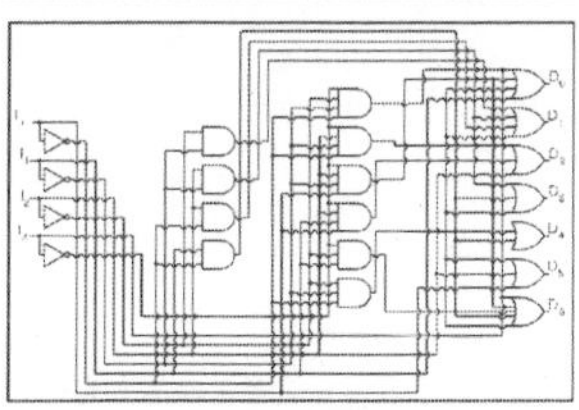

Deciding what to do with the remaining six entries of the truth table is easier with this circuit. This circuit should not be expected to encode an undefined combination of inputs, so we can leave them as"don't care" when we design the circuit. The boolean equations are,

$$D_0 = I_3 + I_1 + \bar{I}_3\bar{I}_2\bar{I}_1 I_3$$

$$D_1 = I_3 + \bar{I}_2\bar{I}_1 + \bar{I}_2\bar{I}_1 + I_2\bar{I}_0$$

$$D_2 = I_2 + \bar{I}_3\bar{I}_2\bar{I}_1\bar{I}_0 + \bar{I}_3 I_2 I_1 I_0$$

$$D_3 = I_3 + I_1\bar{I}_0 + I_2\bar{I}_1$$

$$D_4 = I_1\bar{I}_0 + \bar{I}_2\bar{I}_1\bar{I}_0$$

$$D_5 = I_3 + I_2 + I_0$$

$$D_6 = I_3 + I_1\bar{I}_0 + \bar{I}_3\bar{I}_2\bar{I}_1 + \bar{I}_3 I_2\bar{I}_1 I_0$$

and the circuit is,

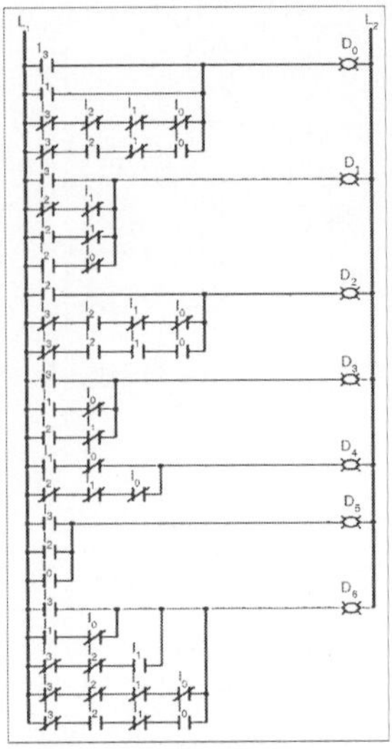

MULTIPLEXES CIRCUIT

a_2	a_1	a_0	d_7	d_6	d_5	d_4	d_3	d_2	d_1	d_0	|	x
-	-	-	-	-	-	-	-	-	-	-	---	-
0	0	0	-	-	-	-	-	-	-	0	|	0
0	0	0	-	-	-	-	-	-	-	1	|	1
0	0	1	-	-	-	-	-	-	0	-	|	0
0	0	1	-	-	-	-	-	-	1	-	|	1
0	1	0	-	-	-	-	-	0	-	-	|	0
0	1	0	-	-	-	-	-	1	-	-	|	1
0	1	1	-	-	-	-	0	-	-	-	|	0
0	1	1	-	-	-	-	1	-	-	-	|	1
1	0	0	-	-	-	0	-	-	-	-	|	0
1	0	0	-	-	-	1	-	-	-	-	|	1
1	0	1	-	-	0	-	-	-	-	-	|	0
1	0	1	-	-	1	-	-	-	-	-	|	1
1	1	0	-	0	-	-	-	-	-	-	|	0
1	1	0	-	1	-	-	-	-	-	-	|	1
1	1	1	0	-	-	-	-	-	-	-	|	0
1	1	1	1	-	-	-	-	-	-	-	|	1

A multiplexer is a combinatorial circuit that is given a certain number (usually a power of two) data inputs, let us say 2n, and n address inputs used as a binary number to select one of the data inputs. The multiplexer has a single output, which has the same value as the selected data input.

In other words, the multiplexer works like the input selector of a home music system. Only one input is selected at a time, and the selected input is transmitted to the single output. While on the music system, the selection of the input is made manually, the multiplexer chooses its input based on a binary number, the address input.

The truth table for a multiplexer is huge for all but the smallest values of n. We therefore use an abbreviated version of the truth table in which some inputs are replaced by `-' to indicate that the input value does not matter.

Here is such an abbreviated truth table for n = 3. The full truth table would have 2(3 + 23) = 2048rows.

We can abbreviate this table even more by using a letter to indicate the value of the selected input, like this:

a_2	a_1	a_0	d_7	d_6	d_5	d_4	d_3	d_2	d_1	d_0	I	x
-	-	-	-	-	-	-	-	-	-	-	---	-
0	0	0	-	-	-	-	-	-	-	c	I	c
0	0	1	-	-	-	-	-	-	c	-	I	c
0	1	0	-	-	-	-	-	c	-	-	I	c
0	1	1	-	-	-	-	c	-	-	-	I	c
1	0	0	-	-	-	c	-	-	-	-	I	c
1	0	1	-	-	c	-	-	-	-	-	I	c
1	1	0	-	c	-	-	-	-	-	-	I	c
1	1	1	c	-	-	-	-	-	-	-	I	c

The same way we can simplify the truth table for the multiplexer, we can also simplify the corresponding circuit. Indeed, our simple design method would yield a very large circuit. The simplified circuit looks like this:

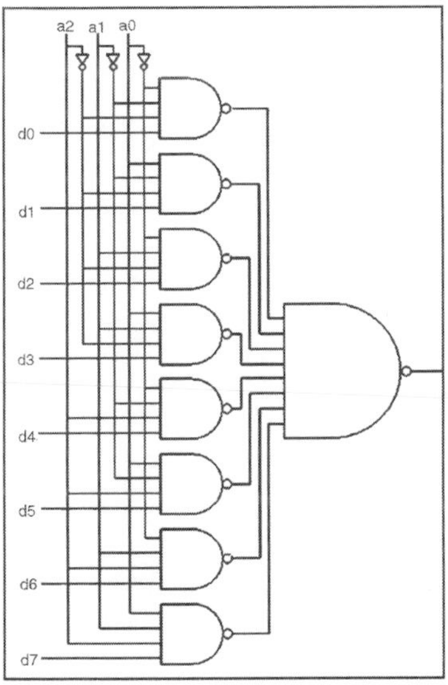

A multiplexer performs the function of selecting the input on any one of'n' input lines and feeding this input to one output line.

Multiplexers are used as one method of reducing the number of integrated circuit packages required by a particular circuit design. This in turn reduces the cost of the system.

Assume that we have four lines, C0, C1, C2 and C3, which are to be multiplexed on a single line, Output (f). The four input lines are also known as the Data Inputs. Since there are four inputs, we will need two additional inputs to the multiplexer, known as the Select Inputs, to select which of the C inputs is to appear at the output. Call these select lines A and B. The gate implementation of a 4-line to 1-line multiplexer is shown below:

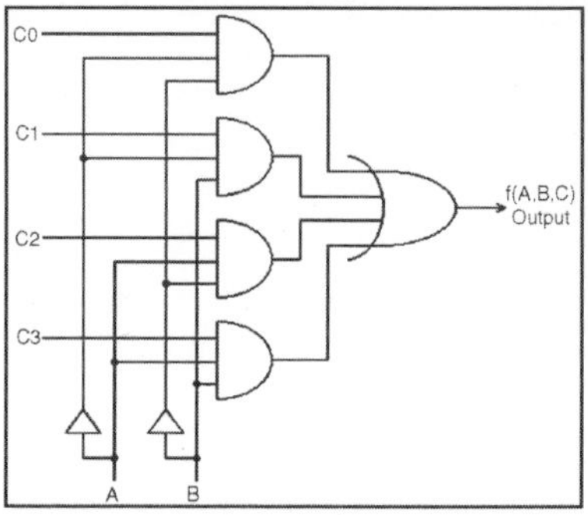

The circuit symbol for the above multiplexer is:

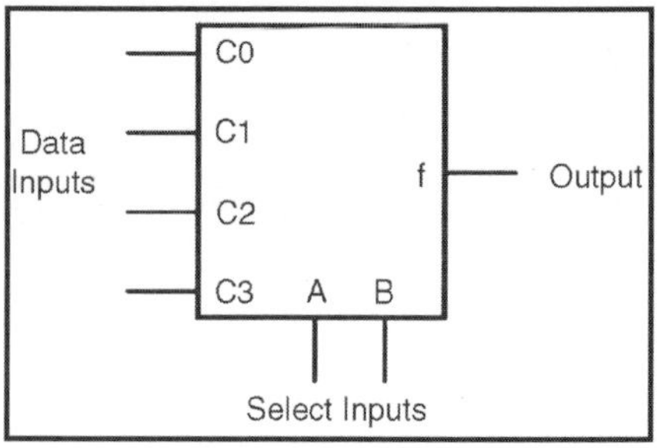

4 INPUT MULTIPLEXER

The multiplexer concept is not limited to two data inputs. If we add a second addressing input, B, we can control as many as four data inputs, as shown to the left. A third and fourth addressing input will allow the multiplexer to control eight or sixteen inputs, respectively.

Inputs A and B are the addressing inputs to this multiplexer. They select which of the four data inputs will be transmitted to the final output, X. If the data inputs are to be multiplexed for transmission to a distant location, the inputs must cycle through all four possible addresses more than twice for each single cycle of each of the data inputs. Otherwise the input data cannot be reconstructed accurately at the receiving end.

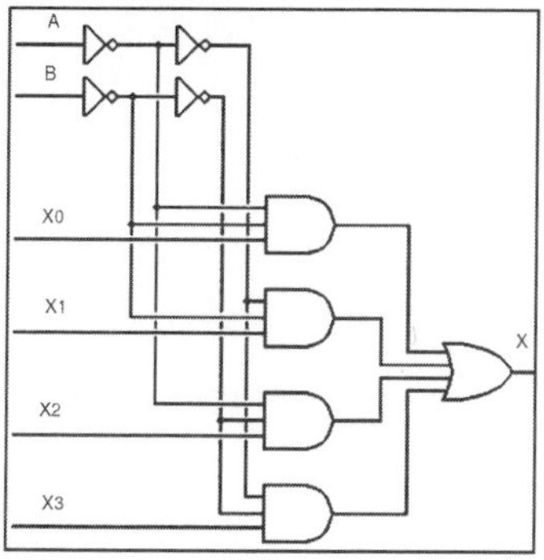

TECHNIQUE OF BINARY MULTIPLIER

Binary multiplication uses the same technique as decimal multiplication. In fact, binary multiplication is much easier because each digit we multiply by is either zero or one. Consider the simple problem of multiplying 1102 by 102. We can use this problem to review some terminology and illustrate the rules for binary multiplication.

- First, we note that 1102 is our multiplicand and 102 is our multiplier.

$$\begin{array}{r} 110 \\ \times\ 10 \\ \hline \end{array}$$

- We begin by multiplying 1102 by the rightmost digit of our multiplier which is 0. Any number times zero is zero, so we just write zeros below.

$$\begin{array}{r} 110 \\ \times 10 \\ \hline 000 \end{array}$$

- Now we multiply the multiplicand by the next digit of our multiplier which is 1. To perform this multiplication, we just need to copy the multiplicand and shift it one column to the left as we do in decimal multiplication.

$$\begin{array}{r} 110 \\ \times 10 \\ \hline 000 \\ 110 \end{array}$$

- Now we add our results together. The product of our multiplication is 1100_2.

$$\begin{array}{r} 110 \\ \times\ 10 \\ \hline 000 \\ 110 \\ \hline 1100 \end{array}$$

When performing binary multiplication, remember the following rules:

- Copy the multiplicand when the multiplier digit is 1. Otherwise, write a row of zeros.
- Shift your results one column to the left as you move to a new multiplier digit.
- Add the results together using binary addition to find the product.

SUBTRACTOR OF BINARY ADDER

HALF ADDER/SUBTRACTOR

This figure shows the configuration for a Half Adder/Subtractor. The first logic gate which is a XOR allows the circuit to do the complement of the binary bit when we want to do a subtraction. In the case that is needed to implement a addition the XOR keep the number the same.

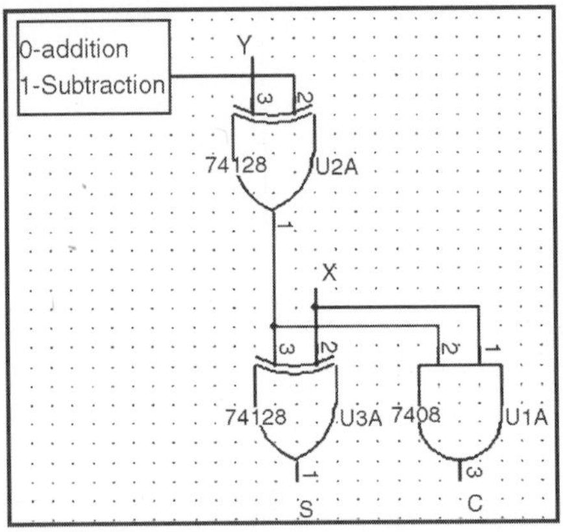

Table. Truth Table.

A	B	C	S
0	0	0	0
0	1	0	1
1	0	0	1
1	1	1	0

FULL ADDER/SUBTRACTOR

Full Adder Circuit

This circuit adds two binary one bit numbers. Also, it manages a carry that could come from another circuit.

This figure shows the configuration for a Full Adder/Subtractor. The box with the name of"Full Adder" has all the logic to do the addition and subtraction depending of the inputs that it receives.

The XOR gate that is outside the box allows the circuit to do the complement of the binary bit when we want to do a subtraction. In the case that is needed to implement an addition the XOR keep the number the same. In the previous graph the internal logic for the box"Full Adder" is shown.

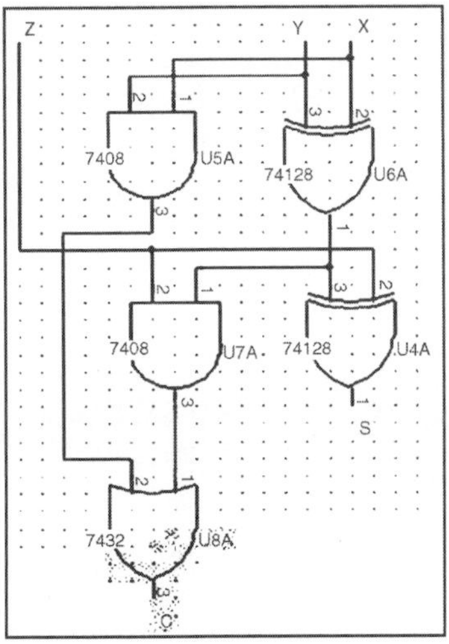

[(AA)'*(BB)']' = A+B

X	Y	Z	C	S
0	0	0	0	0
0	0	1	0	1
0	1	0	0	1
0	1	1	1	0
1	0	0	0	1
1	0	1	1	0
1	1	0	1	0
1	1	1	1	1

3-BIT ADDER/SUBTRACTOR

The 3-bit adder/Subtractor was implemented with three Full adder circuits and three XOR gates outside which implemented the operation (addition/ subtraction) selected by the user. In this circuit, we have three inputs for the first three bits binary number and three inputs for the second three bits binary number.

When the addition is selected the XOR gates keep the binary numbers the same and add them together. On the other hand, when a subtraction is selected the XOR gates complement the second number which is represented with a"Y" and after that, the Full Adder circuit adds both numbers together.

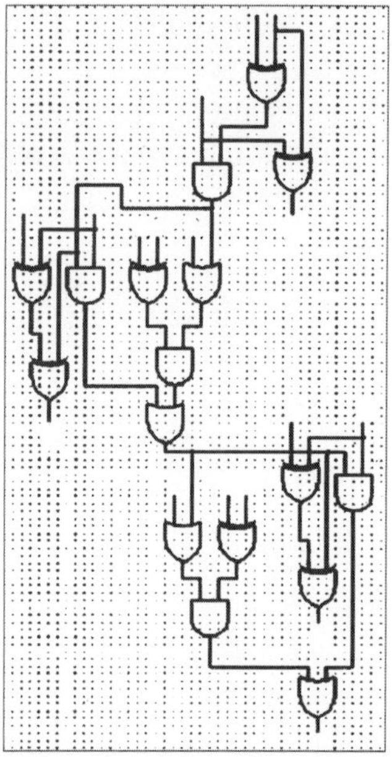

The 3-bit Adder/Subtractor circuit has four outputs. The first three outputs represent the three bits that were the result of the subtraction or addition. The last bit represents a carry.

7

Design of Sequential Circuits

Having seen how to analyse digital circuits, we now investigate how to design digital circuits. We assume that we are given a complete and unambiguous description of the circuit to be designed as a starting point. At this level, most design problems focus on one of two topics: modulo-N counters and sequence detectors.

Here is an overview of the design procedure for a sequential circuit:

- Derive the state diagram and state table for the circuit.
- Count the number of states in the state diagram (call it N) and calculate the umber of flip-flops needed (call it P) by solving the equation 2P-1 < N 2P. This is best solved by guessing the value of P.
- Assign a unique P-bit binary number (state vector) to each state. Often, the first state = 0, the next state = 1, etc.
- Derive the state transition table and the output table.
- Separate the state transition table into P tables, one for each flip-flop.
  WARNING: Things can get messy here; neatness counts.
- Decide on the types of flip-flops to use. When in doubt, use all JK's.
- Derive the input table for each flip-flop using the excitation tables for the type.
- Derive the input equations for each flip-flop based as functions of the input and current state of all flip-flops.
- Summarize the equations by writing them in one place.
- *Draw the circuit diagram*: Most homework assignments will not go this far, as the circuit diagrams are hard to draw neatly.

DESIGN PROBLEM: A MODULO-4 COUNTER

As our first design problem, let's consider a modulo-four counter. When the direction is not specified, we usually intend to build a modulo-four up-counter: 0, 1, 2, 3, 0, 1, 2, 3, etc. We solve these design problems by using the step-wise procedure listed above.

- *Step* 1: Derive the state diagram and state table for the circuit.
  Here is the state diagram. Note that it is quite simple and involves no input.

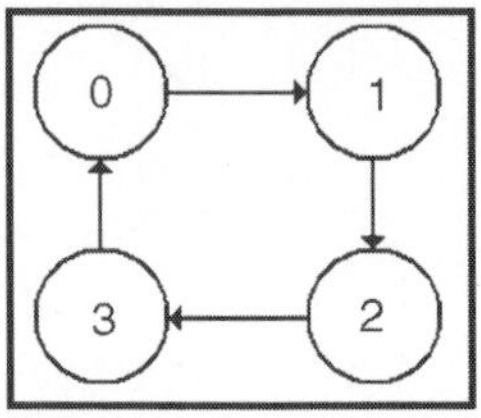

Fig. The State Diagram for a Modulo-4 Up Counter

The state table is simply a rearrangement of the state diagram into a tabular form.

Present State	Next State
0	1
1	2
2	3
3	0

- *Step* 2: Count the number of states in the state diagram (call it N) and calculate the number of flip-flops needed (call it P) by solving the equation 2P–1 < N 2P.
  The number of states on a modulo-N counter is simply N; these are labeled 0 through (N – 1). Specifically, a modulo-4 counter has four states: labeled 0, 1, 2, and 3.
  We solve the equation 2P–1 < 4 2P by noting that 21 = 2 and 22 = 4, so we have determined that 21 < 4 22, hence P = 2. We shall see later that there are valid solutions with more than two flip-flops, but there are none with fewer.
- *Step* 3: Assign a unique P-bit binary number (state vector) to each state.
  Often, the first state = 0, the next state = 1, etc. Some sequential circuits suggest an innovative numbering system, but modulo-N counters never do. We go with the obvious labeling, generated by assigning each decimal number its two-bit binary equivalent as an unsigned integer in the range from 0 to 3.

State	2-bit Vector
0	0 0
1	0 1
2	1 0
3	1 1

- *Step* 4: Derive the state transition table and the output table.
  There is no output table for any modulo-N counter, as the output associated with this type of table is the output on a transition, as is seen in a sequence detector. The transition table is a direct translation of the state table, using the assignments from the previous step.

Present State		Next State
0	00	01
1	01	10
2	10	11
3	11	00

- *Step* 5: Separate the state transition table into P tables, one for each flip-flop.
  Here we separate the state transition table into 2 tables, one for each flip-flop. Note that the flip-flops will be numbered 0 and 1, with flip-flop 0 storing the least significant bit of the state information. Thus, we shall refer to the state information as Y_1Y_0.

Flip-Flop 1			Flip-Flop 0		
Present State		Next State	Present State		Next State
Y_1	Y_0	$Y_1(T+1)$	Y_1	Y_0	$Y_0(T+1)$
0	0	0	0	0	1
0	1	1	0	1	0
1	0	1	1	0	1
1	1	0	1	1	0

- *Step* 6: Decide on the types of flip-flops to use. When in doubt, use all JK's.
  Up to this point, we have made no assumptions about the type of flip-flop to use. In order to proceed any farther with the design, we must now commit to a specific type. In line with this author's preferences, he chooses to use two JK flip-flops.

Q(t)	Q(t+1)	J	K
0	0	0	d
0	1	1	d
1	0	d	1
1	1	d	0

The excitation table for a JK flip-flop is shown at right. Recall that the"d" stands for"Don't Care". For example, if we have Q(t) = 0 and want Q(t+1) = 0, we can use either J = 0, K = 0 or J = 0, K = 1. Similarly either J = 1 and K = 0 or J = 1 and K = 1 will take Q(t) = 0 to Q(t + 1) = 1.

- *Step* 7: Derive the input table for each flip-flop using the excitation tables for the type.
  First look at flip-flop 1, representing the high-order bit. Note that we compare the present state of Y_1 to its next state in order to determine J_1 and K_1.

PS		NS	Input	
Y_1	Y_0	Y_1	J_1	K_1
0	0	0	0	d
0	1	1	1	d
1	0	1	d	0
1	1	0	d	1

Note that in deciding on the input, we must match only the 0's and 1's. We ignore the don't-cares. Note that the"d" for"don't-care" is not a variable to be assigned a value. It is a value that does not need to be matched. At the moment, Y0 is included in the table for future use only. It plays no part in determining the values of J_1 and K_1. Here is the table for Y_0

PS		NS	Input	
Y_1	Y_0	Y_0	J_0	K_0
0	0	1	1	d
0	1	0	d	1
1	0	1	1	d
1	1	0	d	1

Again, Y1 is included in the table for future use only. It plays no part in determining the values of J0 and K0.

- *Step* 8: Derive the input equations for each flip-flop based as functions of the input and current state of all flip-flops.
  At this point, we try to derive an expression that matches each column. Formal methods can be used, but generally are more trouble than they are worth. Here is this author's set of rules to match an expression to a given column.
  - If a column does not have a 0 in it, match it to the constant value 1.
  - If a column does not have a 1 in it, match it to the constant value 0.
  - If the column has both 0's and 1's in it, try to match it to a single variable, which must be part of the present state. Only the 0's and 1's in a column must match the suggested function.
  - If every 0 and 1 in the column is a mismatch, match to the complement of a function.

- If all the above fails, try for simple combinations of the present state.

Let's look at the input table for Y_1.

PS		NS	Input	
Y_1	Y_0	Y_1	J_1	K_1
0	0	0	0	d
0	1	1	1	d
1	0	1	d	0
1	1	0	d	1

Note that the column for J1 has a 0 and a 1 in it as does the column for K1. Each column has two"don't cares" in it, but we ignore these. Because each column has both a 0 and a 1 in it, neither is a match for a constant function. We now try to match J1.

- J_1 does not match Y_1, because Y_1 is 0 in the same row (0 1) as J_1 is 1.
- J1 matches Y_0. In row 0 0, both Y_0 and J_1 are 1. In row 0 1, both Y_0 and J_1 are 1.
- In rows 1 0 and 1 1, J_1 is a"don't care", so we do not need to match it.

Similar logic shows that K1 matches Y_0 also. So now we have the following matches for J_1 and K_1.

PS		NS	Input	
Y_1	Y_0	Y_1	J_1	K_1
0	0	0	0	d
0	1	1	1	d
1	0	1	d	0
1	1	0	d	1

$$J_1 = Y_0$$

$$K_1 = Y_0$$

We now examine Y_0:

PS		NS	Input	
Y_1	Y_0	Y_0	J_0	K_0
0	0	1	1	d
0	1	0	d	1
1	0	1	1	d
1	1	0	d	1

Note that there are no 0's in either the J0 or K0 column. The simplest (and best) match is

$$J_0 = 1$$

and

$$K_0 = 1.$$

- *Step* 9: Summarize the equations by writing them in one place. Here they are:

$$J_1 = Y_0$$
$$K_1 = Y_0$$
$$J_0 = 1$$
$$K_0 = 1$$

This is a counter, so there is no Z output.

- *Step* 10: Draw the circuit diagram.

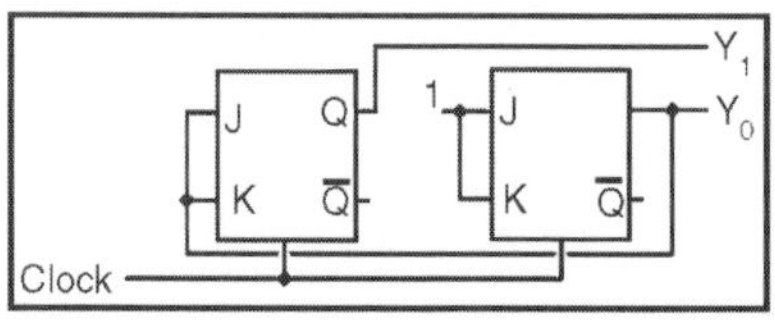

But wait - there is another solution hidden here. Recall that a JK flip-flop can be used to emulate a T flip-flop by setting the J input equal to the K input. Note that the design has the following interesting property.

$$J_1 = K_1 = Y_0$$
$$J_0 = K_0 = 1$$

Given this, we can replace each JK flip-flop with a T flip-flop, arriving at this design.

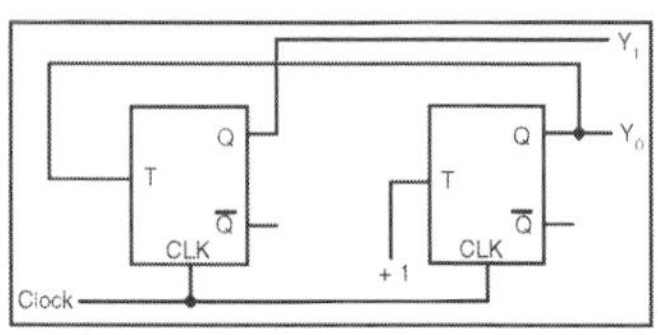

The modulo-4 counter just designed outputs binary codes for the time pulses. Specifically, we assume that it is initialized to $Y_1Y_0 = 00$ and then outputs 01, 10, 11, 00, 01, 10, 11, etc. A more realistic circuit would output discrete pulses corresponding to the decoded output, so that first T0 = 1 and all others are 0, then $T_1 = 1$ and all others are 0, etc. In order to produce the discrete signals T_0, T_1, T_2, and T_3, we need to add a decoding phase to the counter.

Note that the above design is simplified by the fact that the outputs Y_1' and Y_0' are available directly from the flip-flops and do not need to be synthesized using NOT gates.

One can achieve a simpler design at the cost of additional flip-flops. The following design is called a one-hot design, in that it uses a shift register in which exactly one flip-flop at a time is storing a 1. This design also works as a modulo-4 counter and skips the decoder delays.

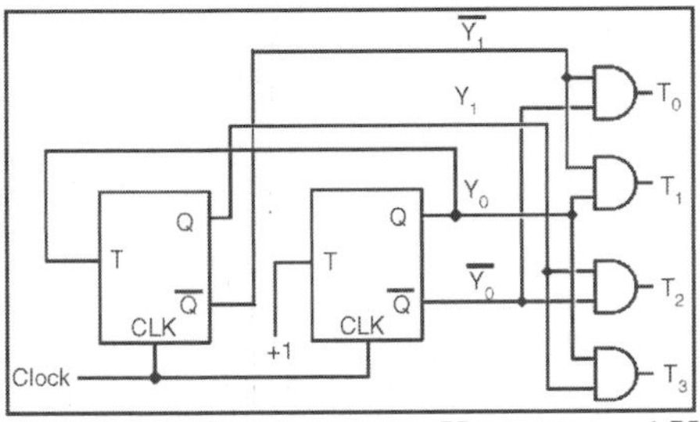

When the counter is initialized, we set $Y_0 = 1$, and $Y_1 = Y_2 = Y_3 = 0$. As the clock ticks, the single 1 is shifted by the shift register, so that the discrete signals become high in sequence.

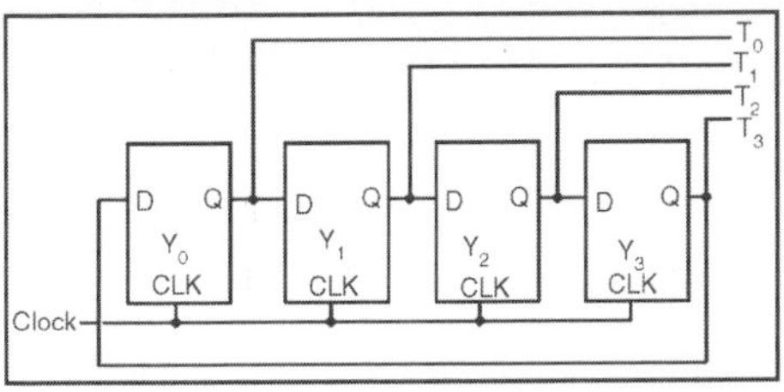

SEQUENTIAL CIRCUITS

Digital electronics is classified into combinational logic and sequential logic. Combinational logic output depends on the inputs levels, whereas sequential logic output depends on stored levels and also the input levels.

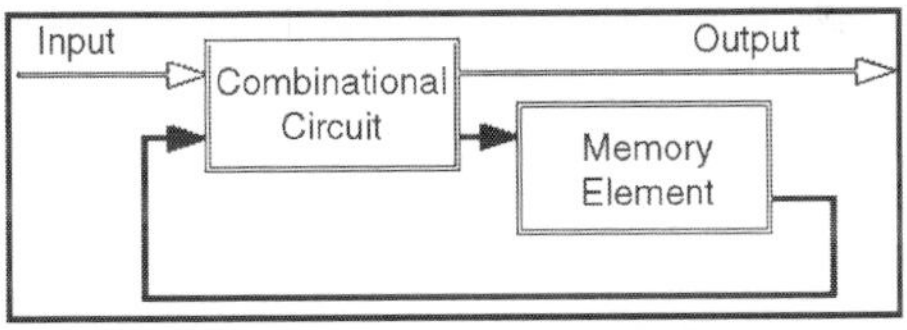

The following figure shows a way to consider sequential circuits.

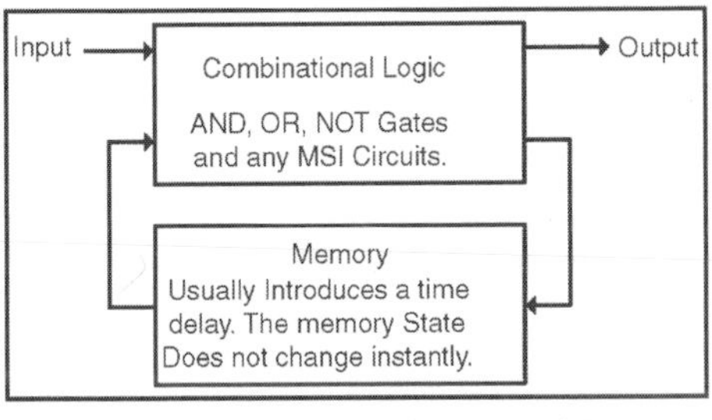

Fig. Sequential Logic Includes Combinational Logic and Memory

Sequential circuits can be characterized into two broad classes - synchronous and asynchronous. As a general rule, asynchronous circuits are faster, but much harder to design. We shall focus totally on synchronous circuits.

By Q(T) we denote the state of a sequential circuit at time T - this is basically its memory. We watch the state of the circuit change from Q(T) to Q(T + 1) as the clock ticks. The constraint on synchronous circuits is that the state of the circuit changes after the input, thus we have a typical sequence as follows:

- At time T, we have input (denoted by X) and state Q(T).
- As a result of the input X and state Q(T), a new state is computed,

This becomes available to the input only at time (T + 1) and so is called Q(T + 1).

The fact that the new state, computed as a result of X and Q(T), is not available to the input of the sequential circuit until the next time step greatly facilitates the design and analysis of the specific circuit.

ANALYSIS OF CLOCKED SEQUENTIAL CIRCUITS

Analysis consists of obtaining a state-table or a state-diagram from a given sequential circuit implementation. In other words analysis closes the loop by forming state-table from a given circuit-implementation. We will show the analysis procedure by deriving the state table of the example circuit we considered in synthesis. The circuit is shown in Figure.

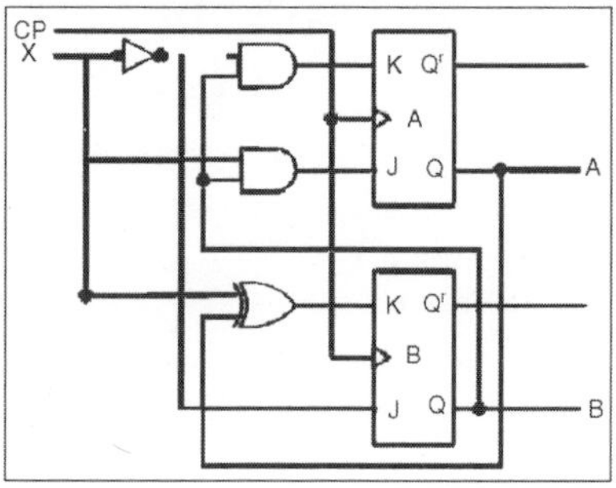

Fig. A Clocked Sequential Circuit

Present State		Next State		Output	
		x = 0	**x = 1**	**x = 0**	**x = 1**
A	B	A	B	y	y
0	0				
0	1				
1	0				
1	1				

The circuit has:

- Clock input, CP.
- One input x

- One output y
- One clocked JK flip-flop
- One clocked D flip-flop (the machine can be in maximum of 4 states)

A State table is representation of sequence of inputs, outputs, and flip-flop states in a tabular form. Two forms of state tables are shown.Analysis is the generation of state table from the given sequential circuit.

The number of rows in the state table is equal to 2 (number of flip-flops+ number of inputs). For the circuit under consideration, number of rows = 2(2+1) = 2(3) = 8

Table. State Table - Form 2.

Present State		Input	NextState	Output
A(t)	B(t)	x	A(t+1)B(t+1)	y
0	0	0		
0	0	1		
0	1	0		
0	1	1		
1	0	0		
1	0	1		
1	1	0		
1	1	1		

In the present case there are two flip-flops and one input, thus a total of 8 rows as shown in the table.

Present State		Input	NextState		Output
A(t)	B(t)	x	A(t+1)	B(t+1)	y
0	0	0	0	0	0
0	0	1	0	1	0
0	1	0	1	0	0
0	1	1	0	1	0
1	0	0	0	0	0
1	0	1	0	0	0
1	1	1	1	1	0
1	1	1	0	1	1

The analysis can start from any arbitrary state. Let us start deriving the state table from the initial state 00.

As a first step, the input equations to the flip-flops and to the combinational circuit must be obtained from the given logic diagram.

These equations are:

$$J_A = BX'$$
$$K_A = BX + B'X'$$
$$D_B = X$$
$$y = ABX$$

The first row of the state-table is obtained as follows:

When input X = 0; and present states A = 0 and B = 0 (as in the first row); then, using the above equations we get:

$$y = 0,$$
$$J_A = 0,$$
$$K_A = 1,$$

and

$$D_B = 0.$$

The resulting state table is exactly same from which we started our design example. Thus analysis is opposite to design and combined they act as a closed loop.

STATE REDUCTION

Ring the input-output relationships. In other words, to reduce the number of states, redundant states should be eliminated. A redundant state Si is a state which is equivalent to another state Sj.

Two states are said to be equivalent if, for each member of the set of inputs, they give exactly the same output and send the circuit either to the same state or to an equivalent state.

Since'm' flip-flops can describe a state machine of up to 2m states, reducing the number of states may (or may not) result in a reduction in the number of flip-flops. For example, if the number of states are reduced from 8 to 5, we still need 3 flip-flops. However, state reduction will result in more don't care states. The increased number of don't care states can help obtain a simplified circuit for the state machine. Consider the shown state diagram.

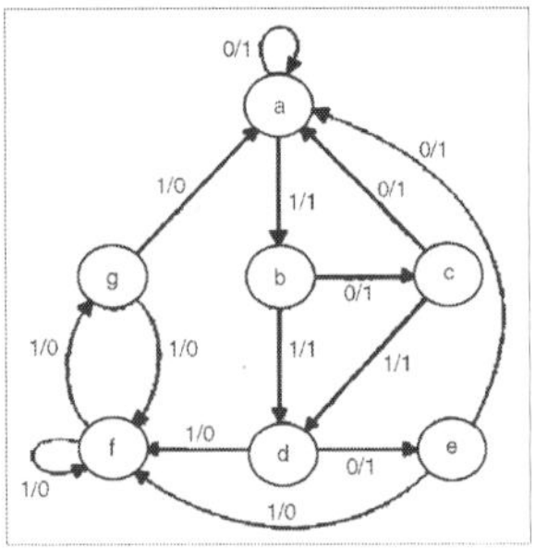

Fig. State Diagram

The state reduction proceeds by first tabulating the information of the state diagram into its equivalent state-table form (as shown in the table) The problem of state reduction requires identifying equivalent states. Each N states is replaced by 1 state.

Consider the following state table.

Table. State Table After Reduction.

Present State	Next State		Output	
	x = 0	x = 1	x = 0	x = 1
a	a	b	1	1
b	c	d	1	1
c	a	d	1	1
d	e	~~f~~ d	1	0
e	a	~~f~~ d	1	0
~~f~~	~~g~~ e	f	1	0
~~g~~	a	f	1	0

States'g' and'e' produce the same outputs, *i.e.*'1' and'0', and take the state machine to same next-states,'a' and'f', on inputs'0' and'1' respectively. Thus, states'g' and'e' are equivalent states.

We can now remove state'g' and replace it with'e' as shown. We next note that the above change has caused the states'd' and'f' to be equivalent. Thus in the next step, we remove state'f' and replace it with'd'. There are no more equivalent states remaining. The reduced state table results in the following reduced state diagram.

STATES ASSIGNMENT

When constructing a state diagram, variable names are used for states as the final number of states is not known a priori. Once the state diagram is constructed, prior to implementation (using gates and flip-flops), we need to perform the step of'state reduction'. The step that follows state reduction is state assignment. In state assignment, binary patterns are assigned to state variables.

Table: Possible State Assignments.

State	Assignment 1	Assignment 2	Assignment 3
a	001	000	000
b	010	010	100
c	011	011	010
d	100	101	101
e	101	111	011

For a given machine, there are several state assignments possible. Different state assignments may result in different combinational circuits of varying complexities. State assignment procedures try to assign binary values to states such that the cost (complexity) of the combinational circuit is reduced.

There are several heuristics that attempt to choose good state assignments (also known as state encoding) that try to reduce the required combinational logic complexity, and hence cost.

As mentioned earlier, for the reduced state machine obtained in the previous example, there can be a number of possible assignments. As an example, three different state assignments are shown in the table for the same machine.

SINGLE BINARY BIT OF FLIP FLOPS

A flip-flop is a"bit box"; it stores a single binary bit. By Q(t), we denote the state of the flip-flop at the present time, or present tick of the clock; either Q(t) = 0 or Q(t) = 1. The student will note that throughout this textbook we make the assumption that all circuit elements function correctly, so that any binary device is assumed to have only two states.

A flip-flop must have an output; this is called either Q or Q(t). This output indicates the current state of the flip-flop, and as such is either a binary 0 or a binary 1.

We shall see that, as a result of the way in which they are constructed, all flip-flops also output $\overline{Q(t)}$, the complement of the current state. Each flip-flop also has, as input, signals that specify how the next state, Q(t + 1), is to relate to the present state, Q(t).

Every flip-flop also has an input derived from the system clock, which allows it to function as a synchronous circuit. It also has connections to power and ground.

Here are the symbols for SR and D flip-flops, which we have not yet defined. Note the small triangles on the clock input; this says"edge triggered".

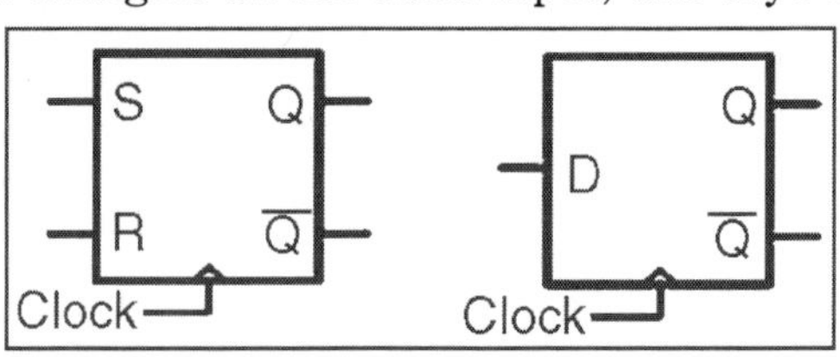

One should note that the more common symbols for flip-flops show the clock input coming in"from the side" as shown in the next figure.

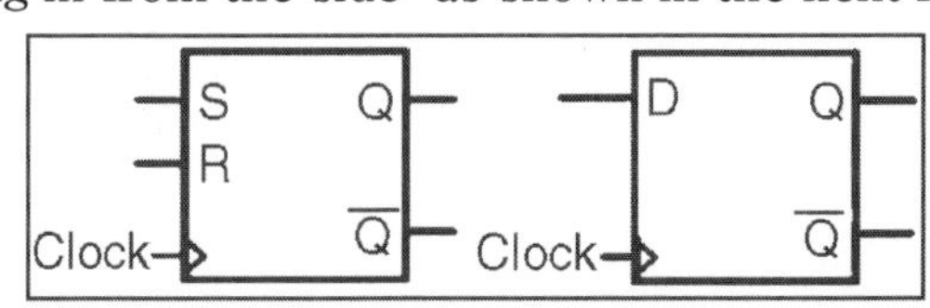

THE CLOCK

The most fundamental characteristic of synchronous sequential circuits is a system clock. This is an electronic circuit that produces a repetitive train of logic 1 and logic 0 at a regular rate, called the clock frequency. Most computer systems have a number of clocks, usually operating at related frequencies; for example – 2 GHz, 1GHz, 500MHz, and 125MHz. The inverse of the clock frequency is the clock cycle time. As an example, we consider a clock with a frequency of 2 GHz (2•109 Hertz). The cycle time is $1.0/(2 \bullet 10^9)$ seconds, or 0.5•10-9 seconds = 0.500 nanoseconds = 500 picoseconds.

Synchronous sequential circuits are sequential circuits that use a clock input to order events. Asynchronous sequential circuits do not use a common clock and, as hinted at above, are much harder to design and test. As we shall focus only on synchronous circuits, we immediately launch a discussion of the clock. The following figure illustrates some of the terms commonly used for a clock.

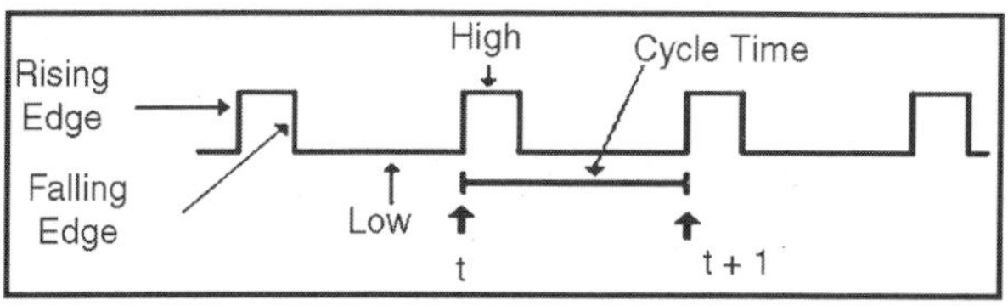

The clock input is very important to the concept of a sequential circuit. At each"tick" of the clock the output of a sequential circuit is determined by its input and by its state. We now provide a common definition of a"clock tick" - it occurs at the rising edge of each pulse.

We use t to represent the time at a clock tick and (t + 1) to denote the time at the next clock tick - the difference between the two is the clock cycle time. Suppose a 2 GHz clock, which corresponds to a clock cycle time of 0.5 nanosecond. Strictly speaking, we should label our timings in nanoseconds: 1.0, 1.5. 2.0. 2.5, etc. The convention is just to count the ticks, referring to the present clock pulse as occurring at time t and the next one at time (t + 1).

DESCRIPTION

By definition, a flip-flop is an edge-triggered latch. We must now show how achieve edge triggering. In essence, what we shall do is take a level triggered latch and give it a clock pulse with a very short positive (logical 1) phase.

The key component of an edge-triggered flip-flop is a pulse generator that we studied in a previous chapter. This could be based on the gate delay of a NOT gate. More modern devices likely use a different circuit so as to generate a shorter pulse.

Now that we have a method to generate a short pulse, we can build an edge-triggered device. The following is a diagram of the components of two typical edge-triggered flip-flops. The top one is a SR flip-flop and the bottom one is a D flip-flop.

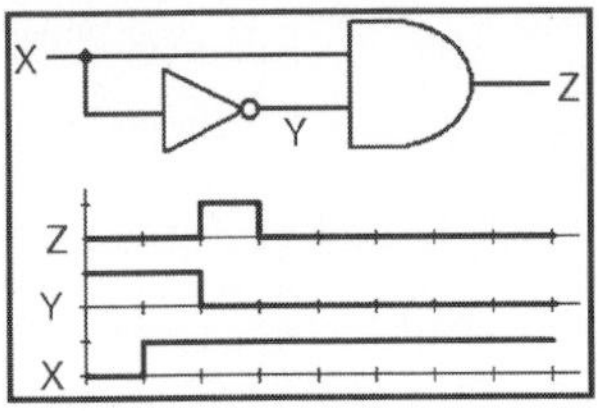

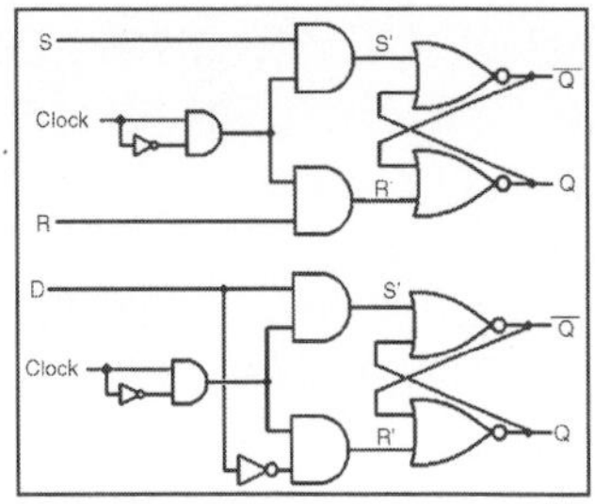

TYPES OF FLIP-FLOPS

In order to avoid the simple example, we shall examine the general case. Here is the circuit for consideration. We postulate a total of gate delays (including that of the D flip-flop) to be the time interval signified by. Thus at time after the D latch is first sensitive to its input, the circuitry has output a new value to become D. But the flip-flop is activated by a short pulse, of time length.

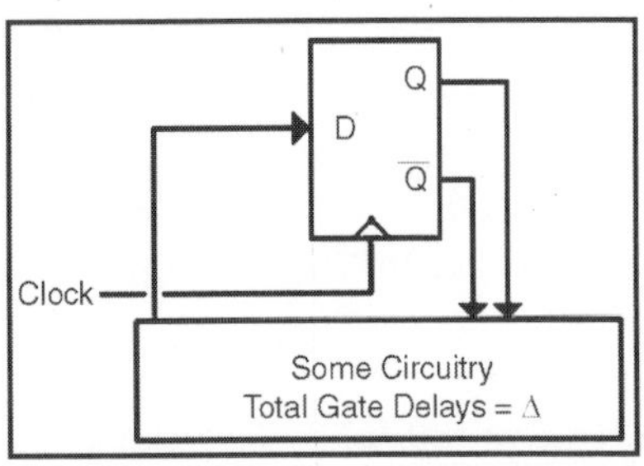

The timing diagram below shows the interruption of the uncontrolled feedback loop. By the time that the output of the circuitry has changed, the D flip-flop is no longer sensitive to input. The input will not become effective until the beginning of the next clock cycle.

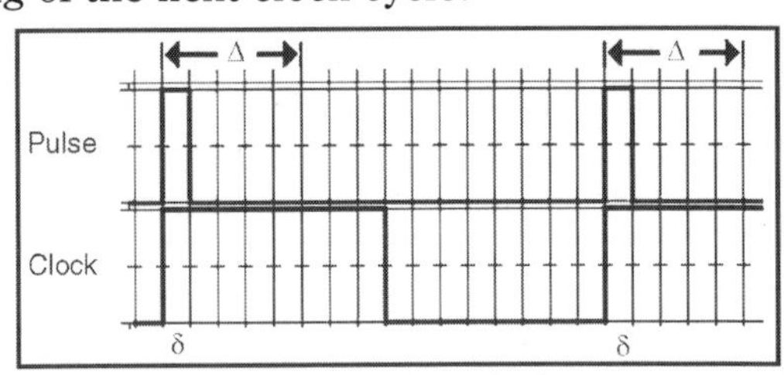

How can one insure that the relative timings of two circuits, the D flip-flop and the rest of the CPU circuitry, operate with the correct timings? This is one of the issues of central importance in the design of a CPU; since we have stumbled into it, let's talk about it.

To be more precise, define two total gate delays: Δ_{MIN} and Δ_{MAX}. Δ_{MIN} is the total time delay for the fastest CPU operation and Δ_{MAX}. the delay for the slowest. We must have MIN >, or the circuit would occasionally display uncontrolled feedback. Conservatively, we might say Δ_{MIN} 1.5• δ. The next criterion is a bit more difficult to state precisely, but it might be stated something like T ≥ 1.5• Δ_{MAX}, where T denotes the clock period. What we say here is that the clock period must be long enough for the CPU output to"settle". Note, however, that a value of T much larger than MAX is just wasted time.

Put another way, the value of Δ_{MAX} determines the fastest clock that can reasonably be applied to the CPU. A good part of the art of CPU design is based on this issue. When we study the RISC (Reduced Instruction Set Computer) movement, we shall see that one of the issues was to remove the more complex instructions, thus reducing Δ_{MAX} and allowing for a faster clock. There is a trade-off here that we shall explore in later chapters.

To be honest, the sum of CPU gate delays is not always the limiting factor in the clock speed. Some recent CPUs have been designed with a"hot clock" that is hot in both ways.

It is very fast, being of the order of 4 to 5 GHz. It is also hot in the literal sense, in that the CPU, operating at that clock rate, emits so much heat that it overheats itself. The art of CPU design is always bumping up against those messy laws of physics.

The SR Flip-Flop

We have completely defined the SR flip-flop, based on the idea of an SR latch. While we could just leave the topic and proceed, it is appropriate at this time to review the subject and state, in one place, what the student is expected to know. We begin with a depiction of the SR flip-flop that will be used in all future discussions.

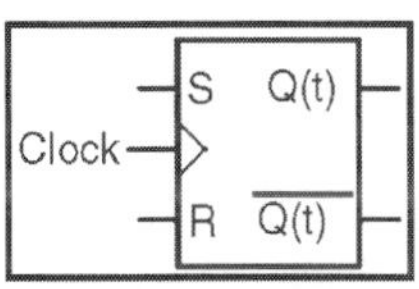

At this point we are no longer interested in the internal construction of the flip-flop, but on its operational characteristics. These are given in the characteristic table.

S	R	Q(t + 1)
0	0	Q(t)
0	1	0
1	0	1
1	1	ERROR

We next address an issue that commonly arises in the use of flip-flops in circuit design. We have a number of scenarios. For each, we know Q(t) and what Q(t + 1) should be. The question is how to achieve that change. For example, if Q(t) = 0 and we want Q(t + 1) = 0, we have two choices: either S = 0 and R = 0, or S = 0 and R = 1. The first option, keeps the state unchanged at 0; the second forces it to 0. As S = 0 is sufficient to do this without regard to the value of R, we say that the input is S = 0 and R = d; the d standing for"don't care".

On the other hand, if Q(t) = 0 and Q(t + 1) = 1, only S = 1 and R = 0 will do. This is the only combination that will give a next state of 1 when the present state is 0.

If we have Q(t) = 1 and want Q(t + 1) = 0, then our choice is simple:

S = 0 and R = 1.

If we have Q(t) = 1 and want Q(t + 1) = 1, then we can choose either S = 0 and R = 0, or S = 1 and R = 0. This is denoted as S = d and R = 0. The above discussions lead to the excitation table for the SR flip-flop.

Q(t)	Q(t + 1)	S	R
0	0	0	d
0	1	1	0
1	0	0	1
1	1	d	0

The JK Flip-Flop: Enhancing the SR Flip-Flop

Recall the characteristic table of the SR flip-flop. We repeat the table here for emphasis.

S	R	Q(t + 1)
0	0	Q(t)
0	1	0
1	0	1
1	1	ERROR

The theoretician examining this table would note two facts immediately.

- The input S = 1 and R = 1 is disallowed; we would like to do something with it.
- The values for Q(t + 1) are three of the possible four Boolean functions of 1 variable.

Considering Q as a Boolean variable, we now show that there are exactly four Boolean functions of this Boolean variable. These are f(Q) = 0, f(Q) = 1, f(Q) = Q, and f(Q) = $\overline{Q}$.

We do this by showing the truth table for each of these functions and noting that there are only four different ways to put 0's and 1's into the two row entries for a function.

Table. The Four Boolean Functions of Boolean Variable Q.

Q	0	Q	$\overline{Q}$	1
0	0	0	1	1
1	0	1	0	1

Given this, our enhanced SR flip-flop would have Q(t + 1) = $\overline{Q}$ (t) as one possible output. For maximal compatibility with the existing SR, we would want to leave the existing valid inputs alone and just make good use of the invalid one. What we get is a JK flip-flop. Recalling that an SR flip-flop is so called because it is a Set-Reset device, we may ask for the meaning of JK. One is tempted to make up some story related to names in German, but the basic answer is that"I don't know". In any case, we present the characteristic table for the JK flip-flop and then ask how one might modify an SR flip to achieve that goal. Here is the desired characteristic table.

J	K	Q(t + 1)
0	0	Q(t)
0	1	0
1	0	1
1	1	$\overline{Q}$ (t)

Viewing this as a modification of the SR flip-flop, we ask how to generate each of S and R from the inputs J and K under the following constraints;

- Except when J = 1 and K = 1, we want to have S = J and R = K. Under these circumstances, the behaviour is identical.
- When J = 1, K = 1, and Q = 0, we want S = 1 and R = 0. This makes Q(t + 1) = 1.
- When J = 1, K = 1, and Q = 1, we want S = 0 and R = 1. This makes Q(t + 1) = 0.

When in doubt about how to create a circuit, we make a truth table. The inputs to the truth table are J, K, and Q (the present state). The outputs are S and R.

Row	Q	J	K	S	R
0	0	0	0	0	0
1	0	0	1	0	1
2	0	1	0	1	0
3	0	1	1	1	0
4	1	0	0	0	0
5	1	0	1	0	1
6	1	1	0	1	0
7	1	1	1	0	1

We have two patterns that almost work: S = J• $\overline{Q}$ and R = K•Q. Let's examine each case. S = J• $\overline{Q}$ produces the expected result for all rows except row 6. In that row we have,

- Q = 1, J = 1, and K = 0. We want Q(t + 1) to be 1. But S = 0 and R = 0 will cause.
- Q(t + 1) = Q(t) = 1, exactly what we want. So this simple formula causes no trouble.
- R = K•Q produces the expected result for all rows except row 1. In that row we have
- Q = 0, J = 0, and K = 1. We want Q(t + 1) to be 0. But S = 0 and R = 0 will cause
- Q(t + 1) = Q(t) = 0, exactly what we want. So again we have no trouble.

Here is the basic detailed circuit for the JK flip-flop.

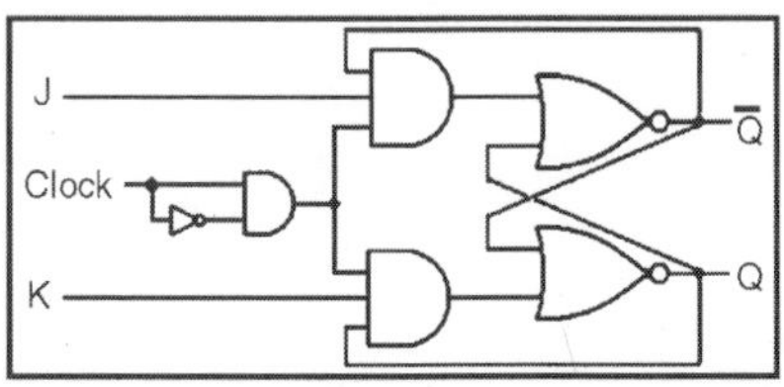

We now give the standard representation of the JK flip-flop as will be used in future discussions. Again, note the triangle symbol on the clock input, indicating edge triggering.

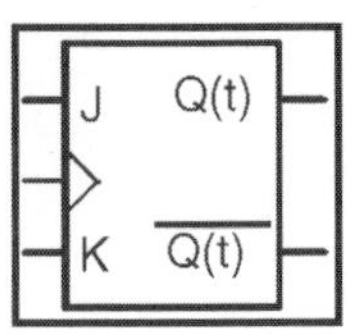

We have already presented the characteristic table for the JK flip-flop. Here it is again.

J	K	Q(t + 1)
0	0	Q(t)
0	1	0
1	0	1
1	1	$\overline{Q}$ (t)

We now create the excitation table for the JK.

- If Q(t) = 0 and Q(t + 1) is to be 0, we can use either J = 0 and K = 0, or J = 0 and K = 1.
- If Q(t) = 0 and Q(t + 1) is to be 1, we can use either J = 1 and K = 0, or J = 1 and K = 1.

Note that this is a new option, not available for an SR flip-flop.

- If Q(t) = 1 and Q(t + 1) is to be 0, we can use either J = 0 and K = 1, or J = 1 and K = 1.

Note that this also is a new option, not available for an SR flip-flop.

- If Q(t) = 1 and Q(t + 1) is to be 1, we can use either J = 0 and K = 0, or J = 1 and K = 0.

This gives the excitation table for the JK flip-flop.

Q(t)	Q(t + 1)	J	K
0	0	0	d
0	1	1	d
1	0	d	1
1	1	d	0

THE D FLIP-FLOP

We have already examined the D latch. The input is labeled"D" for"Data". The D flip-flop has a characteristic table identical to that of a D latch.

D	Q(t + 1)
0	0
1	1

The device is so simple that it does not require an excitation equation. We just use an excitation equation, which simply states"Give it what you want".

$$D = Q(t + 1)$$

Let's compare timing diagrams for a normal D latch versus one that is edge-triggered:

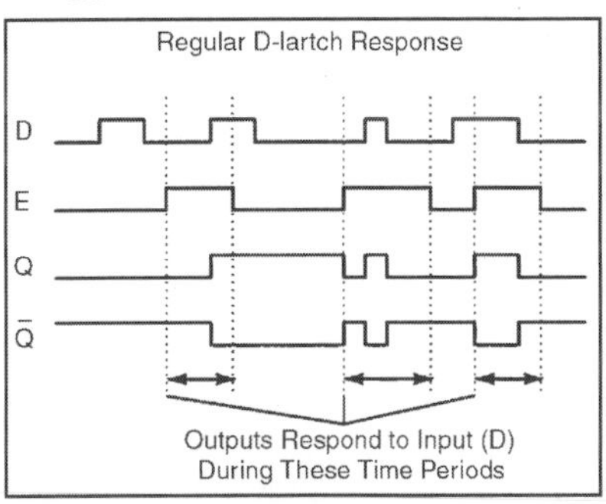

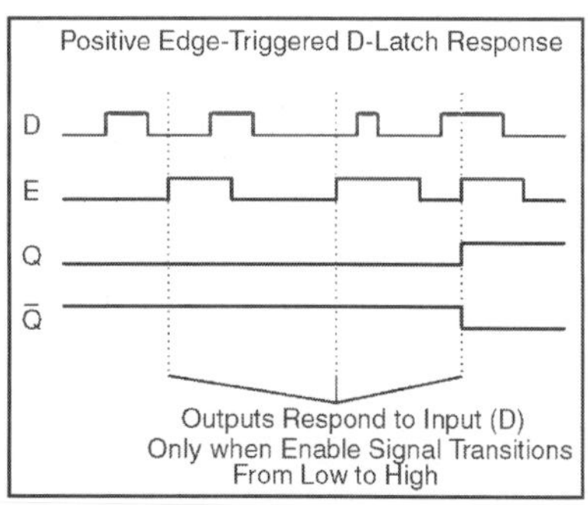

In the first timing diagram, the outputs respond to input D whenever the enable (E) input is high, for however long it remains high. When the enable signal falls back to a low state, the circuit remains latched. In the second timing diagram, we note a distinctly different response in the circuit output(s): it only responds to the D input during that brief moment of time when the enable signal *changes*, or *transitions*, from low to high. This is known as *positive* edge-triggering.

There is such a thing as *negative* edge triggering as well, and it produces the following response to the same input signals:

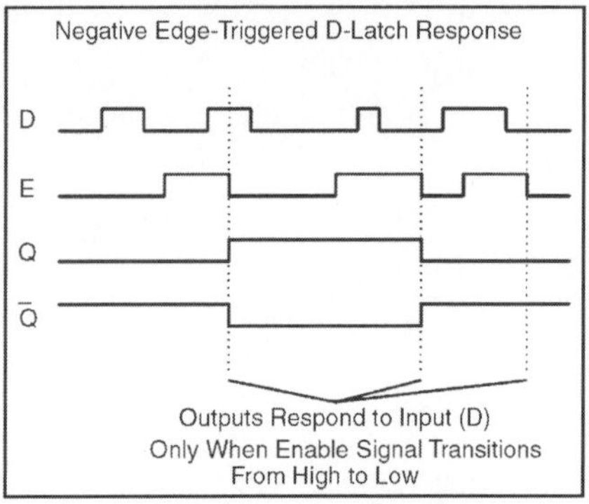

Whenever we enable a multivibrator circuit on the transitional edge of a square-wave enable signal, we call it a *flip-flop*instead of a *latch*.

Consequently, and edge-triggered S-R circuit is more properly known as an S-R flip-flop, and an edge-triggered D circuit as a D flip-flop.

The enable signal is renamed to be the *clock* signal. Also, we refer to the data inputs (S, R, and D, respectively) of these flip-flops as *synchronous* inputs, because they have effect only at the time of the clock pulse edge (transition), thereby synchronizing any output changes with that clock pulse, rather than at the whim of the data inputs.

But, how do we actually accomplish this edge-triggering? To create a "gated" S-R latch from a regular S-R latch is easy enough with a couple of AND gates, but how do we implement logic that only pays attention to the *rising or falling edge* of a changing digital signal?

What we need is a digital circuit that outputs a brief pulse whenever the input is activated for an arbitrary period of time, and we can use the output of this circuit to briefly enable the latch.

We're getting a little ahead of ourselves here, but this is actually a kind of monostable multivibrator, which for now we'll call a *pulse detector*.

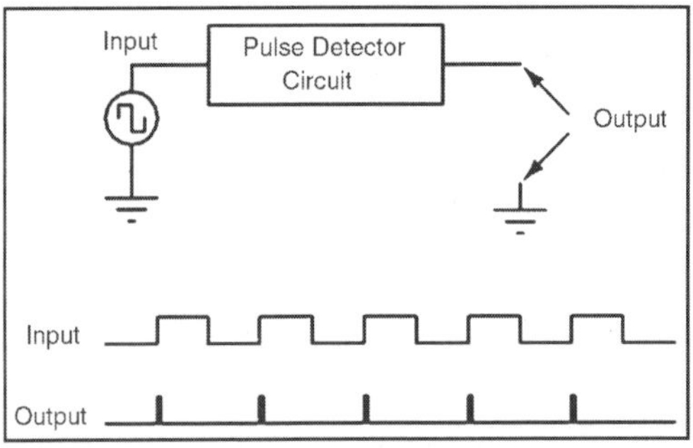

The duration of each output pulse is set by components in the pulse circuit itself. In ladder logic, this can be accomplished quite easily through the use of a time-delay relay with a very short delay time:

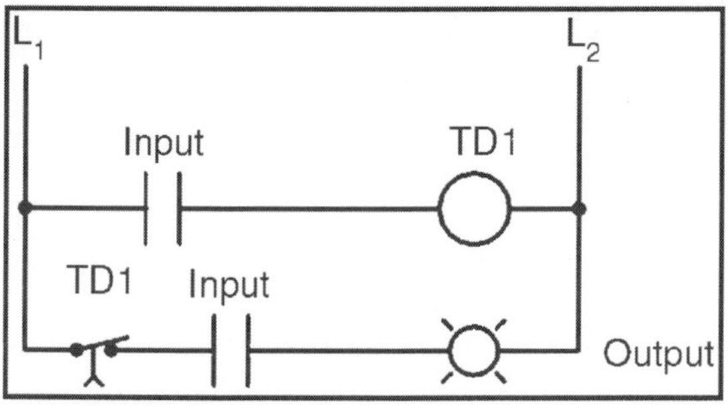

Implementing this timing function with semiconductor components is actually quite easy, as it exploits the inherent time delay within every logic gate (known as *propagation delay*).

What we do is take an input signal and split it up two ways, then place a gate or a series of gates in one of those signal paths just to delay it a bit, then have both the original signal and its delayed counterpart enter into a two-input gate that outputs a high signal for the brief moment of time that the delayed signal has not yet caught up to the low-to-high change in the non-delayed signal.

An example circuit for producing a clock pulse on a low-to-high input signal transition is shown here:

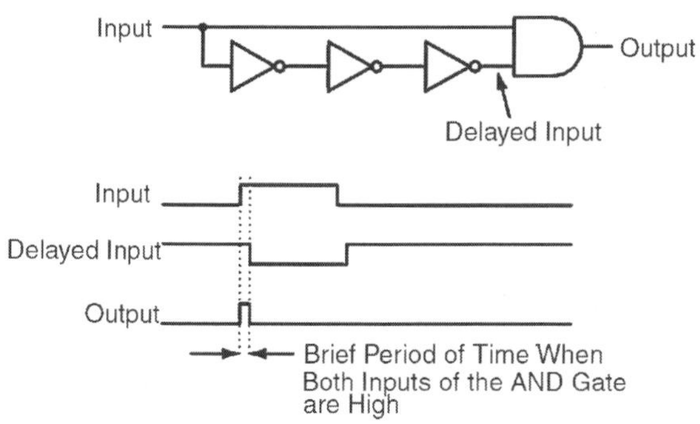

This circuit may be converted into a negative-edge pulse detector circuit with only a change of the final gate from AND to NOR:

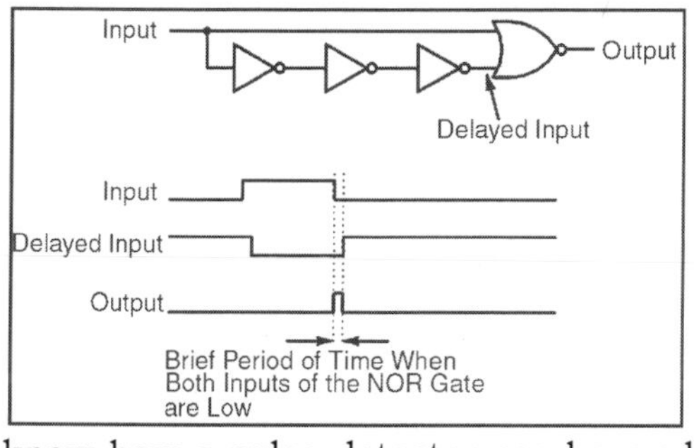

Now that we know how a pulse detector can be made, we can show it attached to the enable input of a latch to turn it into a flip-flop. In this case, the circuit is a S-R flip-flop:

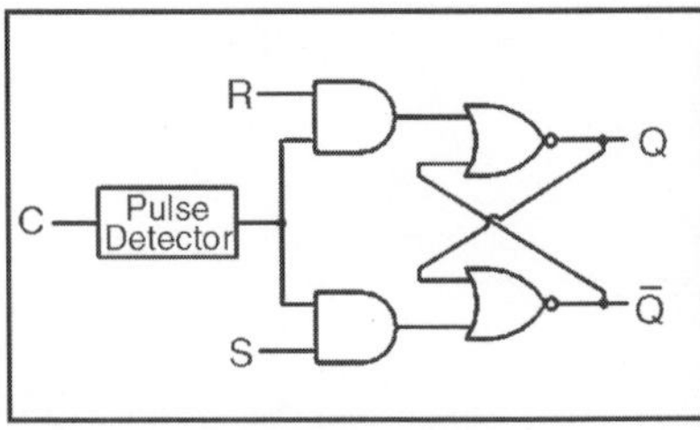

C	S	R	Q	$\bar{Q}$
↑	0	0	Latch	Latch
↑	0	1	0	1
↑	1	0	1	0
↑	1	1	0	0
X	0	0	Latch	Latch
X	0	1	Latch	Latch
X	1	0	Latch	Latch
X	1	1	Latch	Latch

Only when the clock signal (C) is transitioning from low to high is the circuit responsive to the S and R inputs. For any other condition of the clock signal ("x") the circuit will be latched.

A ladder logic version of the S-R flip-flop is shown here:

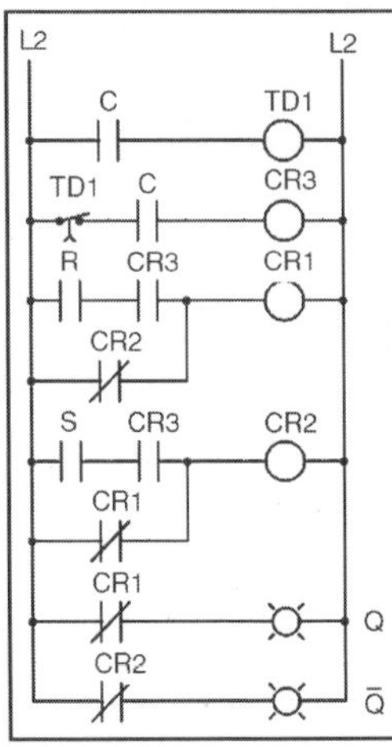

C	S	R	Q	$\bar{Q}$
↑	0	0	Latch	Latch
↑	0	1	0	1
↑	1	0	1	0
↑	1	1	0	0
X	0	0	Latch	Latch
X	0	1	Latch	Latch
X	1	0	Latch	Latch
X	1	1	Latch	Latch

Relay contact CR_3 in the ladder diagram takes the place of the old E contact in the S-R latch circuit, and is closed only during the short time that both C is closed and time-delay contact TR_1 is closed. In either case (gate or ladder circuit), we see that the inputs S and R have no effect unless C is transitioning from a low (0) to a high (1) state. Otherwise, the flip-flop's outputs latch in their previous states.

It is important to note that the invalid state for the S-R flip-flop is maintained only for the short period of time that the pulse detector circuit allows the latch to be enabled. After that brief time period has elapsed, the outputs will latch into either the set or the reset state.

Once again, the problem of a *race condition* manifests itself. With no enable signal, an invalid output state cannot be maintained. However, the valid "latched" states of the multivibrator — set and reset — are mutually exclusive to one another. Therefore, the two gates of the multivibrator circuit will "race" each other for supremacy, and whichever one attains a high output state first will "win."

The block symbols for flip-flops are slightly different from that of their respective latch counterparts:

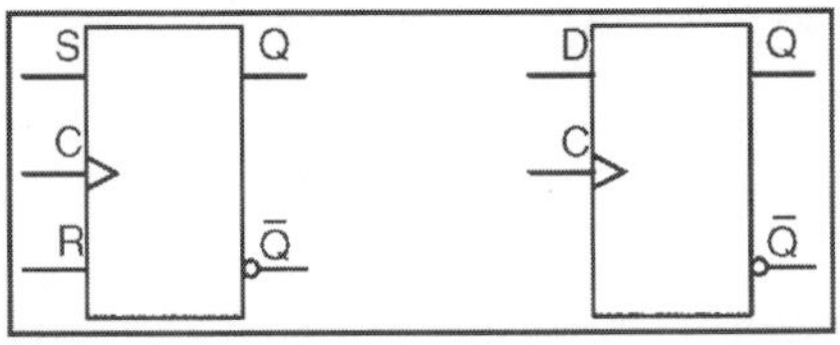

The triangle symbol next to the clock inputs tells us that these are edge-triggered devices, and consequently that these are flip-flops rather than latches. The symbols above are positive edge-triggered: that is, they "clock" on the rising edge (low-to-high transition) of the clock signal. Negative edge-triggered devices are symbolized with a bubble on the clock input line:

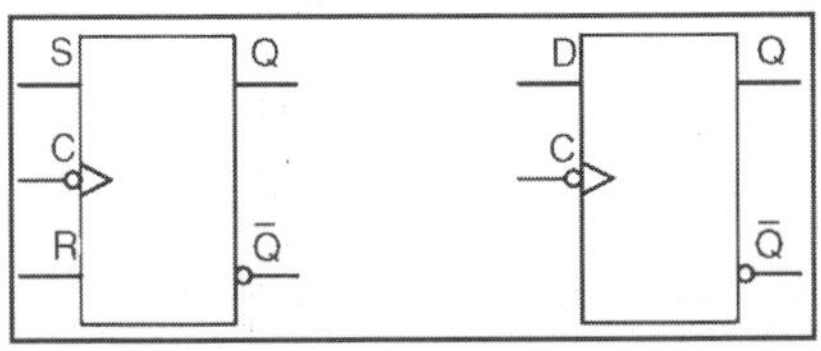

Both of the above flip-flops will "clock" on the falling edge (high-to-low transition) of the clock signal.

The T Flip-Flop

This is the fourth and last of the major types of flip-flops. The input to this flip-flop is labeled"T" for toggle. When T = 0, the state remains the same.

When T = 1, the flip-flop changes state. This gives rise to the following characteristic table.

T	Q(t + 1)
0	Q(t)
1	$\overline{Q}$ (t)

The excitation table for this flip-flop is almost obvious.

Q(t)	Q(t + 1)	T
0	0	0
0	1	1
1	0	1
1	1	0

This gives rise to an excitation equation.

$$T = Q(t) \oplus Q(t + 1)$$

THE JK AS A GENERAL FLIP-FLOP

We now take notice that the JK can be configured to function as any of the other 3 flip-flop types. In this, it is the most general type of flip-flop.

Another variation on a theme of bistable multivibrators is the J-K flip-flop. Essentially, this is a modified version of an S-R flip-flop with no "invalid" or "illegal" output state. Look closely at the following diagram to see how this is accomplished:

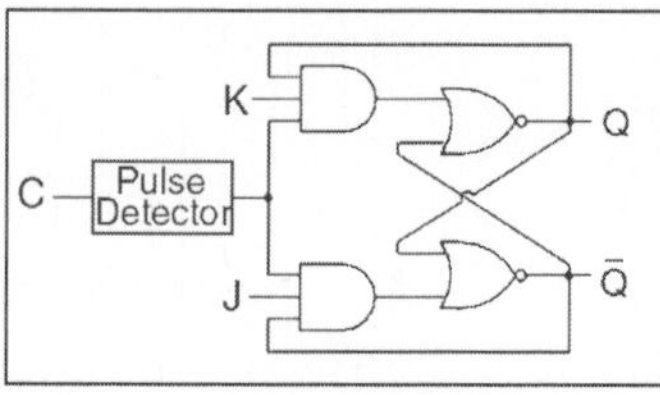

C	J	K	Q	$\bar{Q}$
┌	0	0	Latch	Latch
┌	0	1	0	1
┌	1	0	1	0
┌	1	1	Toggle	Toggle
X	0	0	Latch	Latch
X	0	1	Latch	Latch
X	1	0	Latch	Latch
X	1	1	Latch	Latch

What used to be the S and R inputs are now called the J and K inputs, respectively. The old two-input AND gates have been replaced with 3-input AND gates, and the third input of each gate receives feedback from the Q and not-Q outputs. What this does for us is permit the J input to have effect only when the circuit is reset, and permit the K input to have effect only when the circuit is set.

In other words, the two inputs are *interlocked*, to use a relay logic term, so that they cannot both be activated simultaneously. If the circuit is "set," the J input is inhibited by the 0 status of not-Q through the lower AND gate; if the circuit is "reset," the K input is inhibited by the 0 status of Q through the upper AND gate.

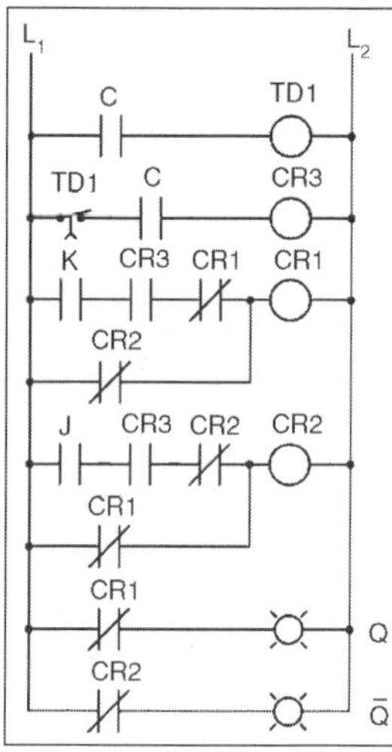

C	J	K	Q	$\bar{Q}$
┌	0	0	Latch	Latch
┌	0	1	0	1
┌	1	0	1	0
┌	1	1	Toggle	Toggle
X	0	0	Latch	Latch
X	0	1	Latch	Latch
X	1	0	Latch	Latch
X	1	1	Latch	Latch

When both J and K inputs are 1, however, something unique happens. Because of the selective inhibiting action of those 3-input AND gates, a "set" state inhibits input J so that the flip-flop acts as if J=0 while K=1 when in fact both are 1. On the next clock pulse, the outputs will switch ("toggle") from set (Q=1 and not-Q=0) to reset (Q=0 and not-Q=1). Conversely, a "reset" state inhibits input K so that the flip-flop acts as if J=1 and K=0 when in fact both are 1. The next clock pulse toggles the circuit again from reset to set.

See if you can follow this logical sequence with the ladder logic equivalent of the J-K flip-flop:

The end result is that the S-R flip-flop's "invalid" state is eliminated (along with the race condition it engendered) and we get a useful feature as a bonus: the ability to toggle between the two (bistable) output states with every transition of the clock input signal.

There is no such thing as a J-K latch, only J-K flip-flops. Without the edge-triggering of the clock input, the circuit would continuously toggle between its two output states when both J and K were held high (1), making it an astable device instead of a bistable device in that circumstance.

If we want to preserve bistable operation for all combinations of input states, we *must* use edge-triggering so that it toggles only when we tell it to, one step (clock pulse) at a time.

The block symbol for a J-K flip-flop is a whole lot less frightening than its internal circuitry, and just like the S-R and D flip-flops, J-K flip-flops come in two clock varieties (negative and positive edge-triggered):

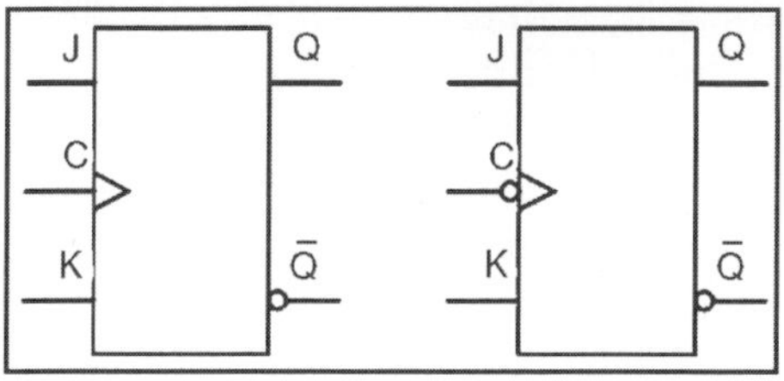

LATCHES AND CIRCUIT ELEMENTS

When showing either latches or flip-flops as circuit elements, it is undesirable to show the"internals" of the device.

Here are the symbols for SR and D latches.

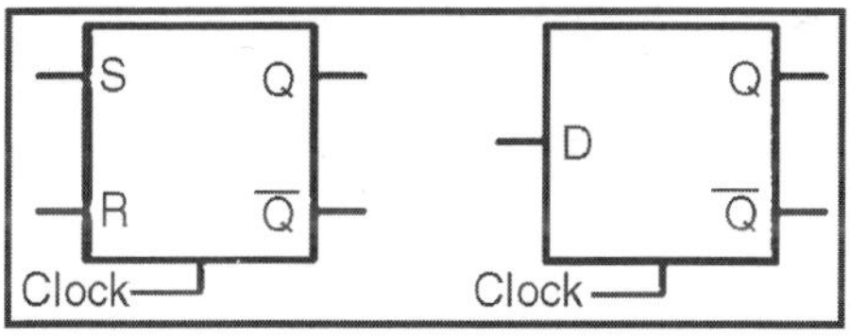

THE S-R LATCH

A bistable multivibrator has two stable states, as indicated by the prefix bi in its name.

Typically, one state is referred to asset and the other as reset. The simplest bistable device, therefore, is known as a set-reset, or S-R, latch.

To create an S-R latch, we can wire two NOR gates in such a way that the output of one feeds back to the input of another, and visa-versa.

The Q and not-Q outputs are supposed to be in opposite states. I say"supposed to" because making both the S and R inputs equal to 1 results in both Q and not-Q being 0. For this reason, having both S and R equal to 1 is called an invalid orillegal state for the S-R multivibrator.

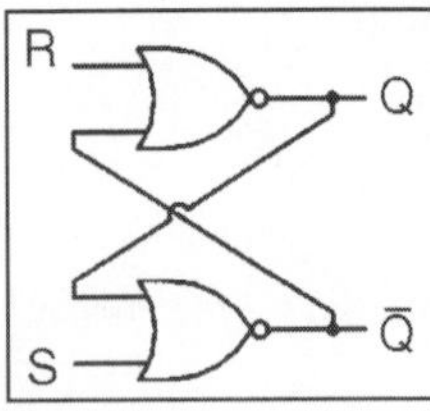

S	R	Q	$\bar{Q}$
0	0	Latch	Latch
0	1	0	1
1	0	1	0
1	1	0	0

Otherwise, making S=1 and R=0"sets" the multivibrator so that Q=1 and not-Q=0. Conversely, making R=1 and S=0"resets" the multivibrator in the opposite state. When S and R are both equal to 0, the multivibrator's outputs"latch" in their prior states.

Note how the same multivibrator function can be implemented in ladder logic, with the same results:

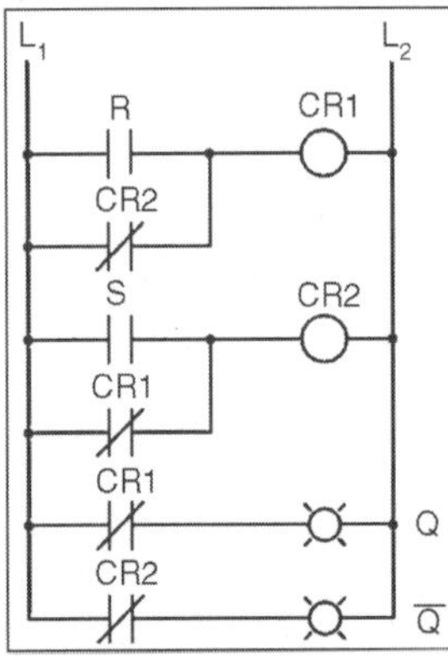

S	R	Q	$\bar{Q}$
0	0	Latch	Latch
0	1	0	1
1	0	1	0
1	1	0	0

By definition, a condition of Q=1 and not-Q=0 is set. A condition of Q=0 and not-Q=1 is reset. These terms are universal in describing the output states of any multivibrator circuit.

The astute observer will note that the initial power-up condition of either the gate or ladder variety of S-R latch is such that both gates (coils) start in the de-energized mode. As such, one would expect that the circuit will start up in an invalid condition, with both Q and not-Q outputs being in the same state.

Actually, this is true! However, the invalid condition is unstable with both S and R inputs inactive, and the circuit will quickly stabilize in either the set or reset condition because one gate (or relay) is bound to react a little faster than the other.

If both gates (or coils) were precisely identical, they would oscillate between high and low like an astable multivibrator upon power-up without ever reaching a point of stability! Fortunately for cases like this, such a precise match of components is a rare possibility. It must be noted that although an astable (continually oscillating) condition would be extremely rare, there will most likely be a cycle or two of oscillation in the above circuit, and the final state of the circuit (set or reset) after power-up would be unpredictable. The root of the problem is a race condition between the two relays CR_1 and CR_2.

A race condition occurs when two mutually-exclusive events are simultaneously initiated through different circuit elements by a single cause In this case, the circuit elements are relays CR_1 and CR_2, and their de-energized states are mutually exclusive due to the normally-closed interlocking contacts. If one relay coil is de-energized, its normally-closed contact will keep the other coil energized, thus maintaining the circuit in one of two states (set or reset).

Interlocking prevents bothrelays from latching. However, if both relay coils start in their de-energized states (such as after the whole circuit has been de-energized and is then powered up) both relays will"race" to become latched on as they receive power (the"single cause") through the normally-closed contact of the other relay.

One of those relays will inevitably reach that condition before the other, thus opening its normally-closed interlocking contact and de-energizing the other relay coil. Which relay"wins" this race is dependent on the physical characteristics of the relays and not the circuit design, so the designer cannot ensure which state the circuit will fall into after power-up.

Race conditions should be avoided in circuit design primarily for the unpredictability that will be created. One way to avoid such a condition is to insert a time-delay relay into the circuit to disable one of the competing relays for a short time, giving the other one a clear advantage. In other words, by purposely slowing down the de-energization of one relay, we ensure that the other relay will always"win" and the race results will always be predictable. Here is an example of how a time-delay relay might be applied to the above circuit to avoid the race condition:

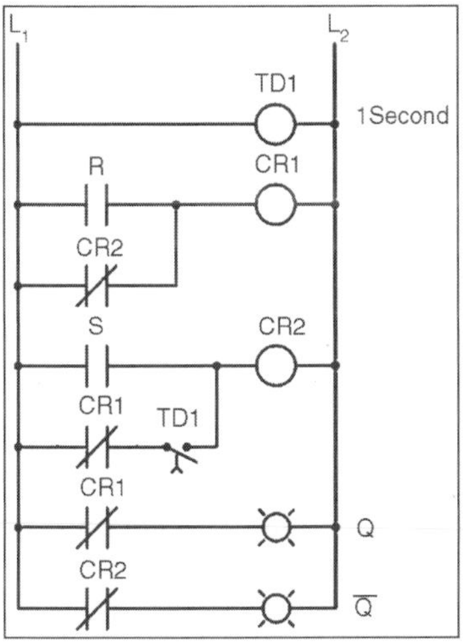

When the circuit powers up, time-delay relay contact TD1 in the fifth rung down will delay closing for 1 second. Having that contact open for 1 second prevents relay CR2 from energizing through contact CR1 in its normally-closed state after power-up.

Therefore, relay CR1 will be allowed to energize first (with a 1-second head start), thus opening the normally-closed CR1contact in the fifth rung, preventing CR2 from being energized without the S input going active. The end result is that the circuit powers up cleanly and predictably in the reset state with S=0 and R=0.

It should be mentioned that race conditions are not restricted to relay circuits. Solid-state logic gate circuits may also suffer from the ill effects of race conditions if improperly designed. Complex computer programmes, for that matter, may also incur race problems if improperly designed. Race problems are a possibility for any sequential system, and may not be discovered until some time after initial testing of the system. They can be very difficult problems to detect and eliminate.

A practical application of an S-R latch circuit might be for starting and stopping a motor, using normally-open, momentary pushbutton switch contacts for both start (S) and stop (R) switches, then energizing a motor contactor with either a CR1 or CR2 contact (or using a contactor in place of CR1 or CR2). Normally, a much simpler ladder logic circuit is employed, such as this:

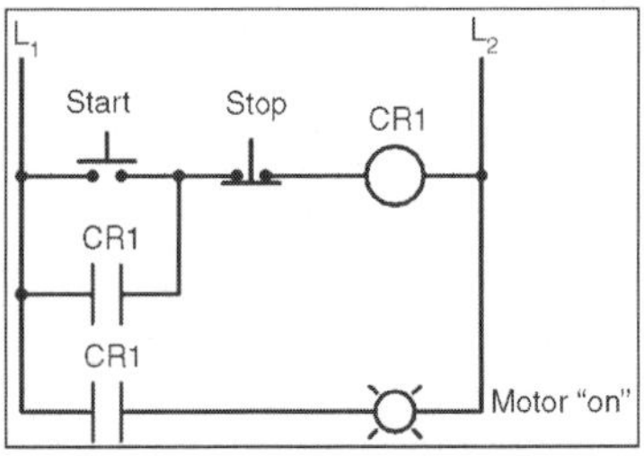

In the above motor start/stop circuit, the CR1 contact in parallel with the start switch contact is referred to as a"seal-in" contact, because it"seals" or latches control relay CR1 in the energized state after the start switch has been released. To break the"sèal," or to"unlatch" or"reset" the circuit, the stop pushbutton is pressed, which de-energizes CR1 and restores the seal-in contact to its normally open status. Notice, however, that this circuit performs much the same function as the S-R latch. Also note that this circuit has no inherent instability problem (if even a remote possibility) as does the double-relay S-R latch design.

In semiconductor form, S-R latches come in prepackaged units so that you don't have to build them from individual gates. They are symbolized as such:

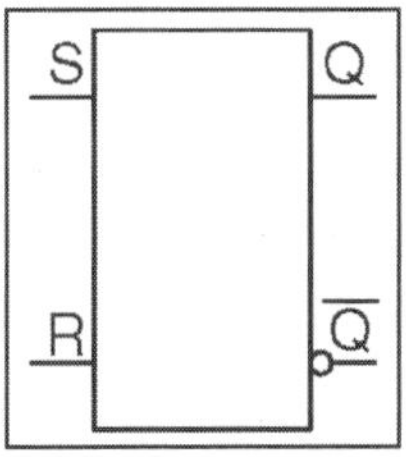

The Gated S-R Latch

It is sometimes useful in logic circuits to have a multivibrator which changes state only when certain conditions are met, regardless of its S and R input states. The conditional input is called the enable, and is symbolized by the letter E. Study the following example to see how this works:

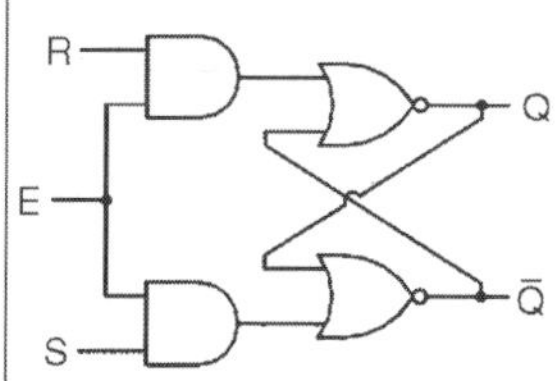

E	S	R	Q	$\bar{Q}$
0	0	0	Latch	Latch
0	0	1	Latch	Latch
0	1	0	Latch	Latch
0	1	1	Latch	Latch
1	0	0	Latch	Latch
1	0	1	0	1
1	1	0	1	0
1	1	1	0	0

When the E=0, the outputs of the two AND gates are forced to 0, regardless of the states of either S or R. Consequently, the circuit behaves as though S and R were both 0, latching the Q and not-Q outputs in their last states. Only when the enable input is activated (1) will the latch respond to the S and R inputs. Note the identical function in ladder logic:

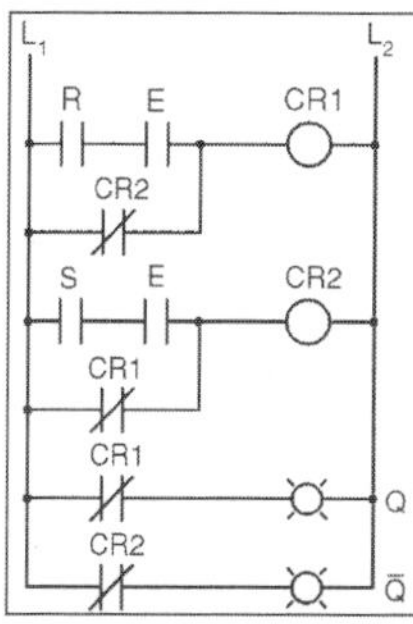

E	S	R	Q	$\bar{Q}$
0	0	0	Latch	Latch
0	0	1	Latch	Latch
0	1	0	Latch	Latch
0	1	1	Latch	Latch
1	0	0	Latch	Latch
1	0	1	0	1
1	1	0	1	0
1	1	1	0	0

A practical application of this might be the same motor control circuit (with two normally-open pushbutton switches for startand stop), except with the addition of a master lockout input (E) that disables both pushbuttons from having control over the motor when it's low (0). Once again, these multivibrator circuits are available as prepackaged semiconductor devices, and are symbolized as such:

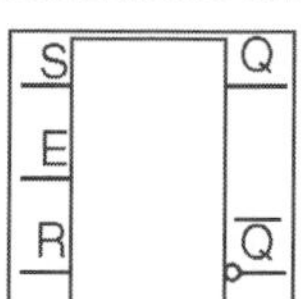

It is also common to see the enable input designated by the letters"EN" instead of just"E."

THE D LATCH

Since the enable input on a gated S-R latch provides a way to latch the Q and not-Q outputs without regard to the status of S or R, we can eliminate one of those inputs to create a multivibrator latch circuit with no"illegal" input states. Such a circuit is called a D latch, and its internal logic looks like this:

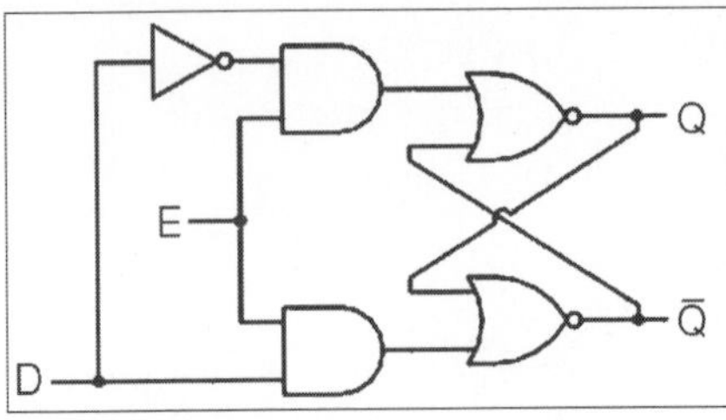

E	D	Q	$\overline{Q}$
0	0	Latch	Latch
0	1	Latch	Latch
1	0	0	1
1	1	1	0

Note that the R input has been replaced with the complement (inversion) of the old S input, and the S input has been renamed to D.

As with the gated S-R latch, the D latch will not respond to a signal input if the enable input is 0 -- it simply stays latched in its last state. When the enable input is 1, however, the Q output follows the D input.

Since the R input of the S-R circuitry has been done away with, this latch has no"invalid" or"illegal" state. Q and not-Q arealways opposite of one another. If the above diagram is confusing at all, the next diagram should make the concept simpler:

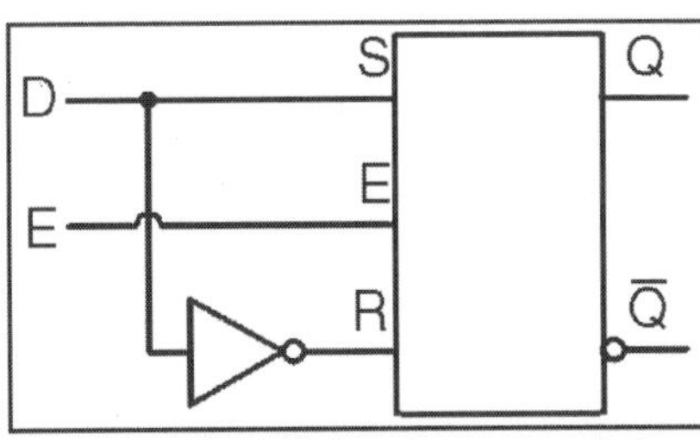

Like both the S-R and gated S-R latches, the D latch circuit may be found as its own prepackaged circuit, complete with a standard symbol:

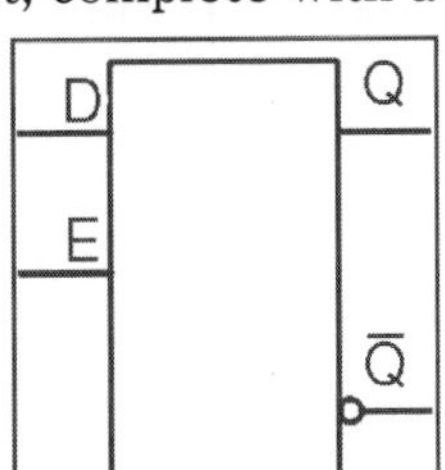

The D latch is nothing more than a gated S-R latch with an inverter added to make R the complement (inverse) of S. Let's explore the ladder logic equivalent of a D latch, modified from the basic ladder diagram of an S-R latch:

An application for the D latch is a 1-bit memory circuit. You can"write" (store) a 0 or 1 bit in this latch circuit by making the enable input high (1) and setting D to whatever you want the stored bit to be. When the enable input is made low (0), the latch ignores the status of the D input and merrily holds the stored bit value, outputting at the stored value at Q, and its inverse on output not-Q.

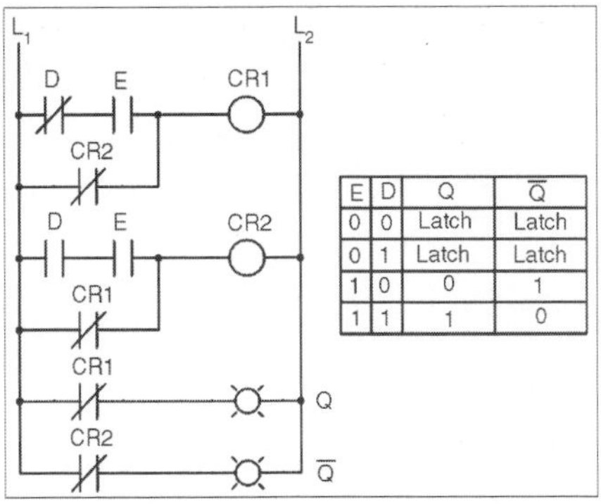

DESIGN PROBLEM

THE MODULO-4 UP-DOWN COUNTER

For the next design, we introduce a problem that uses input. This is a modulo-4 up-down counter. The input X is used to control the direction of counting.

If X = 0, the device counts up: 0, 1, 2, 3, 0, 1, 2, 3, etc.

If X = 1, the device counts down: 0, 3, 2, 1, 0, 3, 2, 1, etc.

- *Step* 1: Derive the state diagram and state table for the circuit.

The state diagram for the modulo-4 up-down counter is shown at right. Notice that the X input is used to determine the counting direction. Again, this type of circuit does not have any output associated with the transitions; the output just reflects which of the four states the machine finds itself in at the moment.

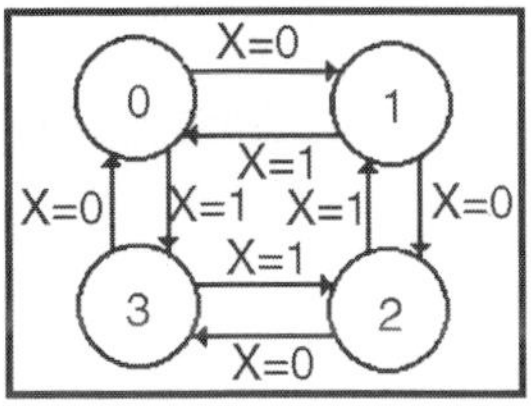

We now produce the sate table by translating the state diagram. As an aside, some students might prefer to begin the design process with the state table and omit the state diagram. That is certainly acceptable practice; whatever works should be used. Here, the state table depends on X - the input used to specify the counting direction.

Present State	Next State	
	X = 0	X = 1
0	1	3
1	2	0
2	3	1
3	0	2

- *Step* 2: Count the number of states in the state diagram (call it N) and calculate the number of flip-flops needed (call it P) by solving the equation $2^{P-1} < N £ 2^P$.

The number of states on a modulo-N counter is simply N; these are labeled 0 through (N – 1). Specifically, a modulo-4 counter has four states: labeled 0, 1, 2, and 3.

We solve the equation $2P\text{-}1 < 4 \leq 2^P$ by noting that $2^1 = 2$ and $2^2 = 4$, so we have determined that $2^1 < 4 \leq 2^2$, hence P = 2. We shall see later that there are valid solutions with more than two flip-flops, but there are none with fewer.

- *Step* 3 Assign a unique P-bit binary number (state vector) to each state.

Often, the first state = 0, the next state = 1, etc.

Some sequential circuits suggest an innovative numbering system, but modulo-N counters never do. We go with the obvious labeling, generated by assigning each decimal number its two-bit binary equivalent as an unsigned integer in the range from 0 to 3.

State	2-bit	Vector
0	0	0
1	0	1
2	1	0
3	1	1

- *Step* 4: Derive the state transition table and the output table.

There is no output table for any modulo-N counter, as the output associated with this type of table is the output on a transition, as is seen in a sequence detector. The transition table is a direct translation of the state table, using the assignments from the previous step. The transition table for the modulo-4 up-down counter is as follows.

Present State		Next State	
		X = 0	X = 1
0	00	01	11
1	01	10	00
2	10	11	01
3	11	00	10

- *Step* 5: Separate the state transition table into P tables, one for each flip-flop.

Here we separate the state transition table into 2 tables, one for each flip-flop. Note that the flip-flops will be numbered 0 and 1, with flip-flop 0 storing the least significant bit of the state information. Thus, we shall refer to the state information as Y_1Y_0.

Flip-Flop 1			Flip-Flop 0		
PS	Next State		PS	Next State	
Y_1Y_0	Y_1, X = 0	Y_1, X = 1	Y_1Y_0	Y_0, X = 0	Y_0, X = 1
0 0	0	1	0 0	1	1
0 1	1	0	0 1	0	0
1 0	1	0	1 0	1	1
1 1	0	1	1 1	0	0

The student who is paying attention at this point will notice an interesting feature concerning flip-flop 0; specifically that its next state does not depend on X.

This is due to the fact that in considering a modulo-N counter, one moves from odd numbers to even numbers and from even numbers to odd numbers in both counting up and counting down.

Step 6: Decide on the types of flip-flops to use. When in doubt, use all JK's.

Up to this point, we have made no assumptions about the type of flip-flop to use. In order to proceed any farther with the design, we must now commit to a specific type. In line with this author's preferences, he chooses to use two JK flip-flops.

Q(t)	Q(t+1)	J	K
0	0	0	d
0	1	1	d
1	0	d	1
1	1	d	0

The excitation table for a JK flip-flop is shown at right. Recall that the"d" stands for"Don't Care". For example, if we have Q(t) = 0 and want Q(t+1) = 0, we can use either J = 0, K = 0 or J = 0, K = 1. Similarly either J = 1 and K = 0 or J = 1 and K = 1 will take Q(t) = 0 to Q(t + 1) = 1.

- *Step* 7 Derive the input table for each flip-flop using the excitation tables for the type.

Here is the input table for flip-flop 1. Note that the arrangement of the table has been altered to reflect the fact that we now have a binary input.

	X = 0			X = 1		
Y_1Y_0	Y_1	J_1	K_1	Y_1	J_1	K_1
0 0	0	0	d	1	1	d
0 1	1	1	d	0	0	d
1 0	1	d	0	0	d	1
1 1	0	d	1	1	d	0

Here is the input table for flip-flop 0.

	X = 0			X = 1		
Y_1Y_0	Y_0	J_0	K_0	Y_1	J_0	K_0
0 0	1	1	d	1	1	d
0 1	0	d	1	0	d	1
1 0	1	1	d	1	1	d
1 1	0	d	1	0	d	1

- *Step* 8: Derive the input equations for each flip-flop based as functions of the input and current state of all flip-flops.

At this point, we try to derive an expression that matches each column. Formal methods can be used, but generally are more trouble than they are worth. Here is this author's set of rules to match an expression to a given column.

- If a column does not have a 0 in it, match it to the constant value 1.
- If a column does not have a 1 in it, match it to the constant value 0.
- If the column has both 0's and 1's in it, try to match it to a single variable, which must be part of the present state. Only the 0's and 1's in a column must match the suggested function.
- If every 0 and 1 in the column is a mismatch, match to the complement of a function.
- If all the above fails, try for simple combinations of the present state.

The reader will note that there are two columns for each variable for which an equation is desired; one column for X = 0 and one column for X = 1. For example, consider the table for flip-flop 0, just above. If we work on a column-by column basis, we shall arrive at four equations.

- One for J_0 when X = 0,
- One for K_0 when X = 0,
- One for J_0 when X = 1, and
- One for K_0 when X = 1.

However, we need a single equation for J_0 and a single equation for K_0.

8

Networks Design of Binary Multiplier

Binary multiplication requires only shifting and adding. For example, multiplication 1310 by 0510 in binary:

```
      Multiplicand    →1 1 0 1  (13)
      Multiplier      →0 1 0 1  (05)
Partial Product     →   1 1 0 1
                      0 0 0 0
                      0 1 1 0 1
                    1 1 0 1
                    1 0 0 0 0 0 1
                   0 0 0 0
                  0 1 0 0 0 0 0 1  (65)
```

Note that each partial product is either the multiplicand (1101) shifted over by the appropriate number of places or zero. Multiplication of two 4-bit numbers requires a 4-bit multiplicand register, a 4-bit multiplier register, a 4-bit full adder, and an 8-bit register for the product. The product register serves as an accumulator to accumulate the sum of the partial products. If the multiplicand were shifted left each time before it was added to the accumulator, as was done in the previous example, an 8-bit adder would be needed. So it is better to shift the contents of the product register to the right each time, as shown in the block diagram as shown in figure below:

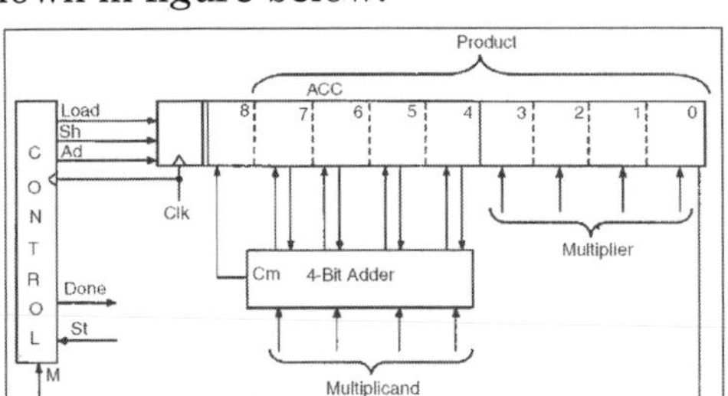

Fig. Block Diagram for Binary Multiplier

This type of multiplier is sometimes referred to as a serial - parallel multiplier, since the multiplier bits are processed serially, but the addition takes place in parallel. As indicated by the arrows on the diagram, 4 bits from the accumulator (ACC) and 4 bits from the multiplicand register are connected to the adder inputs; the 4 sum bits and the carry output from the adder are connected back to the accumulator. When an add signal (Ad) occurs, the adder outputs are transferred to the accumulator by the next clock pulse, thus causing the multiplicand to be added to the accumulator. An extra bit at the left end of the product register temporarily stores any carry that is generated when the multiplicand is added to the accumulator.

When a shift signal (Sh) occurs, all 9 bits of ACC are shifted right by the next clock pulse. Since the lower 4 bits of the product register are initially unused, we will store the multiplier in this location instead of in a separate register. As each multiplier bit is used, it is shifted out the right end of the register to make room for additional product bits. A shift signal (Sh) causes the contents of the product register (including the multiplier) to be shifted right one place when the next clock pulse occurs. The control circuit puts out the proper sequence of add and shift signals after a start signal (St=1) has been received. If the current multiplier bit (M) is 1, the multiplicand is added to the accumulator followed by a right shift; if the multiplier bit is 0, the addition is skipped, and only the right shift occurs. The multiplication example (13X05) is reworked below showing the location of the bits in the registers at each clock time.

Multiplication Steps:	
Initial Contents of product Register	0 0 0 0 0 0 1 0 1 ← M (5)
(Add Multiplicand Since M=1)	1 1 0 1
After Addition	0 1 1 0 1 0 1 0 1
After Shift	0 0 1 1 0 1 0 1 0 ←M
(Skip Addition Since M=0)	0 0 0 0
After Addition	0 0 1 1 0 1 0
After Shift	0 0 0 1 1 0 1 0 1 ← M
(Add Multiplicand Since M=1)	1 1 0 1
After Addition	1 0 0 0 0 0 1 0 1
After Shift	0 1 0 0 0 0 0 1 0
(Skip Addition Since M=0)	
After Shift (Final Answer)	0 0 1 0 0 0 0 0 1

The control circuit must be designed to output the proper sequence of add and shift signals. Figure below shows a state graph for the control circuit.

In this table, S0 is the reset state, and the network stays in S0 until a start signal (St=1) is received. This generates a load signal, which causes the multiplier to be loaded into the lower 4 bits of the Accumulator (ACC) and the upper 5 bits of the accumulator to be cleared.

In state S1, the low-order bit of the multiplier (M) is tested. If M=1, an add signal is generated, and if M=0, a shift signal is generated. Similarly, in states S3, S5 and S7, the current multiplier bit (M) is teted to determine whether to generate an add or shift signal. A shift signal is always generated at the next clock time following an add signal (states S2, S4, S6, and S8). After four shifts have been generated, the control network goes to S9, and a done signal is generated before returning to S0.

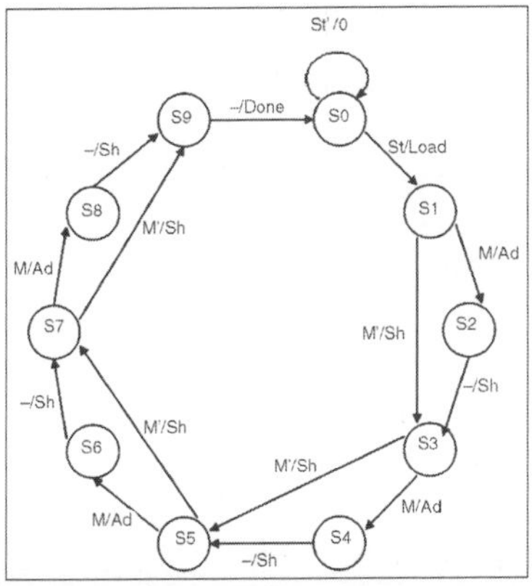

The behavioural VHDL model (Figure below) corresponds directly to the state graph. Since there are 10 states, we have declared an integer range 0 to 9 for the state signal. The signal ACC represents the 9-bit accumulator output.

The statement alias M: bit is ACC(0); allows us to use the name M in place of ACC(0) The done signal needs to be turned on only in state 9. If we had used the statement when 9 => State ⇐ 0; Done ⇐'1', Done would be turned on at the same time the State changed to 0. This is too late, since we want Done to turn on when the state becomes 9.

Behavioural Model for 4×4 Binary Multiplier

```
    – This is behavioural model of a multiplier for unsigned
    - Binary numbers. It multiplies a 4-bit
    multiplicand
    - By a 4-bit multiplier to give an 8-bit
  product.
    - The maxiumum number of clock cycles needed
    - Multiply is 10.
    Library Bitlib;
    use bitlib.bit_pack. all;
    entity mult 4×4 is
          Port (Clk, st: in bit)
             Mplier,McandL:in bit_vector (3 downto 0);
             Done: out bit
end mult4×4;
    architecture bej- This is behavioural model of
  Mult 4×4 is
       Signal State: integer range 0 to 9;
          Signal Acc: bit_vector range 0 to 9;     --
          accumlator
```

```
            Alias M: Bit is Acc (0); --M is bit 0 of
               Acc
      bigin
         process
         begin
            wait until CIK = '1'; --executes on rising
      edge of clock
            Case state is
               when 0⇒ --initial state
               if st= '1' then
                  Acc(8downto 4) ⇐ "00000"; --Begin Cycle
                     Acc(3 downto 0) ⇐ Mplier; --Load the
Multiplier
            state1 ⇐1; end if;
      when 1|3|5|7 ⇒ -- "add/shift" state
         if M = '1' then -- Add Multiplicand
            Acc(8downto 4)⇐ add 4 (Acc(7downto
               4),Mcand, '0');
         state ⇐ State+1; else
         Acc ⇐'0'and Acc (8 downto 1);--shift
            accumulator right
            State ⇐state + 2;
            end if;
         when if;
      when 2|4|6|8  ⇒ -- "shift" state
      Acc ⇐ '0'and Acc(8 downto 1); --Right Shift
      State ⇐ state + 1;
   when 9 ⇒ --End of cycle
      State ⇐0;
   end case;
 end process;
      Done ⇐ '1'when state = 9else '0';
   end behavel;
```

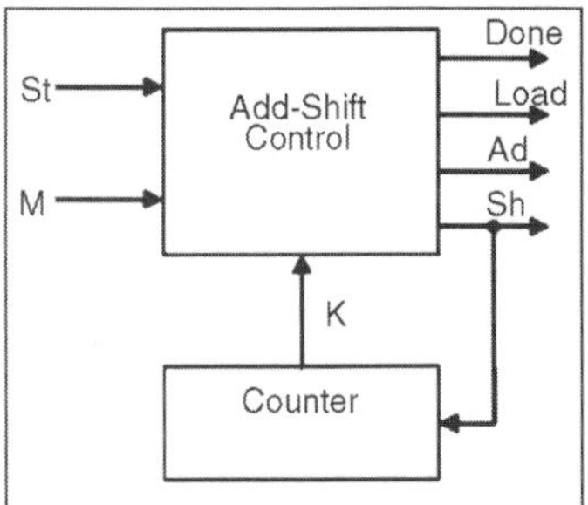

(a) Multiplier Control

Fig. Multiplier Control

As the state graph for the multiplier indicates, the control performs two functions generating add or shift signals as needed and counting the number of shifts. If the number of bits is large, it is convenient to divide the control network into a counter and an add-shift control, as shown in Figure (a).

First, we will derive a state graph for the add-shift control that tests St and M and outputs the proper sequence of add and shift signals (Figure (b)).

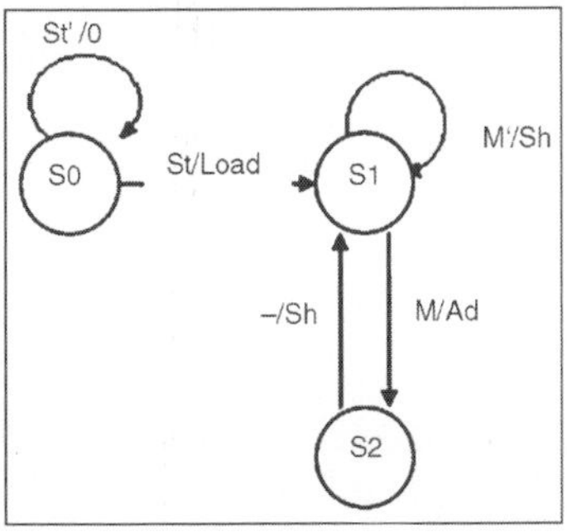

Fig. State-graph for Add-shift Control

Then we will add a completion signal (K) from the counter that stops the multiplier after the proper number of shifts have been completed.

Starting in S0 in Figure above, when a start signal St=1 is received, a load signal is generated and the network goes to state S2; if M=0, a shift signal is generated and the network stays in S1. In S2, a shift signal is generated since a shift always follows an add.

The graph of Figure shown in (b) will generate the proper sequence of add and shift signals, but it has no provision for stopping the multiplier.

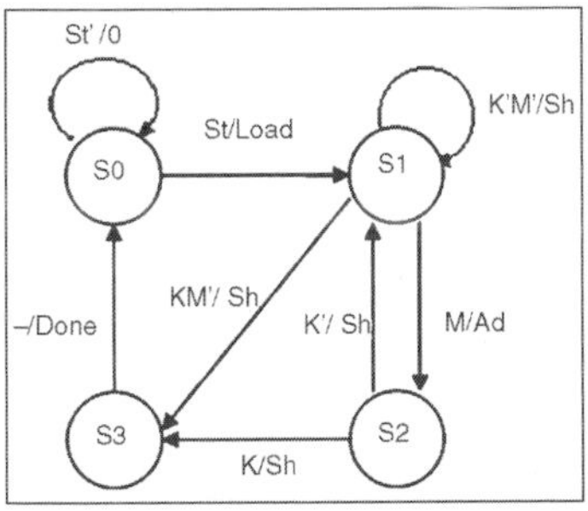

Fig. Final state Graph for add -shift Control

In order to determine when the multiplication is completed, the counter is incremented each time a shift signal is generated.

If the multiplier is n bits, n shifts are required. We will design the counter so that a completion signal (K) is generated after n-1 shifts have occurred. When K=1, the network should perform one more addition id necessary and then do the final shift. The control operation in above Figure (c) is the same as Figure 6(b) as long as K=0. In state S1, if K=1, we test M as usual. If M =0, we output the final shift signal and go to the done state (S3); however if M=1, we add before shifting and go to state S2. In state S2, if K=1, we output one more shift signal and then go to S3.

The last shift signal will increment the counter to 0 at the same time the add-shift control goes to the done state. As an example, consider the multiplier of figure, but replace the control network with

Figure (a). Since n=4, a 2-bit counter is needed to count the 4 shifts, and K=1 when the counter is in state 3(112). Table below shows the operation of the multiplier when 1101 is multiplied by 1011. S0, S1, S2, and S3 represent states of the control circuit (Figure (c)).

Time	State	Counter	Product Register	St	M	K	Load	Ad	sh	Down
t_0	SO	00	000000000	0	0	0	0	0	0	0
t_1	SO	00	000000000	1	0	0	1	0	0	0
t_2	S1	00	000001011	0	1	0	0	1	0	0
t_3	S2	00	011011011	0	1	0	0	0	1	0
t_4	S1	01	001101101	0	1	0	0	0	1	0
t_5	S2	01	100111101	0	1	0	0	0	1	0
t_6	S1	10	010011110	0	0	0	0	0	1	0
t_7	S1	11	001001111	0	1	1	0	1	0	0
t_8	S2	11	100011111	0	1	1	0	0	1	0
t_9	S3	00	010001111	0	1	0	0	0	0	1

STATE MACHINE CHARTS

A state machine is a computational model based on automata theory. The state machine model is widely used to describe discrete systems, where the current behaviour is a function of previously occurring events.

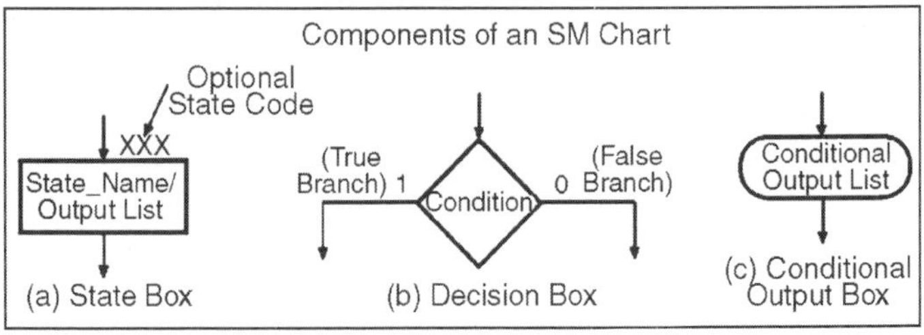

At a given point in time the system is in one of several possible states. The system can change its state depending on input from the environment. As a state change occurs, actions can be performed on the environment.

The arrows in Figure represent transitions, progressions from one state to another. For example, when a seminar is in the Scheduled state, it can either be opened for enrollment or cancelled. he notation for the labels on transitions is in the format event [guard][/method list]. It is mandatory to indicate the event which causes the transition, such as open or cancelled. Guard, conditions that must be true for the transition to be triggered, are optionally indicated.

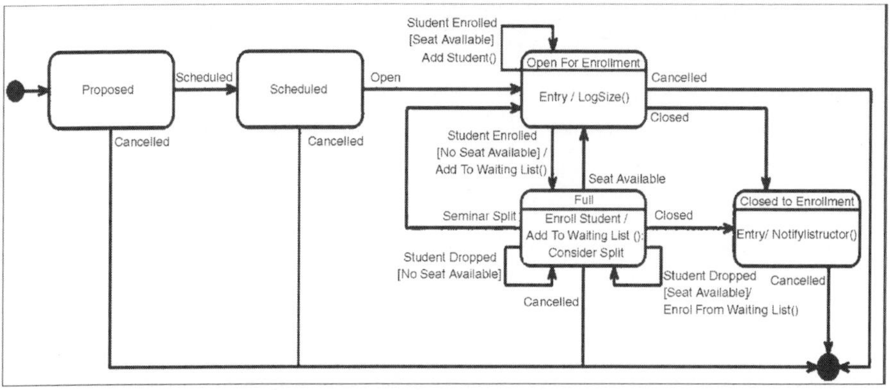

The [not seat available] guard is shown on the student enrolled transition from the Open For Enrollment to the Closed To Enrollment state.

Guards can be described in any manner, including both free form text or formal language - when I'm white boarding I'll use free form text to ensure that it's readable by everyone but with a sophisticated CASE tool I would consider using either a programming language such as Java or a modeling language such as Object Constraint Language (OCL) if the tool has the ability to actually process that information into something useful (such as executable code).

The invocation of methods, such as addToWaitingList() can optionally be indicated on transitions. The order in the listing implying the order in which they are invoked.

States are represented by the values of the attributes of an object. For example, a seminar is in the Open For Enrollment state when it has been flagged as open and seats are available to be filled. It is possible to indicate the invocation of methods within a state, for example, upon entry into the Closed To Enrollment state the methodnotify Instructor() is invoked.

The notation used within the same as that used on transitions, the only difference being that the method list is mandatory and the event is optional.

For example in the Full state the operations add To Waiting List() and consider Split() are invoked whenever a student is enrolled. Had there been no event indicated those methods would be invoked continuously (in a loop) whenever the object is in that state.

I indicate the methods to run during the state when I want to indicate a method is to be run continuously, perhaps a method that polls other objects for information or a method that implements the logic of an important business process.

Methods to be invoked when the object enters the state are indicated by the keyword entry, as you see with both the Open For Enrollment and Closed

To Enrollment states in Figure. Methods to be invoked as the object exits the state are indicated by the keyword exit. The capability to indicate method invocations when you enter and exit a state is useful because it enables you to avoid documenting the same method several times on each of the transitions that enter or exit the state, respectively.

Transitions are the result of the invocation of a method that causes an important change in state. Understanding that not all method invocations will result in transitions is important.

For example, the invocation of a getter method likely wouldn't cause a transition because it isn't changing the state of the object (unless lazy initialization is being applied). Furthermore, Figure indicates an attempt to enroll a student in a full seminar may not result in the object changing state, unless it is determined that the seminar should be split, even though the state of the object changes (another student is added to the waiting list).

You can see that transitions are a reflection of your business rules. For example, you see that you can attempt to enroll a student in a course only when it is open for enrollment or full, and that a seminar may be split (presumably into two seminars) when the waiting list is long enough to justify the split. You can have recursive transitions, also called self transitions, that start and end in the same state. An example of which is the student dropped transition when the seminar is full.

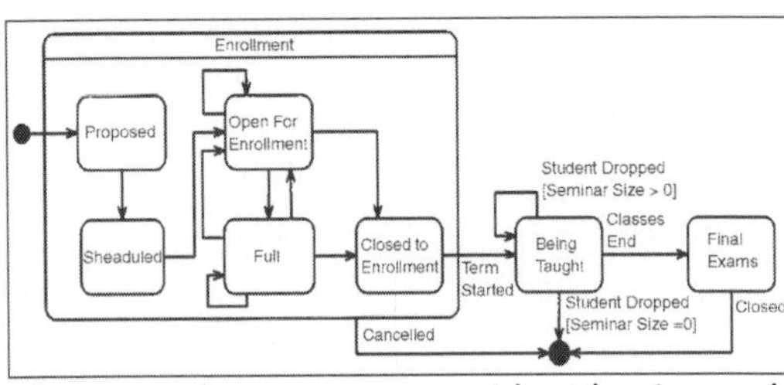

For the sake of convention, we say an object is always in one and only one state, implying transitions are instantaneous. Although we know this is not completely true (every method is going to take some time to run), this makes life a lot easier for us to assume transitions take no time to complete.

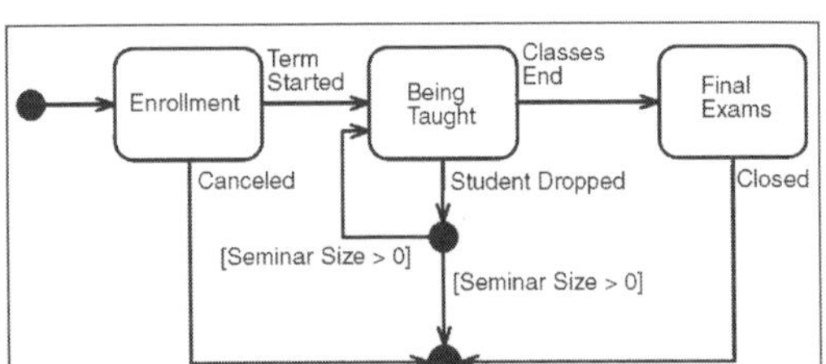

Because the lifecycle of a seminar is so complex Figure 1 only depicts part of it. Figure depicts the entire lifecycle, with Figure 1 shown as a substate of theEnrollment state. I could have included all of the details in Figure shown below but chose not to in order to keep the diagram simple. In fact, instead of creating a diagram such as Figure typically prefer something more along the lines of the high-level view of Figure as seen below.

This approach keeps the diagrams small and easy to understand. *A state machine is generally defined by a set of*: States - A state.

DERIVATION OF SM CHARTS

CONSTRUCTION OF SM CHARTS

The construction of an SM chart for a sequential control network is similar to that used to derive a state graph. Steps are:

- First draw a block diagram of the system that we are controlling.
- Define the required input and output signals to the control network.
- Then construct the chart that tests the input signals and generates the proper sequence of output signals.

REALIZATION OF SM CHARTS

The realization consists of a combinational sub network, together with Flip-flop storing the state of the network. As an example SM chart for binary multiplier is taken up. This SM chart is given in Fig shown below.

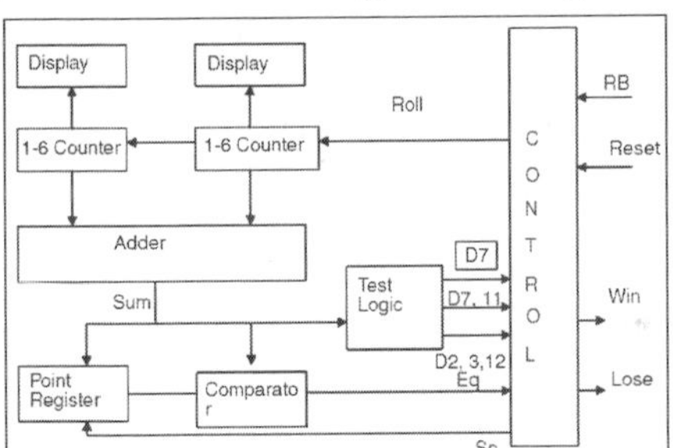

The controller for the multiplier can be realised using a PLA and D flip-flops. The controller chart has 4 states and you need 2 (2 = 4) D flip-flops. The PLA table is given as shown

A	B	St	M	K		A+	B+	Load	Sh	Ad	Done
So	0	0	0	-	-	0	0	0	0	0	0
0	0	1	-	-	0	1	1	0	0	0	
0	1	-	0	1	1	1	0	1	0	0	
S1	0	1	-	0	1	1	1	0	1	0	0
0	1	-	1	-	1	0	0	0	1	0	
1	0	-	-	0	0	1	0	1	0	0	
S2	1	0	-	-	0	0	1	0	1	0	0
S3	1	1	-	-	-	0	0	0	0	0	1

In this the PLA has 5 inputs and 6 outputs. Each row in the table corresponds to one of the link paths in the SM chart. Since So has two exit paths the table has two rows for present state So. The first row corresponds to the St = 0 exit path. So the next state and output are 0. In the second row, St = 1, so the next state is a don't care in the corresponding rows. The outputs for each row can be filled in by tracing the corresponding link paths on the SM chart. For example, the link path from S1 to S2 passes through conditional output Ad, so Ad = 1 in this row. Since S2 has a Moore output Sh, Sh = 1 in both of the rows for which AB = 10. By inspection of the PLA table, the logic equations for the multiplier control are: A+ = A'BM'K + A'BM + AB'K; B+ = A'B'St + A'BM' + AB'; Load = A'B'St

Sh = A'B'St = AB'; Ad = A'BM'; Done = AB

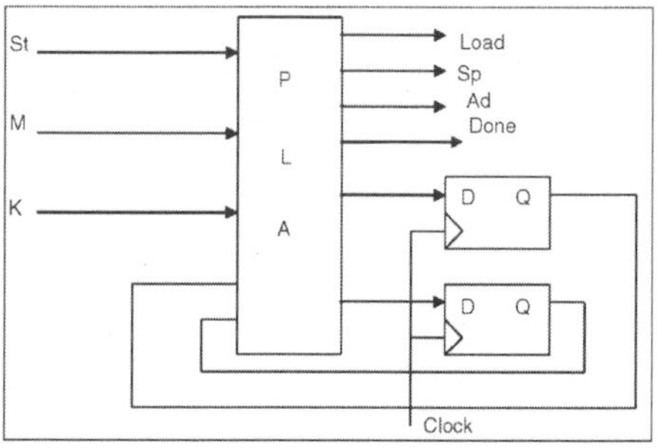

Fig. PLA Realization of Multiplier Control

IMPLEMENTATION OF DICE GAME

This is a game based on a dice, which will have 6 faces with numbers 1,2,3,4,5,6. In this game two dices will be thrown, and depending on the sum of the numbers seen on the faces of the dices, the result is decided. One can come out with many conditions to decide whether a player has won or not. The purpose of this example is to design a circuit that will simulate the above said idea.

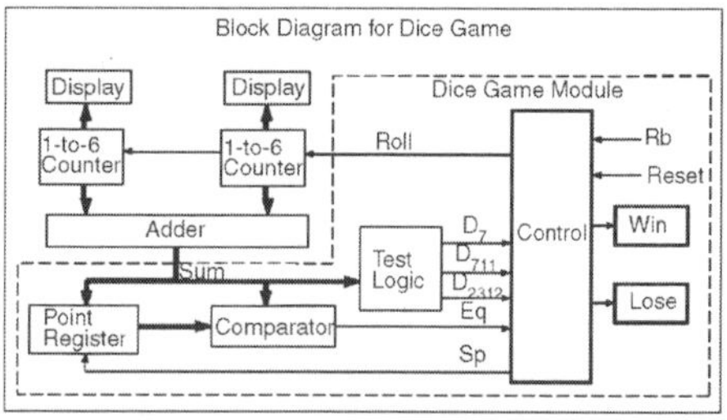

Fig. Above shows the Block Diagram for the Dice Game and Below Shows the SM Chart of Game.

Two counters are used to simulate the ROLL of DICE. Each counter counts in the sequence 1,2,3,4,5,6,1,.........Thus after the ROLL of the DICE the SUM of the values in the two counters will be in the range 2 through 12.

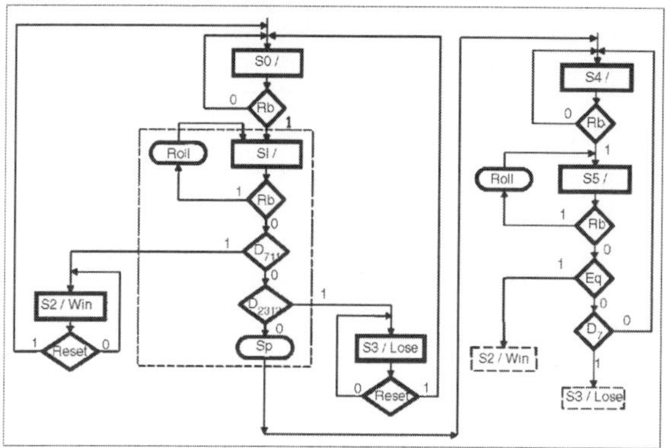

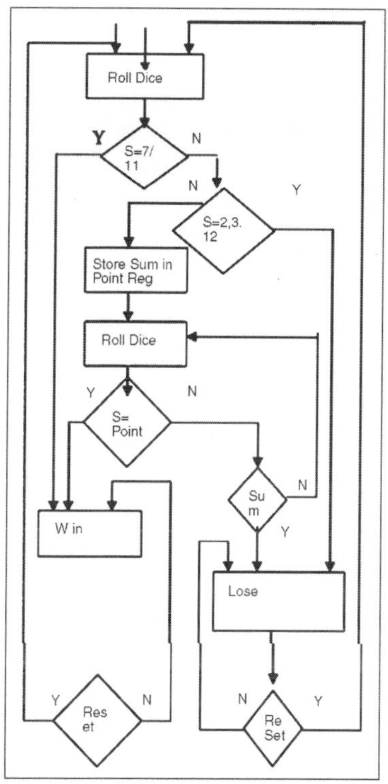

The rules of the game are as follows:

- After the first ROLL of the DICE, the player wins if the SUM is 7 or 11.Th player
  LOSES if the SUM is 2,3 or 12. Otherwise the SUM the player obtained on the first ROLL is referred to as a POINT, and the player must ROLL the DICE again.
- On the second or subsequent ROLL of the DICE, the player WINS if the SUM is 7.
- Otherwise, the player must ROLL again until he or she finally WINS or LOSES or RESETS (starts a new game).

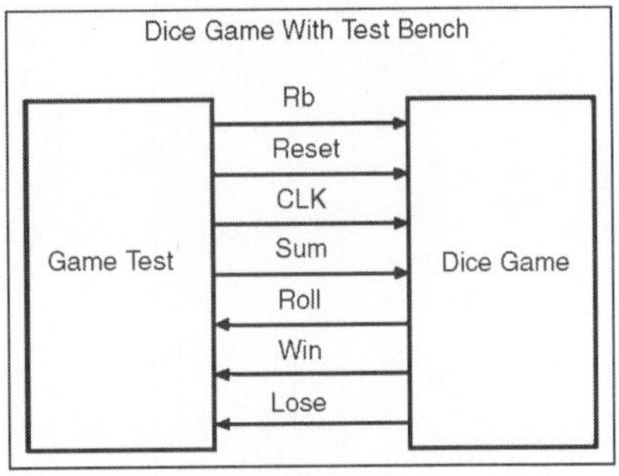

BOOLEAN ALGEBRAIC IDENTITIES

In mathematics, an *identity* is a statement true for all possible values of its variable or variables. The algebraic identity of $x + 0 = x$ tells us that anything (x) added to zero equals the original "anything," no matter what value that "anything" (x) may be. Like ordinary algebra, Boolean algebra has its own unique identities based on the bivalent states of Boolean variables.

The first Boolean identity is that the sum of anything and zero is the same as the original "anything." This identity is no different from its real-number algebraic equivalent:

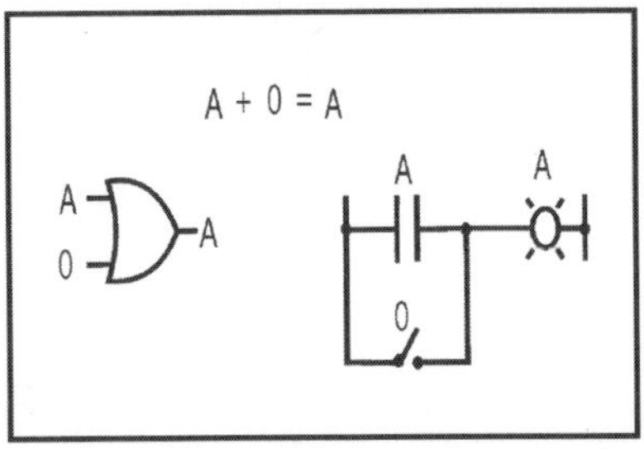

No matter what the value of A, the output will always be the same: when A=1, the output will also be 1; when A=0, the output will also be 0. The next identity is most definitely *different* from any seen in normal algebra. Here we discover that the sum of anything and one is one:

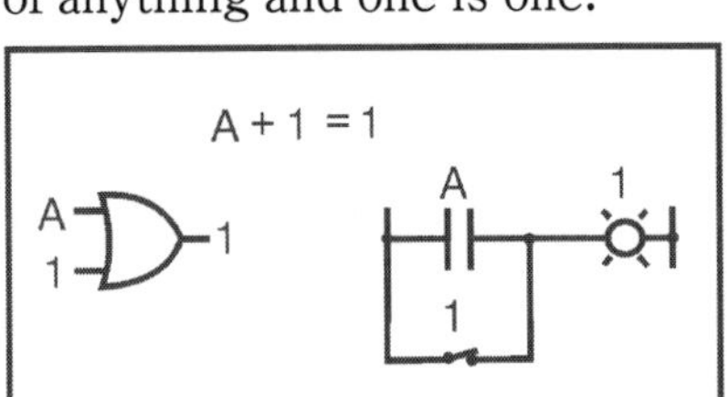

No matter what the value of A, the sum of A and 1 will always be 1. In a sense, the "1" signal *overrides*the effect of A on the logic circuit, leaving the output fixed at a logic level of 1.Next, we examine the effect of adding A and A together, which is the same as connecting both inputs of an OR gate to each other and activating them with the same signal:

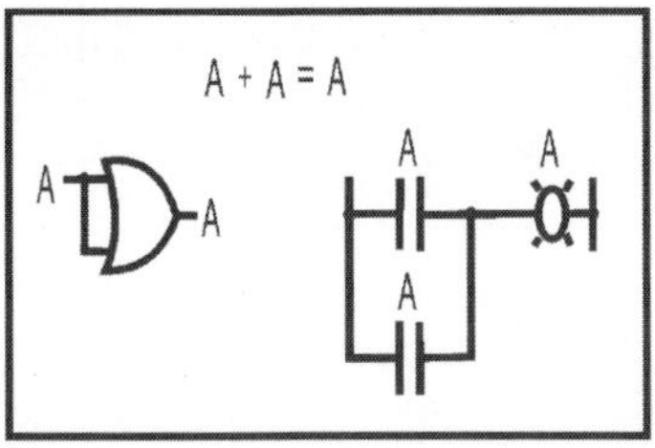

In real-number algebra, the sum of two identical variables is twice the original variable's value (x + x = 2x), but remember that there is no concept of "2" in the world of Boolean math, only 1 and 0, so we cannot say that A + A = 2A. Thus, when we add a Boolean quantity to itself, the sum is equal the original quantity:

$$0 + 0 = 0,$$

and

$$1 + 1 = 1.$$

Introducing the uniquely Boolean concept of complementation into an additive identity, we find an interesting effect. Since there must be one "1" value between any variable and its complement, and since the sum of any Boolean quantity and 1 is 1, the sum of a variable and its complement must be 1:

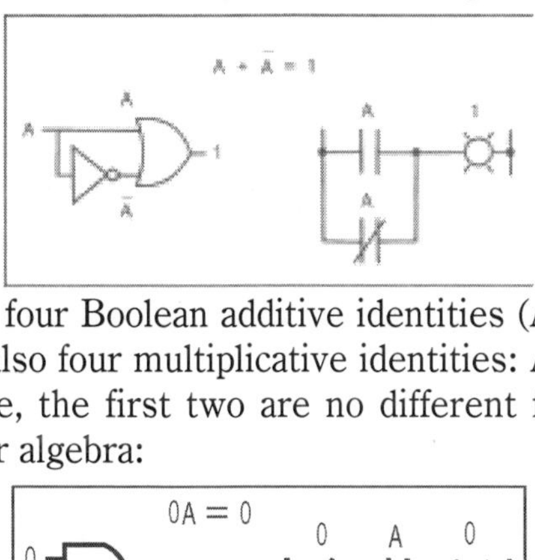

Just as there are four Boolean additive identities (A+0, A+1, A+A, and A+A'), so there are also four multiplicative identities: A × 0, A × 1, A × A, and A × A'. Of these, the first two are no different from their equivalent expressions in regular algebra:

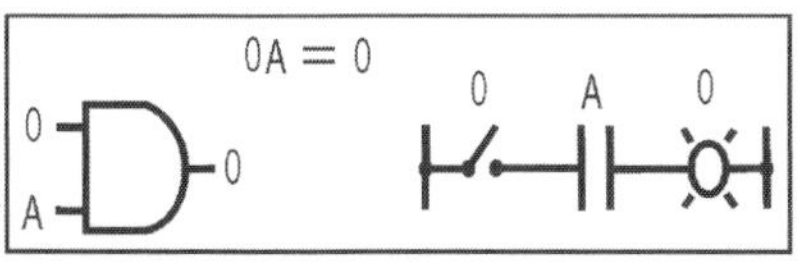

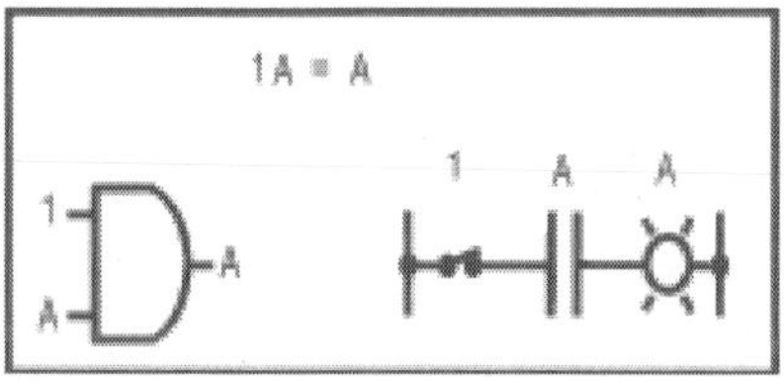

The third multiplicative identity expresses the result of a Boolean quantity multiplied by itself. In normal algebra, the product of a variable and itself is the *square* of that variable ($3 \times 3 = 3^2 = 9$). However, the concept of "square"

implies a quantity of 2, which has no meaning in Boolean algebra, so we cannot say that A × A = A^2. Instead, we find that the product of a Boolean quantity and itself is the original quantity, since 0 × 0 = 0 and 1 × 1 = 1:

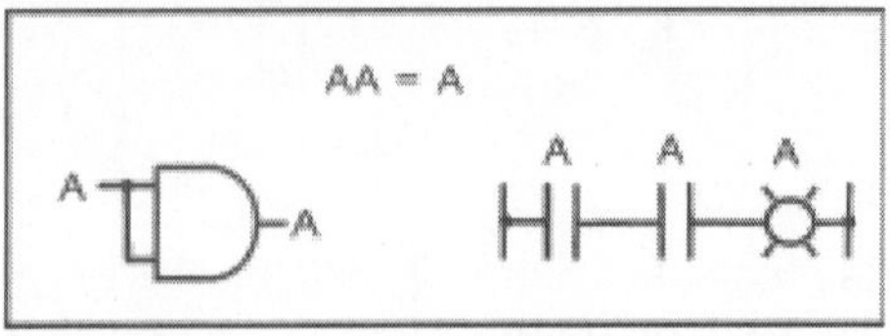

The fourth multiplicative identity has no equivalent in regular algebra because it uses the complement of a variable, a concept unique to Boolean mathematics. Since there must be one "0" value between any variable and its complement, and since the product of any Boolean quantity and 0 is 0, the product of a variable and its complement must be 0:

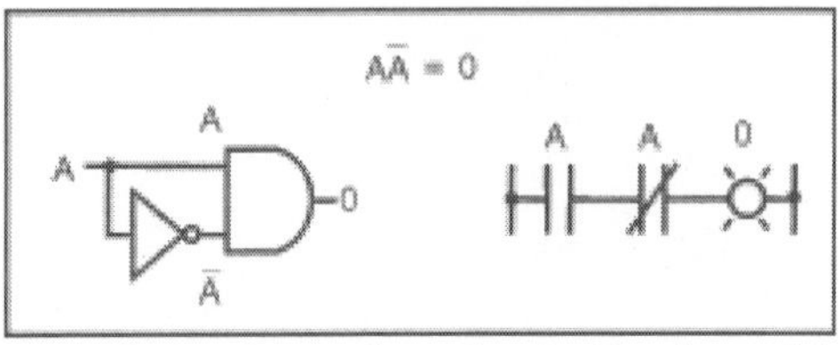

To summarize, then, we have four basic Boolean identities for addition and four for multiplication:

Table. Basic Boolean Algebraic Identities.

Additive	Multiplicative
A + 0 = A	0A = 0
A + 1 = 1	1A = A
A + A = A	AA = A
A + $\bar{A}$ = 1	$A\bar{A}$ = 0

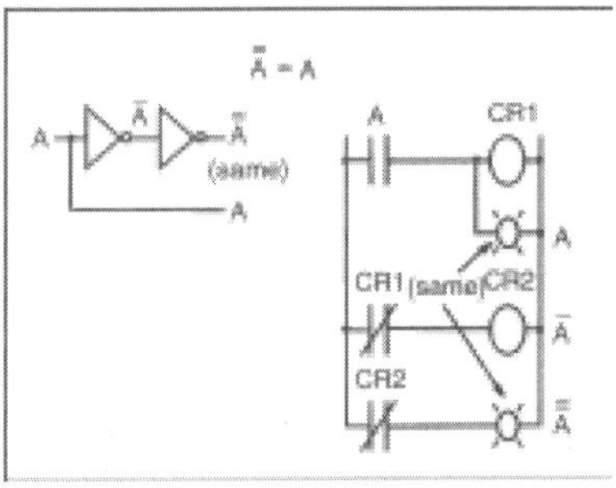

Another identity having to do with complementation is that of the *double complement*: a variable inverted twice. Complementing a variable twice (or any even number of times) results in the original Boolean value. This is analogous to negating (multiplying by -1) in real-number algebra: an even number of negations cancel to leave the original value:

BOOLEAN ALGEBRA PROPERTIES

Another type of mathematical identity, called a "property" or a "law," describes how differing variables relate to each other in a system of numbers. One of these properties is known as the *commutative property*, and it applies equally to addition and multiplication. In essence, the commutative property tells us we can reverse the order of variables that are either added together or multiplied together without changing the truth of the expression:

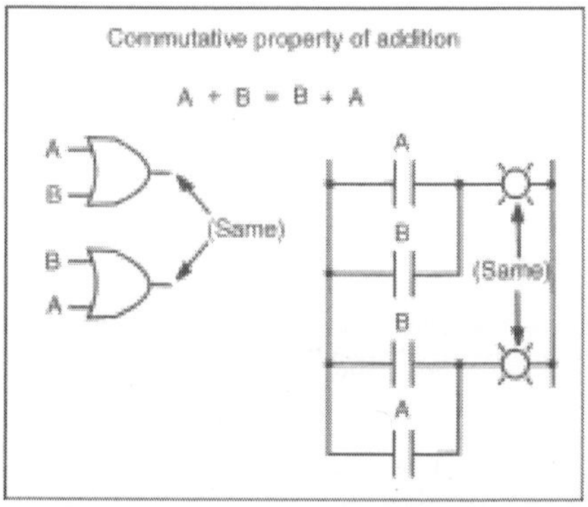

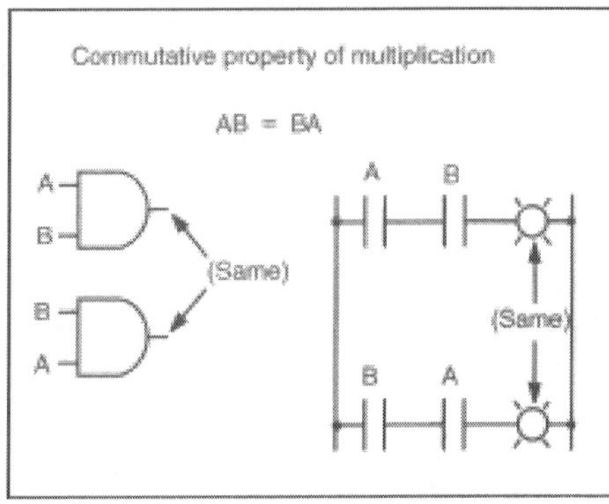

Along with the commutative properties of addition and multiplication, we have the *associative property*, again applying equally well to addition and multiplication. This property tells us we can associate groups of added or multiplied variables together with parentheses without altering the truth of the equations.

SIMPLIFICATION RULES

Boolean algebra finds its most practical use in the simplification of logic circuits. If we translate a logic circuit's function into symbolic (Boolean) form, and apply certain algebraic rules to the resulting equation to reduce the number of terms and/or arithmetic operations, the simplified equation may be translated back into circuit form for a logic circuit performing the same function with fewer components. If equivalent function may be achieved with fewer components, the result will be increased reliability and decreased cost of manufacture.

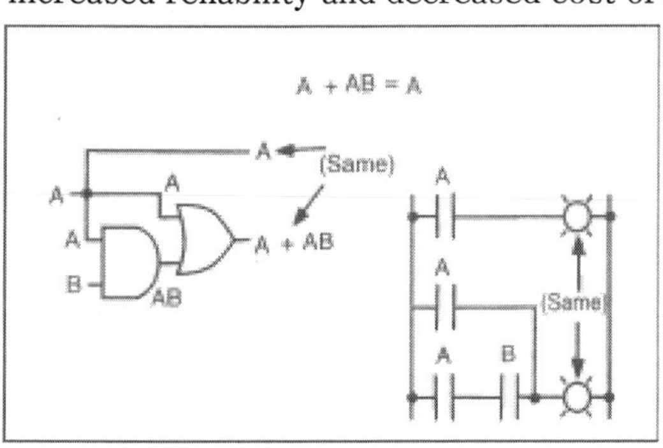

The identities and properties already reviewed in this chapter are very useful in Boolean simplification, and for the most part bear similarity to many identities and properties of "normal" algebra. However, the rules shown in this section are all unique to Boolean mathematics. This rule may be proven symbolically by factoring an "A" out of the two terms, then applying the rules of A + 1 = 1 and 1A = A to achieve the final result:

$A + AB$

↓ Factoring A Out of Both Terms

$A\,(1 + B)$

↓ Applying identity A + 1 = 1

$A\,(1)$

↓ Applying identity 1A = A

A

Please note how the rule A + 1 = 1 was used to reduce the (B + 1) term to 1. When a rule like "A + 1 = 1" is expressed using the letter "A", it doesn't mean it only applies to expressions containing "A". What the "A" stands for in a rule like A + 1 = 1 is *any* Boolean variable or collection of variables. This is perhaps the most difficult concept for new students to master in Boolean simplification: applying standardized identities, properties, and rules to expressions not in standard form. For instance, the Boolean expression ABC + 1 also reduces to 1 by means of the "A + 1 = 1" identity. In this case, we recognize that the "A" term in the identity's standard form can represent the entire "ABC" term in the original expression.

The next rule looks similar to the first one shown in this section, but is actually quite different and requires a more clever proof:

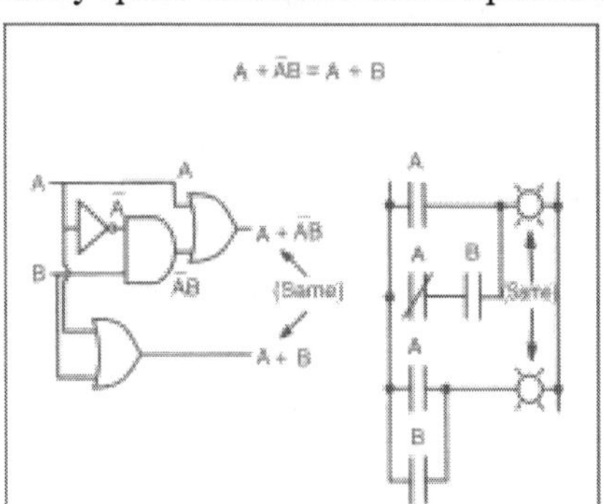

$A + \bar{A}B$

↓ Applying the Previous Rule to Expand A Term: A + AB = A

$A + AB + \bar{A}B$

↓ Factoring B Out of 2nd and 3rd Terms

$A + B\,(A + \bar{A})$

↓ Applying Identity $A + \bar{A} = 1$

$A + B\,(1)$

↓ Applying Identity 1A = A

$A + B$

Note how the last rule (A + AB = A) is used to "un-simplify" the first "A" term in the expression, changing the "A" into an "A + AB".

Knowing when to take such a step and when not to is part of the art-form of algebra, just as a victory in a game of chess almost always requires calculated sacrifices. Another rule involves the simplification of a product-of-sums expression:

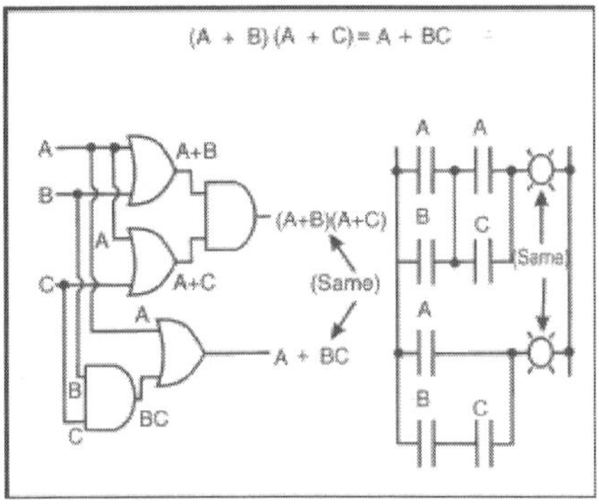

And, the corresponding proof:

(B + C) (A + C)

↓ Distributing Terms

AA + AC + AB + BC

↓ Applying Identity AA = A

A + AC + AB + BC

↓ Applying Rule A + AB = A to the A + AC term

A + AB + BC

↓ Applying Rule A + AB = A to the A + AB term

A + BC

Useful Boolean Rules:

$$A + AB = A$$

$$A + \overline{A}B = A + B$$

$$(A + B)(A + C) = A + BC$$

BOOLEAN ANALYSIS OF SWITCHING NETWORKS

A switching device is one where the non-linearity is utilised to produce two opposite states. CMOS devices in digital circuits, for instance, have their output connected to either the positive or the negative supply rail and are never found at anything in between except during a transient period when the device is actually switching.

Here the non-linearity is designed to be extreme, and the analyst can actually take advantage of that fact. These kinds of networks can be analysed using Boolean algebra by assigning the two states ("on"/"off", "positive"/"negative" or whatever states are being used) to the boolean constants "0" and "1".

The transients are ignored in this analysis, along with any slight discrepancy between the actual state of the device and the nominal state assigned to a boolean value. For instance, boolean "1" may be assigned to the state of +5V. The output of the device may actually be +4.5V but the analyst still considers this to be boolean "1". Device manufacturers will usually specify a range of values in their data sheets that are to be considered undefined (ie the result will be unpredictable).

The transients are not entirely uninteresting to the analyst. The maximum rate of switching is determined by the speed of transition from one state to the other. Happily for the analyst, for many devices most of the transition occurs in the linear portion of the devices transfer function and linear analysis can be applied to obtain at least an approximate answer. It is mathematically possible to derive boolean algebras which have more than two states. There is not too much use found for these in electronics, although three-state devices are passingly common.

Separation of Bias and Signal Analyses

This technique is used where the operation of the circuit is to be essentially linear, but the devices used to implement it are non-linear. A transistor amplifier is an example of this kind of network. The essence of this technique is to separate the analysis in to two parts. Firstly, the dc biases are analysed using some non-linear method. This establishes the quiescent operating point of the circuit. Secondly, the small signal characteristics of the circuit are analysed using linear network analysis. Examples of methods that can be used for both these stages are given below.

Graphical Method of dc Analysis

In a great many circuit designs, the dc bias is fed to a non-linear component via a resistor (or possibly a network of resistors). Since resistors are linear components, it is particularly easy to determine the quiescent operating point of the non-linear device from a graph of its transfer function. The method is as follows: from linear network analysis the output transfer function (that is output voltage against output current) is calculated for the network of resistor(s) and the generator driving them. This will be a straight line and can readily be superimposed on the transfer function plot of the non-linear device. The point where the lines cross is the quiescent operating point.

Perhaps the easiest practical method is to calculate the (linear) network open circuit voltage and short circuit current and plot these on the transfer function of the non-linear device.

The straight line joining these two point is the transfer function of the network. In reality, the designer of the circuit would proceed in the reverse direction to that described. Starting from a plot provided in the manufacturers data sheet for the non-linear device, the designer would choose the desired operating point and then calculate the linear component values required to achieve it.

It is still possible to use this method if the device being biased has its bias fed through another device which is itself non-linear - a diode for instance. In this case however, the plot of the network transfer function onto the device being biased would no longer be a straight line and is consequently more tedious to do.

SMALL SIGNAL EQUIVALENT CIRCUIT

This method can be used where the deviation of the input and output signals in a network stay within a substantially linear portion of the non-linear devices transfer function, or else are so small that the curve of the transfer function can be considered linear. Under a set of these specific conditions, the non-linear device can be represented by an equivalent linear network. It must be remembered that this equivalent circuit is entirely notional and only valid for the small signal deviations. It is entirely inapplicable to the dc biasing of the device.

For a simple two-terminal device, the small signal equivalent circuit may be no more than two components. A resistance equal to the slope of the v/i curve at the operating point (called the dynamic resistance), and tangent to the curve. A generator, because this tangent will not, in general, pass through the origin. With more terminals, more complicated equivalent circuits are required.

A popular form of specifying the small signal equivalent circuit amongst transistor manufacturers is to use the two-port network parameters known as [h] parameters. These are a matrix of four parameters as with the [z] parameters but in the case of the [h] parameters they are a hybrid mixture of impedances, admittances, current gains and voltage gains. In this model the three terminal transistor is considered to be a two port network, one of its terminals being common to both ports. The [h] parameters are quite different depending on which terminal is chosen as the common one. The most important parameter for transistors is usually the forward current gain, h_{21}, in the common emitter configuration. This is designated h_{fe} on data sheets.

The small signal equivalent circuit in terms of two-port parameters leads to the concept of dependent generators.

That is, the value of a voltage or current generator depends linearly on a voltage or current elsewhere in the circuit.

There will always be dependent generators in a two-port parameter equivalent circuit. This applies to the [h] parameters as well as to the [z] and any other kind. These dependencies must be preserved when developing the equations in a larger linear network analysis.

PIECEWISE LINEAR METHOD

In this method, the transfer function of the non-linear device is broken up into regions. Each of these regions is approximated by a straight line. Thus, the transfer function will be linear up to a particular point where there will be a discontinuity. Past this point the transfer function will again be linear but with a different slope.

A well known application of this method is the approximation of the transfer function of a pn junction diode. The actual transfer function of an ideal diode has been given at the top of this (non-linear) section. However, this formula is rarely used in network analysis, a piecewise approximation being used instead. It can be seen that the diode current rapidly diminishes to $-I_0$ as the voltage falls. This current, for most purposes, is so small it can be ignored. With increasing voltage, the current increases exponentially. The diode is modelled as an open circuit up to the knee of the exponential curve, then past this point as a resistor equal to the bulk resistance of the semiconducting material.

The commonly accepted values for the transition point voltage are 0.7V for silicon devices and 0.3V for germanium devices. An even simpler model of the diode, sometimes used in switching applications, is short circuit for forward voltages and open circuit for reverse voltages. The model of a forward biased pn junction having an approximately constant 0.7V is also a much used approximation for transistor base-emitter junction voltage in amplifier design.

The piecewise method is similar to the small signal method in that linear network analysis techniques can only be applied if the signal stays within certain bounds. If the signal crosses a discontinuity point then the model is no longer valid for linear analysis purposes. The model does have the advantage over small signal however, in that it is equally applicable to signal and dc bias. These can therefore both be analysed in the same operations and will be linearly superimposable.

9

Signals Circuit Theory

SIGNAL (ELECTRICAL ENGINEERING)

In the fields of communications, signal processing, and in electrical engineering more generally, a signal is any time-varying or spatial-varying quantity. In the physical world, any quantity measurable through time or over space can be taken as a signal. Within a complex society, any set of human information or machine data can also be taken as a signal.

Such information or machine data (for example, the dots on a screen, the ink making up text on a paper page, or the words now flowing into the reader's mind) must all be part of systems existing in the physical world – either living or non-living. Despite the complexity and even mystery – in the case of the reader's mind – of such systems, their outputs and inputs can often be represented with great fidelity as simple quantities measurable through time or across space.

In the latter half of the 20th century, electrical engineering itself separated into several disciplines, specializing in the design and analysis of physical signals and systems, on the one hand, and in the functional behaviour and conceptual structure of the complex human and machine systems, on the other. These engineering disciplines have led the way in the design, study, and implementation of systems that take advantage of signals as simple measurable quantities in order to facilitate the transmission, storage, and manipulation of information.

SOME DEFINITIONS

Definitions specific to subfields are common. For example, in information theory, a *signal* is a codified message, that is, the sequence of states in a communication channel that encodes a message. In a *communication system*, a *transmitter* encodes a *message* into a signal, which is carried to a *receiver* by the communications *channel*. For example, the words "Mary had a little lamb" might be the message spoken into a telephone. The telephone transmitter converts the sounds into an electrical voltage signal.

The signal is transmitted to the receiving telephone by wires; and at the receiver it is reconverted into sounds. Signals can be categorized in various ways. The most common distinction is between discrete and continuous spaces that the functions are defined over, for example discrete and continuous time domains.

Discrete-time signals are often referred to as *time series* in other fields. Continuous-time signals are often referred to as *continuous signals* even when the signal functions are not continuous; an example is a square-wave signal. A second important distinction is between discrete-valued and continuous-valued. Digital signals are discrete-valued, but are often derived from an underlying continuous-valued physical process.

Discrete-time and Continuous-time Signals

If for a signal, the quantities are defined only on a discrete set of times, we call it a discrete-time signal. In other words, a discrete-time real (or complex) signal can be seen as a function from the set of integers to the set of real (or complex) numbers. A continuous-time real (or complex) signal is any real-valued (or complex-valued) function which is defined for all time t in an interval, most commonly an infinite interval.

Analog and Digital Signals

Less formally than the theoretical distinctions mentioned above, two main types of signals encountered in practice are *analog* and *digital*. In short, the difference between them is that digital signals are *discrete* and *quantized*, as defined below, while analog signals possess neither property.

DISCRETIZATION

One of the fundamental distinctions between different types of signals is between continuous and discrete time. In the mathematical abstraction, the domain of a continuous-time (CT) signal is the set of real numbers (or some interval thereof), whereas the domain of a discrete-time (DT) signal is the set of integers (or some interval). What these integers represent depends on the nature of the signal.

DT signals often arise via sampling of CT signals. An audio signal, for example consists of a continually fluxuating voltage on a line that can be digitized by an ADC circuit, wherein the circuit will read the voltage level on the line, say, every 50 μs. The resulting stream of numbers is stored as digital data on a discrete-time signal. Computers and other digital devices are restricted to discrete time.

A discrete signal or discrete-time signal is a time series, perhaps a signal that has been sampled from a continuous-time signal. Unlike a continuous-time signal, a discrete-time signal is not a function of a continuous-time argument, but is a sequence of quantities; that is, a function over a domain of discrete integers. Each value in the sequence is called a sample.

When a discrete-time signal is a sequence corresponding to uniformly spaced times, it has an associated sampling rate; the sampling rate is not apparent in the data sequence, so may be associated as a separate data item.

DIGITAL SIGNALS

A digital signal is a discrete-time signal that takes on only a discrete set of values. It typically derives from a discrete signal that has been quantized.

Common practical digital signals are represented as 8-bit (256 levels), 16-bit (65,536 levels), 32-bit (4.3 billion levels), and so on, though any number of quantization levels is possible, not just powers of two.

QUANTIZATION

If a signal is to be represented as a sequence of numbers, it is impossible to maintain arbitrarily high precision - each number in the sequence must have a finite number of digits. As a result, the values of such a signal are restricted to belong to a finite set; in other words, it is quantized. In digital signal processing, quantization is the process of approximating a continuous range of values (or a very large set of possible discrete values) by a relatively small set of discrete symbols or integer values.

More specifically, a signal can be multi-dimensional and quantization need not be applied to all dimensions. Discrete signals (a common mathematical model) need not be quantized, which can be a point of confusion. A common use of quantization is in the conversion of a discrete signal (a sampled continuous signal) into a digital signal by quantizing. Both of these steps (sampling and quantizing) are performed in analog-to-digital converters with the quantization level specified in bits.

A specific example would be compact disc (CD) audio which is sampled at 44,100 Hz and quantized with 16 bits (2 bytes) which can be one of 65,536 (*i.e.* 2) possible values per sample. In electronics, adaptive quantization is a quantization process that varies the step size based on the changes of the input signal, as a means of efficient compression.Two approaches commonly used are forward adaptive quantization and backward adaptive quantization.

Mathematical Description

The simplest and best-known form of quantization is referred to as scalar quantization, since it operates on scalar (as opposed to multi-dimensional vector) input data. The integer-valued quantization index i is the representation that is typically stored or transmitted, and then the final interpretation is constructed using $g(i)$ when the data is later interpreted.

In computer audio and most other applications, a method known as *uniform quantization* is the most common. There are two common variations of uniform quantization, called *mid-rise* and *mid-tread* uniform quantizers.

In this case the $f(x)$ and $g(i)$ operators are just multiplying scale factors (one multiplier being the inverse of the other) along with an offset in $g(i)$ function to place the representation value in the middle of the input region for each quantization index.

The value $2^{-(M-)}$ is often referred to as the *quantization step size*. Using this quantization law and assuming that quantization noise is approximately uniformly distributed over the quantization step size (an assumption typically accurate for rapidly varying x or high M) and further assuming that the input signal x to be quantized is approximately uniformly distributed over the entire interval from -1 to 1, the signal to noise ratio (SNR) of the quantization.

From this equation, it is often said that the SNR is approximately 6 dB per bit. For mid-tread uniform quantization, the offset of 0.5 would be added within the floor function instead of outside of it. Sometimes, mid-rise quantization is used without adding the offset of 0.5. This reduces the signal to noise ratio by approximately 6.02 dB, but may be acceptable for the sake of simplicity when the step size is small.

In digital telephony, two popular quantization schemes are the 'A-law' (dominant in Europe) and 'ì-law' (dominant in North America and Japan). These schemes map discrete analog values to an 8-bit scale that is nearly linear for small values and then increases logarithmically as amplitude grows. Because the human ear's perception of loudness is roughly logarithmic, this provides a higher signal to noise ratio over the range of audible sound intensities for a given number of bits.

QUANTIZATION AND DATA COMPRESSION

Quantization plays a major part in lossy data compression. In many cases, quantization can be viewed as the fundamental element that distinguishes lossy data compression from lossless data compression, and the use of quantization is nearly always motivated by the need to reduce the amount of data needed to represent a signal. In some compression schemes, like MP3 or Vorbis, compression is also achieved by selectively discarding some data, an action that can be analyzed as a quantization process (*e.g.*, a vector quantization process) or can be considered a different kind of lossy process.

One example of a lossy compression scheme that uses quantization is JPEG image compression. During JPEG encoding, the data representing an image (typically 8-bits for each of three colour components per pixel) is processed using a discrete cosine transform and is then quantized and entropy coded. By reducing the precision of the transformed values using quantization, the number of bits needed to represent the image can be reduced substantially. For example, images can often be represented with acceptable quality using JPEG at less than 3 bits per pixel (as opposed to the typical 24 bits per pixel needed prior to JPEG compression). Even the original representation using 24 bits per pixel requires quantization for its PCM sampling structure.

In modern compression technology, the entropy of the output of a quantizer matters more than the number of possible values of its output (the number of values being 2^M in the above example). In order to determine how many bits are necessary to effect a given precision, algorithms are used. Suppose, for example, that it is necessary to record six significant digits, that is to say, millionths. The number of values that can be expressed by N bits is equal to two to the Nth power.

To express six decimal digits, the required number of bits is determined by rounding (6/log 2)—where log refers to the base ten, or common, logarithm—up to the nearest integer. Since the logarithm of 2, base ten, is approximately 0.30102, the required number of bits is then given by (6/0.30102), or 19.932, rounded up to the nearest integer, *viz.*, 20 bits. This type of quantization—where a set of binary digits, *e.g.*, an arithmetic register in a CPU, are used to represent a quantity—is called Vernier quantization. It is also possible, although rather less efficient, to rely upon equally spaced quantization levels.

This is only practical when a small range of values is expected to be captured: for example, a set of eight possible values requires eight equally spaced quantization levels—which is not unreasonable, although obviously less efficient than a mere trio of binary digits (bits)—but a set of, say, sixty-four possible values, requiring sixty-four equally spaced quantization levels, can be expressed using only six bits, which is obviously far more efficient.

RELATION TO QUANTIZATION IN NATURE

At the most fundamental level, some physical quantities are quantized. This is a result of quantum mechanics). Signals may be treated as continuous for mathematical simplicity by considering the small quantizations as negligible. In any practical application, this inherent quantization is irrelevant for two reasons. First, it is overshadowed by signal noise, the intrusion of extraneous phenomena present in the system upon the signal of interest.

The second, which appears only in measurement applications, is the inaccuracy of instruments. Thus, although all physical signals are intrinsically quantized, the error introduced by modeling them as continuous is vanishingly small.

EXAMPLES OF SIGNALS

- *Motion*. The motion of a particle through some space can be considered to be a signal, or can be represented by a signal. The domain of a motion signal is one-dimensional (time), and the range is generally three-dimensional. Position is thus a 3-vector signal; position and orientation is a 6-vector signal.
- *Sound*. Since a sound is a vibration of a medium (such as air), a sound signal associates a pressure value to every value of time and three space coordinates. A microphone converts sound pressure at some

place to just a function of time, using a voltage signal as an analog of the sound signal.

- *Compact discs* (CDs). CDs contain discrete signals representing sound, recorded at 44,100 samples per second. Each sample contains data for a left and right channel, which may be considered to be a 2-vector (since CDs are recorded in stereo).
- *Pictures*. A picture assigns a colour value to each of a set of points. Since the points lie on a plane, the domain is two-dimensional. If the picture is a physical object, such as a painting, it's a continuous signal. If the picture is a digital image, it's a discrete signal. It's often convenient to represent colour as the sum of the intensities of three primary colours, so that the signal is vector-valued with dimension three.
- *Videos*. A video signal is a sequence of images. A point in a video is identified by its position (two-dimensional) and by the time at which it occurs, so a video signal has a three-dimensional domain. Analog video has one continuous domain dimension (across a scan line) and two discrete dimensions (frame and line).
- *Biological membrane potentials*. The value of the signal is a straightforward electric potential ("voltage"). The domain is more difficult to establish. Some cells or organelles have the same membrane potential throughout; neurons generally have different potentials at different points. These signals have very low energies, but are enough to make nervous systems work; they can be measured in aggregate by the techniques of electrophysiology.

FREQUENCY ANALYSIS

Signals are often analyzed or modeled in terms of their frequency spectrum. Frequency domain techniques are applicable to all signals, both continuous-time and discrete-time. If a signal is passed through an LTI system, the frequency spectrum of the resulting output signal is the product of the frequency spectrum of the original input signal and the frequency response of the system.

Frequency domain is a term used to describe the analysis of mathematical functions or signals with respect to frequency. Speaking non-technically, a time-domain graph shows how a signal changes over time, whereas a frequency-domain graph shows how much of the signal lies within each given frequency band over a range of frequencies.

A frequency-domain representation can also include information on the phase shift that must be applied to each sinusoid in order to be able to recombine the frequency components to recover the original time signal. The frequency domain relates to the Fourier transform or Fourier series by decomposing a function into an infinite or finite number of frequencies.

This is based on the concept of Fourier series that any waveform can be expressed as a sum of sinusoids (sometimes infinitely many.) A spectrum analyzer is the tool commonly used to visualize real-world signals in the frequency domain.

MAGNITUDE AND PHASE

In using the Laplace, Z-, or Fourier transforms, the frequency spectrum is complex, describing the magnitude and phase of a signal, or of the response of a system, as a function of frequency. In many applications, phase information is not important. By discarding the phase information it is possible to simplify the information in a frequency domain representation to generate a frequency spectrum or spectral density.

A spectrum analyzer is a device that displays the spectrum. The power spectral density is a frequency-domain description that can be applied to a large class of signals that are neither periodic nor square-integrable; to have a power spectral density a signal needs only to be the output of a wide-sense stationary random process.

PARTIAL FREQUENCY-DOMAIN EXAMPLE

Due to popular simplifications of the hearing process and titles such as Plomp's "The Ear as a Frequency Analyzer," the inner ear is often thought of as converting time-domain sound waveforms to frequency-domain spectra. The frequency domain is not actually a very accurate or useful model for hearing, but a time/frequency space or time/place space can be a useful description.

BRIDGE CIRCUIT

A bridge circuit is a type of electrical circuit in which the current in a conductor splits into two parallel paths and then recombines into a single conductor, thereby enclosing a loop. It was originally used for measurement purposes, but can also be used in power supplies.

The best-known bridge circuit, the Wheatstone bridge, was invented by Samuel Hunter Christie and popularized by Charles Wheatstone, and is used for measuring resistance. It is constructed from four resistors, one of which has an unknown value (R_x), one of which is variable (R_2), and two of which are fixed and equal (R_1 and R_3), connected as the sides of a square. Two opposite corners of the square are connected to a source of electrical current, such as a battery. A galvanometer is connected across the other two opposite corners. The variable resistor is adjusted until the galvanometer reads zero.

It is then known that the ratio between the variable resistor and its neighbour is equal to the ratio between the unknown resistor and its neighbour, and this enables the value of the unknown resistor to be calculated. The Wheatstone bridge has also been developed to measure impedance in AC circuits, resulting in designs such as the Wien bridge, the Maxwell bridge and the Heaviside bridge. All are based on the same principle, which is to compare the output of two potentiometers sharing a common source.

In power supply design, a bridge circuit or bridge rectifier is an arrangement of diodes or similar devices used to rectify an electric current, *i.e.* to convert it from an unknown or alternating polarity to a direct current of known polarity. In some motor controllers, a H-bridge is used to control the direction the motor turns.

DIGITAL AND ANALOG SIGNALS

Signals carry information and are defined as any physical quantity that varies with time, space, or any other independent variable. For example, a sine wave whose amplitude varies with respect to time or the motion of a particle with respect to space can be considered as signals.

A system can be defined as a physical device that performs an operation on a signal. For example, an amplifier is used to amplify the input signal amplitude. In this case, the amplifier performs some operations on the signal, which has the effect of increasing the amplitude of the desired information-bearing signal. Signals can be categorised in various ways; for example discrete and continuous time domains. Discrete-time signals are defined only on a discrete set of times. Continuous-time signals are often referred to as continuous signals even when the signal functions are not continuous; an example is a square-wave signal.

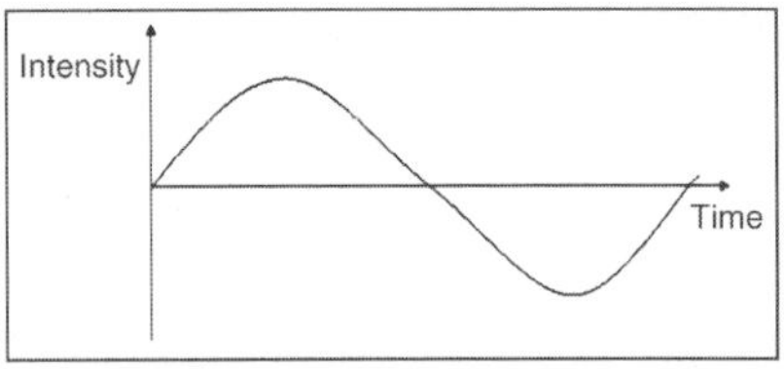

Fig. Analog Signal

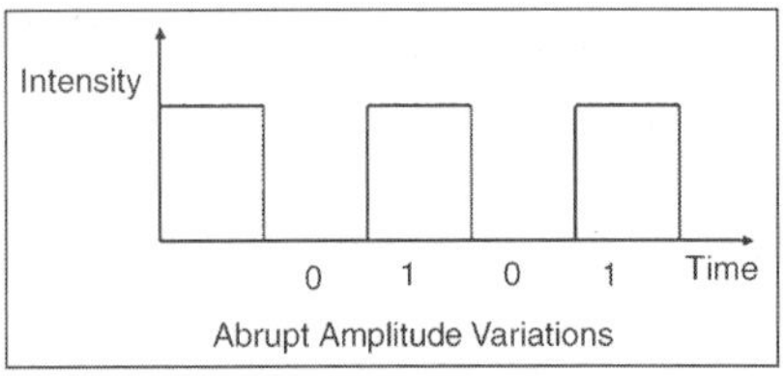

Fig. Digital Signal

Another category of signals is discrete-valued and continuous-valued or otherwise known as digital and analog signals. Digital signals are discrete-valued and analog signals are continuous electrical signals that vary in time as shown in Figures. Analog devices and systems process signals whose voltages or other quantities vary in a continuous manner. They can take on any value across a continuous range of voltage, current, or other metric. The analog signals can have an infinite number of values. Analog systems can be called wave systems.

They have a value that changes steadily over time and can have any one of an infinite set of values in a range. Analog signals represent some physical quantity and they can be a model of the real quantity. Most of the time, the variations corresponds to that of the non-electric (original) signal. For example, the telephone transmitter converts the sounds into an electrical voltage signal. The intensity of the voice causes electric current variations. Therefore, the two are analogous hence the name analog. At the receiving end, the signal is reproduced in the same proportion. Hence the electric current is a model and is an electrical representation of one's voice. Not all analog signals vary as smoothly as the waveform shown in Figure. Digital signals are non-continuous, they change in individual steps. They consist of pulses or digits with discrete levels or values. The value of each pulse is constant, but there is an abrupt change from one digit to the next. Digital signals have two amplitude levels. The value of which are specified as one of two possibilities such as 1 or 0, HIGH or LOW, TRUE or FALSE and so on. In reality, the values are anywhere within specific ranges and we define values within a given range.

A digital system is the one that handles only discrete values or signals. Any set that is restricted to a finite number of elements contains discrete information. The word digital describes any system based on discontinuous data or events. Digital is the method of storing, processing and transmitting information through the use of distinct electronic pulses that represent the binary digits 0 and 1. Examples of discrete sets are the 10 decimal digits, the 26 letters of the alphabet, etc. A digital system would be to flick the light switch on and off. There's no 'in between' values.

ADVANTAGES OF DIGITAL SIGNALS

The usual advantages of digital circuits when compared to analog circuits are:

- *Noise Margin (resistance to noise/robustness):* Digital circuits are less affected by noise. If the noise is below a certain level (the noise margin), a digital circuit behaves as if there was no noise at all. The stream of bits can be reconstructed into a perfect replica of the original source. However, if the noise exceeds this level, the digital circuit cannot give correct results.
- *Error Correction and Detection:* Digital signals can be regenerated to achieve lossless data transmission, within certain limits. Analog signal transmission and processing, by contrast, always introduces noise.
- *Easily Programmable:* Digital systems interface well with computers and are easy to control with software. It is often possible to add new features to a digital system without changing hardware, and to do this remotely, just by uploading new software. Design errors or bugs can be worked-around with a software upgrade, after the product is in customer hands. A digital system is often preferred because of (re-)programmability and

ease of upgrading without requiring hardware changes.

- Cheap Electronic Circuits: More digital circuitry can be fabricated per square millimeter of integrated-circuit material. Information storage can be much easier in digital systems than in analog ones. In particular, the great noise-immunity of digital systems makes it possible to store data and retrieve it later without degradation. In an analog system, aging and wear and tear will degrade the information in storage, but in a digital system, as long as the wear and tear is below a certain level, the information can be recovered perfectly. Theoretically, there is no data-loss when copying digital data. This is a great advantage over analog systems, which faithfully reproduce every bit of noise that makes its way into the signal.

DISADVANTAGES

The world in which we live is analog, and signals from this world such as light, temperature, sound, electrical conductivity, electric and magnetic fields, and phenomena such as the flow of time, are for most practical purposes continuous and thus analog quantities rather than discrete digital ones. For a digital system to do useful things in the real world, translation from the continuous realm to the discrete digital realm must occur, resulting in quantisation errors. This problem can usually be mitigated by designing the system to store enough digital data to represent the signal to the desired degree of fidelity. The Nyquist-Shannon sampling theorem provides an important guideline as to how much digital data is needed to accurately portray a given analog signal.

Digital systems can be fragile, in that if a single piece of digital data is lost or misinterpreted, the meaning of large blocks of related data can completely change. This problem can be diminished by designing the digital system for robustness. For example, a parity bit or other error-detecting or error-correcting code can be inserted into the signal path so that minor data corruptions can be detected and possibly corrected. Digital circuits use more energy than analog circuits to accomplish the same calculations and signal processing tasks, thus producing more heat as well. In portable or battery-powered systems this can be a major limiting factor. Digital circuits are made from analog components, and care has to be taken to all noise and timing margins, to parasitic inductances and capacitances, to proper filtering of power and ground connections, to electromagnetic coupling amongst data lines. Inattention to these can cause problems such as "glitches", pulses do not reach valid switching (threshold) voltages, or unexpected ("undecoded") combinations of logic states. A corollary of the fact that digital circuits are made from analog components is the fact that digital circuits are slower to perform calculations than analog circuits that occupy a similar amount of physical space and consume the same amount of power. However, the digital circuit will perform the calculation with much better repeatability, due to the high noise immunity of digital circuitry.

10

Circuits and Transformers

The circuit will light three automobile headlight bulbs in parallel. If these bulbs draw 5 amperes each, the current labeled I2 will, of course, be 15 amperes a-c. But the value of I1 is not so obvious. For any transformer the ratio of the input to output current is the inverse of the ratio of the input to output voltage. If the input voltage of the transformer is twice the output voltage, the input current will always be half the output current. This is expressed as

$$\frac{V_{\text{input}}}{V_{\text{output}}} = \frac{I_{\text{output}}}{I_{\text{input}}}$$

Note that this equation can also be expressed as

$$\text{Vinput} \times \text{Iinput} = \text{Voutput} \times \text{Ioutput}$$

(*i.e.*)

$$\text{Input watts} = \text{output watts}$$

In other words, if a certain amount of power is coming out of a transformer, the same amount will be going in. This makes sense as long as the transformer itself does not use up much power. Most transformers operate so efficiently that errors introduced by using these equations are negligible. In the example, we are given the output current (15 amperes), the output voltage (6 volts), and the input voltage (115 volts). Therefore

$$I_{\text{input}} = \frac{15 \times 6}{115} = 0.78 \text{ ampere}$$

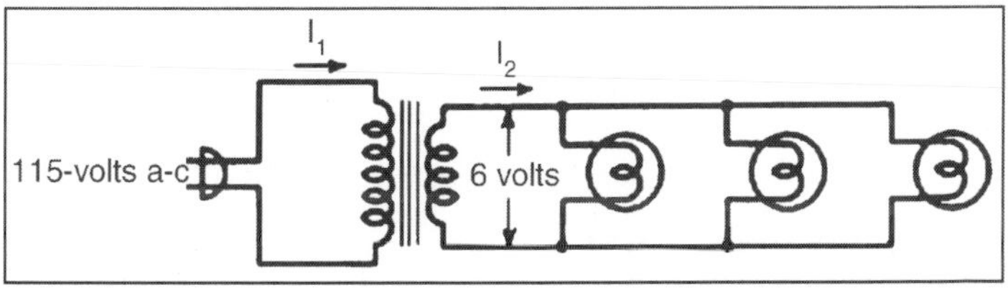

Fig. Circuit to Light Three Automobile Headlight Bulbs in Parallel. I_1, is the Input Current and I_2 t he Output Current of the Transformer.

To extend this treatment, consider what happens when there is no connection (an infinite resistance) between the two output terminals of the transformer. The output voltage will still be 6 volts, but now the output current is zero. The output power is zero watts, and therefore the input power must also be zero watts. Since the input voltage is not zero, the input current must be zero. In other words, according to this equation, a transformer plugged into the wall will not draw any current in its primary as long as nothing is connected across its secondary. This conclusion looks peculiar but is true for all practical purposes.

ESTIMATING THE RATINGS

Transformers which have no ratings engraved on them can usually be found in the dark corners of the laboratory. This section will describe several ways in which you may start to classify them. First of all, if what you think might be a transformer has only two wires coming out of it, it is a choke rather than a transformer. A choke looks like a transformer but it consists of only a single coil of wire on an iron core. If there are either three of four terminals on the unknown transformer and they are labeled with letters such as *b*, *p*, or *g*, it is an output transformer and should not be used except in certain electronic applications which will not be discussed here. But if the transformer has at least four terminals or wires, and if they are not labeled with the letters, *b*, *p*, or *g*, it may be usable.

The easiest way to determine its output voltage is to apply 115 volts, 60 cycles per second, to the primary coil and measure the voltages resulting in the secondary or secondaries. On some transformers the primary is labeled "primary," or "input," or "115 volts a-c," and then this procedure is straightforward, but usually there are no such markings and you have to guess which wires connect to the input coil. When there are only four wires or terminals and none is labeled, the four must be divided into two pairs, one pair from the input and the other from the output coil. If the pairing is not obvious from the geography of the transformer, it can be determined simply by using an ohmmeter to find out which pairs of wires are connected together. Now examine the leads themselves.

If one pair consists of fairly heavy wires, it is very likely that the unit is a step-down transformer and that the heavy pair are the secondary leads. If one pair of terminals is mounted on a large ceramic holder, that pair is probably the output of a very-high-voltage transformer (*e.g.*, 5000 volts or more). If neither is true, either coil can be used as the primary and the other as the secondary. In one case you will have a step-up transformer, and in the other a step-down transformer. To find out the ratio of input to output voltage, one pair of terminals may be connected across the 115-volt wall voltage, and the voltage between the others measured with an a-c voltmeter. However, since it is possible that the application of 115 volts to one of the coils will damage it, a safer procedure is to increase the input voltage gradually from zero to 115 volts, stopping if the transformer begins to heat up.

If there are five wires on an unknown transformer, it is probable that the fifth wire is a centre tap on the output coil. If so, an ohmmeter will indicate that there is one pair and one trio of connected wires. The output voltages (two of them) can then be measured by applying a known a-c voltage across the pair and measuring the voltages between pairs of the three remaining wires. When there are more than five wires or terminals on a transformer, the job becomes a lot more difficult. In general, the best procedure is to find out with an ohmmeter which wires are connected to each other, and then try to reconstruct the circuit diagram of the transformer from this information.

Suppose you find a transformer with eight wires coming out of its casing, and an ohmmeter indicates that there are one pair and two groups of three. Because primaries rarely have centre taps, the pair is very likely the primary. Then examine the wires themselves. If in one of the groups of three, two of the wires are thick, the chances are very good that the group of three is a low-voltage output coil with a centre tap (probably 6 volts), and that the other group of three is a high-voltage output coil with a centre tap (*e.g.*, 300 volts).

It is very difficult for an unpracticed worker to judge the output current rating of an unknown transformer. The only real way to determine this is to let the transformer draw some current and see if it gets hot. This is done by putting the rated voltage across the input terminals and connecting a variable resistor across the output. Then gradually increase the current by decreasing the resistance until the transformer either gets hot or is delivering the required current. A good criterion for "hot" is this: If you can hold your hand against the transformer, it is not overheating. If you have to pull your hand away, the transformer is overloaded. Any transformer running near its rated load will be hot, but it should not be so hot that you cannot hold your hand against it.

ADJUSTABLE-VOLTAGE TRANSFORMERS

One of the most useful gadgets around a laboratory is the adjustable voltage transformer, often referred to as "Variac." (Variac is a trade name for a particular, widely sold brand of adjustable voltage transformer.) The customary input voltage to this type of transformer is 115 volts, 60 cycles, and the output can be adjusted continuously from zero to 115 or zero to 140 volts by turning a knob on the top.

The theory of operation of this sort of transformer is too complicated to be discussed herein. At any given setting of the slider arm, it acts like a normal, 2-coil transformer except that there is a direct electric connection between the input and the output. This connection is sometimes important. Adjustable voltage transformers are very convenient for changing the intensity of light bulbs, the speed of motors, etc.

They are particularly useful for testing devices the ratings of which are unknown. For example, suppose that you need a relay and you find one in the stockpile, but the coil voltage required to make it operate is unknown.

Connect it across the output of an adjustable-voltage transformer and gradually increase the voltage from zero until it closes firmly. Then read the required voltage with a voltmeter.

(If the relay never closes firmly but just buzzes, it is a d-c relay.) The same procedure can be applied to other devices such as unlabeled motors or transformers: increase the applied voltage gradually until the devices operate properly.

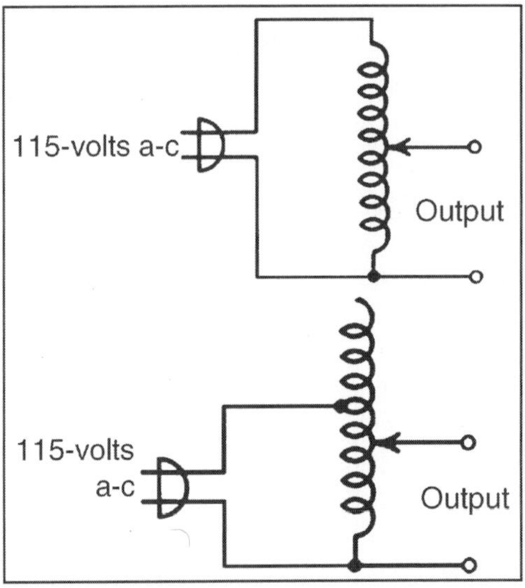

Fig. (*a*) Diagram of a Transformer whose output Voltage is Adjustable from 0 to 115 Volts. (*b*) Diagram of a 0 to 140-volt Adjustable Voltage Transformer.

ISOLATION TRANSFORMERS

There is a class of transformers called isolation transformers which usually have the same number of turns of wire on the primary as on the secondary, so that they neither step up nor step down the input voltage.

They are used when it is desired to isolate a circuit electrically from the source of power. For example, suppose that you constructed the very simple device to deliver a weak electric shock to a human subject. One side of the wall socket is already connected to "ground," that is, to the earth itself, and to the water pipes, radiator, etc. This side of the socket is so labeled.

Now suppose that the subject reaches over and touches some metal object that is in contact with a radiator.

He will be completing the circuit, and will no longer be a useful subject. This possibility can be eliminated by placing an isolation transformer between the adjustable voltage transformer and the wall socket.

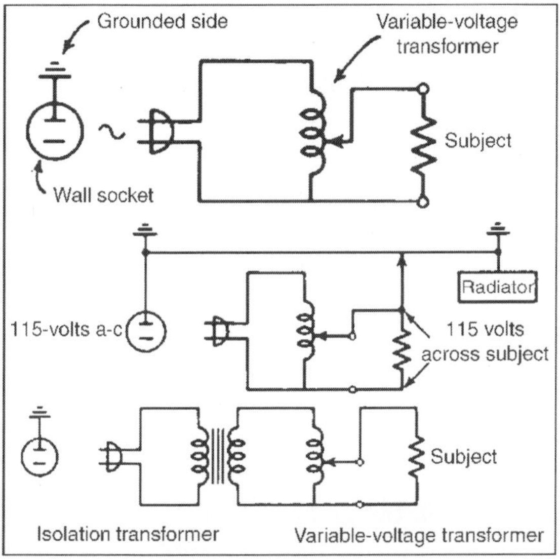

Fig. (*a*) Circuit to Shock Subjects. This Circuit Should not be Used Because of the Serious Shock Hazard. (*b*) Diagram of the Actual Circuit when a Subject, being Shocked by the Circuit Touches any Grounded Object. The Full wall-socket Voltage Appears Across the Subject. (*c*) A Safe Circuit for Shocking Human Subjects. The Isolation Transformer Eliminates the Shock Hazard of the Circuit.

CONSTANT VOLTAGE TRANSFORMERS

The voltage at the wall socket is usually said to be 115 volts. Actually, that voltage varies widely from time to time. It is a common experience to see the room lights dim when the refrigerator goes on. The refrigerator motor, when it first turns on, draws a large current which flows through the wires leading into the house and to the wall socket. The increased current running through the resistance of those wires causes an increased voltage drop in the wires, and less Voltage is left to hear the lights.

In and around the laboratory, and all along the circuit that includes the power station and the laboratory, devices turn on and off repeatedly, so that it is not unusual for the voltage at the wall to vary from 90 to 130 volts during the day. (The voltage can be higher than 115 because, to overcome the average voltage loss in the wires, the power company delivers more than 115 volts to the system. If everyone but you should suddenly turn off everything electrical, all of your light bulbs would burn out.)

For many devices, such changes in voltage do not matter very much (*e.g.*, fan motors, relays), but several types of apparatus are very sensitive to voltage changes and must be protected from them. All laboratory apparatus in which the intensity of a light bulb is important fall into this category because relatively small changes in line voltage will produce large changes in any lamp's brightness. For example, variations in the line voltage can be a source of great variance in tachistoscopic studies.

Devices which take variable a-c inputs and deliver constant-voltage a-c outputs are called voltage regulators or voltage stabilizers and may be ordered at any electronics supply store.

There are a number of different types of voltage regulators, some considerably more costly than others. The cheapest, and one which meets most requirements, is called a constant-voltage transformer. It has no moving parts and is essentially a standard transformer built around a special core.

The molecules of metal in any transformer core are involved in the transfer of energy from the input to the output coil. In a constant-voltage transformer, all of the core molecules are brought into action when the instantaneous input voltage is, say, 90 volts. The coils are wound so that the output voltage is 115 volts under this condition.

Any further increase in the input voltage cannot affect the output voltage appreciably because all of the core molecules are already involved (the core is said to be saturated). This transformer will, therefore, give almost constant output voltage for input voltages ranging from 90 to some higher voltage where the system begins to break down.

Typically, the upper limit is approximately 130 volts. Transformers of this type are built with various output-current capacities from about 1/10 ampere up to 10 or 20 amperes.

The rating is usually stated in units, called volt-amperes, which are very closely related to watts. If you want 115 volts at about 4 amperes, buy a 500-volt-ampere unit. (For a-c circuits, power—in watts—is sometimes different from the simple product of volts times amperes.)

The constant-voltage transformers described above have several defects which may be of consequence in certain applications. First, they only hold the voltage stable to about 1lper cent of its nominal level. For the rare situation in which something better is needed, a more expensive unit must be bought. Second, these transformers usually hum and, because of this, should not be used when absolute quiet is needed. A third defect is that, since the core saturates in the normal range of input voltages, the output wave form is a flat-topped wave that approaches the shape of a square wave. For the rare occasions when a very low harmonic content is required of the supply voltage, regulators which give pure sine-wave outputs can be purchased. Voltage regulators are not to be confused with the very common piece of electronic equipment called a regulated power supply.

The term "power supply," when it is applied to a single unit of equipment, usually refers to a unit that supplies d-c power. A regulated, or stabilized, power supply unit is one that delivers a constant d-c output voltage independent of changes in the line voltage.

TRANSFORMER TYPES

A variety of types of electrical transformer are made for different purposes. Despite their design differences, the various types employ the same basic principle as discovered in 1831 by Michael Faraday, and share several key functional parts.

POWER TRANSFORMERS

Laminated core

This is the most common type of transformer, widely used in appliances to convert mains voltage to low voltage to power electronics

- Widely available in power ratings ranging from mW to MW
- Insulated laminations minimize eddy current losses
- Small appliance and electronic transformers may use a split bobbin, giving a high level of insulation between the windings
- Rectangular core
- Core laminate stampings are usually in EI shape pairs. Other shape pairs are sometimes used.
- Mumetal shields can be fitted to reduce EMI (electromagnetic interference)
- A screen winding is occasionally used between the 2 power windings
- Small appliance and electronics transformers may have a thermal cut out built in
- Occasionally seen in low profile format for use in restricted spaces
- Laminated core made with silicon steel with high permeability

Toroidal

Doughnut shaped toroidal transformers are used to save space compared to EI cores, and sometimes to reduce external magnetic field. These use a ring shaped core, copper windings wrapped round this ring (and thus threaded through the ring during winding), and tape for insulation.

Toroidals compared to EI core transformers:

- Lower external magnetic field
- Smaller for a given power rating
- Higher cost in most cases, as winding requires more complex and slower equipment
- Less robust
- Central fixing is either
  - Bolt, large metal washers and rubber pads
  - Bolt and potting resin
- Overtightening the central fixing bolt may short the windings
- Greater inrush current at switch-on.

Autotransformer

An autotransformer has only a single winding, which is tapped at some point along the winding. AC or pulsed voltage is applied across a portion of the winding, and a higher (or lower) voltage is produced across another portion of the same winding. The higher voltage will be connected to the ends of the winding, and the lower voltage from one end to a tap. For example, a transformer with a tap at the centre of the winding can be used with 230 volts across the entire winding, and 115 volts between one end and the tap.

It can be connected to a 230-volt supply to drive 115-volt equipment, or reversed to drive 230-volt equipment from 115 volts. Since the current in the windings is lower, the transformer is smaller, lighter cheaper and more efficient. For voltage ratios not exceeding about 3:1, an autotransformer is cheaper, lighter, smaller and more efficient than an isolating (two-winding) transformer of the same rating. Large three-phase autotransformers are used in electric power distribution systems, for example, to interconnect 33 kV and 66 kV sub-transmission networks.

In practice, transformer losses mean that autotransformers are not perfectly reversible; one designed for stepping down a voltage will deliver slightly less voltage than required if used to step up. The difference is usually slight enough to allow reversal where the actual voltage level is not critical. This is true of isolated winding transformers too.

Variac

By exposing part of the winding coils of an autotransformer, and making the secondary connection through a sliding carbon brush, an autotransformer with a near-continuously variable turns ratio can be obtained, allowing for wide voltage adjustment in very small increments.

Stray Field Transformer

A Stray field transformer has a significant stray field or a (sometimes adjustable) magnetic bypass in its core. It can act as a transformer with inherent current limitation due to its lower tight coupling between the primary and the secondary winding, which is unwanted in much other cases. The output and input currents are low enough to prevent thermal overload under each load condition - even if the secondary is shortened.

Stray field transformers are used for arc welding and high voltage discharge lamps (cold cathode fluorescent lamps, series connected up to 7,5 kV AC working voltage). It acts both as voltage transformer and magnetic ballast.

Polyphase Transformers

For three-phase power, three separate single-phase transformers can be used, or all three phases can be connected to a single polyphase transformer.

The three primary windings are connected together and the three secondary windings are connected together. The most common connections are Y-Delta, Delta-Y, Delta-Delta and Y-Y. A vector group indicates the configuration of the windings and the phase angle difference between them.

If a winding is connected to earth (grounded), the earth connection point is usually the centre point of a Y winding. If the secondary is a Delta winding, the ground may be connected to a centre tap on one winding (high leg delta) or one phase may be grounded (corner grounded delta). A special purpose polyphase transformer is the zigzag transformer. There are many possible configurations that may involve more or fewer than six windings and various tap connections.

Resonant Transformers

A resonant transformer operates at the resonant frequency of one or more of its coils and (usually) an external capacitor. The resonant coil, usually the secondary, acts as an inductor, and is connected in series with a capacitor. When the primary coil is driven by a periodic source of alternating current, such as a square or sawtooth wave at the resonant frequency, each pulse of current helps to build up an oscillation in the secondary coil.

Due to resonance, a very high voltage can develop across the secondary, until it is limited by some process such as electrical breakdown. These devices are used to generate high alternating voltages, and the current available can be much larger than that from electrostatic machines such as the Van de Graaff generator or Wimshurst machine. Examples:

- Tesla coil
- Oudin coil (or Oudin resonator; named after its inventor Paul Oudin)
- D'Arsonval apparatus
- Ignition coil or induction coil used in the ignition system of a petrol engine
- Flyback transformer of a CRT television set or video monitor.
- Electrical breakdown and insulation testing of high voltage equipment and cables. In the latter case, the transformer's secondary is resonated with the cable's capacitance.

Other applications of resonant transformers are as coupling between stages of a superheterodyne receiver, where the selectivity of the receiver is provided by the tuned transformers of the intermediate-frequency amplifiers.

Constant Voltage Transformer

By arranging particular magnetic properties of a transformer core, and installing a ferro-resonant tank circuit (a capacitor and an additional winding), a transformer can be arranged to automatically keep the secondary winding voltage relatively constant for varying primary supply without additional circuitry or manual adjustment.

CVA transformers run hotter than standard power transformers, because regulating action depends on core saturation, which reduces efficiency somewhat. The output waveform is heavily distorted unless careful measures are taken to prevent this. Saturating transformers provide a simple rugged method to stabilize an AC power supply.

Ferrite Core

Ferrite core power transformers are widely used in switched mode power supplies (SMPSUs). The powder core enables high frequency operation, and hence much smaller size to power ratio than laminated iron transformers. Ferrite transformers are not usable as power transformers at mains frequency.

Planar Transformer

Manufacturers etch spiral patterns on a printed circuit board to form the "windings" of a planar transformer. (Manufacturers literally wind pieces of wire on some core or bobbin to form the windings of other kinds of transformers).

Some planar transformers are commercially sold as discrete components — the transformer is the only thing on that printed circuit board. Other planar transformers are one of many components on one large printed circuit board.

- Much thinner than other transformers, for low-profile applications (even when several PCBs are stacked)
- Almost all use a ferrite planar core

Oil Cooled Transformer

For large transformers used in power distribution or electrical substations, the core and coils of the transformer are immersed in oil which cools and insulates. Oil circulates through ducts in the coil and around the coil and core assembly, moved by convection. The oil is cooled by the outside of the tank in small ratings, and in larger ratings an air-cooled radiator is used.

Where a higher rating is required, or where the transformer is used in a building or underground, oil pumps are used to circulate the oil and an oil-to-water heat exchanger may also be used. Formerly, indoor transformers required to be fire-resistant used PCB liquids; since these are now banned, substitute fire-resistant liquids such as silicone oils are instead used.

Isolating Transformer

Most transformers isolate, meaning the secondary winding is not connected to the primary. But this isn't true of all transformers. However the term 'isolating transformer' is normally applied to mains transformers providing isolation rather than voltage transformation. They are simply 1:1 laminated core transformers. Extra voltage tappings are sometimes included, but to earn the name 'isolating transformer' it is expected that they will usually be used at 1:1 ratio.

INSTRUMENT TRANSFORMERS

CURRENT TRANSFORMERS

A current transformer (CT) is a measurement device designed to provide a current in its secondary coil proportional to the current flowing in its primary. Current transformers are commonly used in metering and protective relaying in the electrical power industry where they facilitate the safe measurement of large currents, often in the presence of high voltages. The current transformer safely isolates measurement and control circuitry from the high voltages typically present on the circuit being measured.

Current transformers are often constructed by passing a single primary turn (either an insulated cable or an uninsulated bus bar) through a well-insulated toroidal core wrapped with many turns of wire. The CT is typically described by its current ratio from primary to secondary. For example, a 4000:5 CT would provide an output current of 5 amperes when the primary was passing 4000 amperes.

The secondary winding can be single ratio or have several tap points to provide a range of ratios. Care must be taken that the secondary winding is not disconnected from its load while current flows in the primary, as this will produce a dangerously high voltage across the open secondary and may permanently affect the accuracy of the transformer.

Specially constructed wideband CTs are also used, usually with an oscilloscope, to measure high frequency waveforms or pulsed currents within pulsed power systems. One type provides a voltage output that is proportional to the measured current; another, called a Rogowski coil, requires an external integrator in order to provide a proportional output.

Voltage Transformers

Voltage transformers (VTs) or potential transformers (PTs) are another type of instrument transformer, used for metering and protection in high-voltage circuits. They are designed to present negligible load to the supply being measured and to have a precise voltage ratio to accurately step down high voltages so that metering and protective relay equipment can be operated at a lower potential.

Typically the secondary of a voltage transformer is rated for 69 or 120 Volts at rated primary voltage, to match the input ratings of protection relays. The transformer winding high-voltage connection points are typically labelled as H1, H2 (sometimes H0 if it is internally grounded) and X1, X2, and sometimes an X3 tap may be present. Sometimes a second isolated winding (Y1, Y2, Y3) may also be available on the same voltage transformer. The high side (primary) may be connected phase to ground or phase to phase. The low side (secondary) is usually phase to ground. The terminal identifications (H1, X1, Y1, etc.) are often referred to as polarity. This applies to current transformers as well. At any instant terminals with the same suffix numeral have the same polarity and phase. Correct identification of terminals and wiring is essential for proper operation of metering and protection relays.

While VTs were formerly used for all voltages greater than 240V primary, modern meters eliminate the need VTs for most secondary service voltages. VTs are typically used in circuits where the system voltage level is above 600 V. Modern meters eliminate the need of VT's since the voltage remains constant and it is measured in the incoming supply.

Pulse Transformers

A pulse transformer is a transformer that is optimised for transmitting rectangular electrical pulses (that is, pulses with fast rise and fall times and a relatively constant amplitude). Small versions called *signal* types are used in digital logic and telecommunications circuits, often for matching logic drivers to transmission lines.

Medium-sized *power* versions are used in power-control circuits such as camera flash controllers. Larger *power* versions are used in the electrical power distribution industry to interface low-voltage control circuitry to the high-voltage gates of power semiconductors. Special high voltage pulse transformers are also used to generate high power pulses for radar, particle accelerators, or other high energy pulsed power applications.

To minimise distortion of the pulse shape, a pulse transformer needs to have low values of leakage inductance and distributed capacitance, and a high open-circuit inductance.

In power-type pulse transformers, a low coupling capacitance (between the primary and secondary) is important to protect the circuitry on the primary side from high-powered transients created by the load.

For the same reason, high insulation resistance and high breakdown voltage are required. A good transient response is necessary to maintain the rectangular pulse shape at the secondary, because a pulse with slow edges would create switching losses in the power semiconductors.

The product of the peak pulse voltage and the duration of the pulse (or more accurately, the voltage-time integral) is often used to characterise pulse transformers.

Generally speaking, the larger this product, the larger and more expensive the transformer. Pulse transformers by definition have a duty cycle of less than 1, whatever energy stored in the coil during the pulse must be "dumped" out before the pulse is fired again.

RF Transformers

There are several types of transformer used in radio frequency (RF) work. Steel laminations are not suitable for RF.

Air-Core Transformers

These are used for high frequency work. The lack of a core means very low inductance. Such transformers may be nothing more than a few turns of wire soldered onto a printed circuit board.

Ferrite-Core Transformers

Widely used in intermediate frequency (IF) stages in superheterodyne radio receivers. These are mostly tuned transformers, containing a threaded ferrite slug that is screwed in or out to adjust IF tuning. The transformers are usually canned for stability and to reduce interference.

Transmission-Line Transformers

For radio frequency use, transformers are sometimes made from configurations of transmission line, sometimes bifilar or coaxial cable, wound around ferrite or other types of core. This style of transformer gives an extremely wide bandwidth but only a limited number of ratios (such as 1:9, 1:4 or 1:2) can be achieved with this technique.

The core material increases the inductance dramatically, thereby raising its Q factor. The cores of such transformers help improve performance at the lower frequency end of the band. RF transformers sometimes used a third coil (called a tickler winding) to inject feedback into an earlier (detector) stage in antique regenerative radio receivers.

Baluns

Baluns are transformers designed specifically to connect between balanced and unbalanced circuits. These are sometimes made from configurations of transmission line and sometimes bifilar or coaxial cable and are similar to transmission line transformers in construction and operation.

Audio Transformers

Audio transformers are usually the factor which limit sound quality when used; electronic circuits with wide frequency response and low distortion are relatively simple to design.

Transformers are also used in DI boxes to convert high-impedance instrument signals (*e.g.* bass guitar) to low impedance signals to enable them to be connected to a microphone input on the mixing console.

A particularly critical component is the output transformer of an audio power amplifier. Valve circuits for quality reproduction have long been produced with no other (inter-stage) audio transformers, but an output transformer is needed to couple the relatively high impedance (up to a few hundred ohms depending upon configuration) of the output valve(s) to the low impedance of a loudspeaker. (The valves can deliver a low current at a high voltage; the speakers require high current at low voltage.) Most solid-state power amplifiers need no output transformer at all.

For good low-frequency response a relatively large iron core is required; high power handling increases the required core size. Good high-frequency response requires carefully designed and implemented windings without excessive leakage inductance or stray capacitance. All this makes for an expensive component.

Early transistor audio power amplifiers often had output transformers, but they were eliminated as designers discovered how to design amplifiers without them.

Loudspeaker Transformers

In the same way that transformers are used to create high voltage power transmission circuits that minimize transmission losses, loudspeaker transformers can be used allow many individual loudspeakers to be powered from a single audio circuit operated at higher-than normal loudspeaker voltages. This application is common in industrial public address applications. Such circuits are commonly referred to as constant voltage speaker systems, although the audio waveform is a changing voltage. Such systems are also known by other terms such as 25-, 70- and 100-volt speaker systems, referring to the nominal voltage of the loudspeaker line.

At the audio amplifier, a large audio transformer may be used to step-up the low impedance, low-voltage output of the amplifier to the designed line voltage of the loudspeaker circuit.

At the distant loudspeaker location, a smaller step-down transformer returns the voltage and impedance to ordinary loudspeaker levels. The loudspeaker transformers commonly have multiple primary taps, allowing the volume at each speaker to be adjusted in discrete steps.

Output Transformer (valve)

Valve (tube) amplifiers almost always use an output transformer to match the high load impedance requirement of the valves (several kilohms) to a low impedance speaker.

Small Signal Transformers

Moving coil phonograph cartridges produce a very small voltage. In order for this to be amplified with a reasonable signal-noise ratio, a transformer is usually used to convert the voltage to the range of the more common moving-magnet cartridges.

Microphones may also be matched to their load with a small transformer, which is mumetal shielded to minimise noise pickup.

These transformers are less widely used today, as transistorized buffers are now cheaper.

'Interstage' and Coupling Transformers

A use for interstage transformers is in the case of push-pull amplifiers where an inverted signal is required.

Here two secondary windings wired in opposite polarities may be used to drive the output devices. These phase splitting transformers are not much used today.

Cast Resin Transformers

Cast-resin power transformers have been widely used for a long time. These transformers have the advantage of easy installation and improved fire behaviour in case of class. This indoor type transformer is totally dry, without cooling oil.

Homemade and Obsolete Transformers

Transformer Kits

Transformers may be wound at home using commercial transformer kits, which contain laminations and bobbin.

Or ready made transformers may be disassembled and rewound. These approaches are occasionally used by home constructors, but are usually avoided where possible due to the number of hours required to hand wind a transformer.

100per cent Homemade

It is possible to make the transformer laminations by hand too. Such transformers are encountered at times in 3rd world countries, using laminations cut from scrap sheet steel, paper slips between the laminations, and string to tie the assembly together. The result works, but is usually noisy due to poor clamping of laminations.

- Picture
- Device in use

Hedgehog

Hedgehog transformers are occasionally encountered in homemade 1920s radios. They are homemade audio interstage coupling transformers. Enamelled copper wire is wound round the central half of the length of a bundle of insulated iron wire (eg florists' wire), to make the windings.

The ends of the iron wires are then bent around the electrical winding to complete the magnetic circuit, and the whole is wrapped with tape or string to hold it together.

These were sometimes used when the cost of a ready made transformer could not be justified. Inductance tends to be on the low side, with consequent loss of bass. With the speakers of the day this was no bad thing.

Variocouplers

Variocouplers are rf transformers with 2 windings and variable coupling between the windings. They were standard equipment in 1920s radio sets. Pancake coil variocouplers were common in 1920s radios for variable rf coupling. The 2 planar coils were arranged to swing away from each other and for the angle between them to increase to 90 degrees, thus giving wide variation in coupling. No core was used.

These were mostly used to control reaction. The pancake structure was a means to minimize stray capacitance. In another design of variocoupler, 2 coils were wound on a 2 circular bands, and housed one inside the other, with provision for rotating the inner coil. Coupling varies as one coil is rotated between 0 and 90 degrees from the other. These had higher stray capacitance than the pancake type.

Not Transformers

Finally there are some items often mistaken for transformers, but which are not always transformers. Wall warts: small power supplies with integral mains plug. These can contain a transformer and other circuitry. Most use a laminated iron transformer, but an increasing number now contain a small SMPSU.

These are smaller and much lighter. Halogen lighting transformers: Toroidal transformers are sometimes used for this task, but most halogen 'transformers' are SMPSUs.

11

MOSFET

DEFINITION

STRUCTURE OPERATION AND V-I CHARACTERISTICS

Mosfets are 4-terminal devices with a gate that controls the conduction of the drain to source channel. The body (or substrate) of the typical n-channel enhancement mosfet is a P-type material with the drain and source formed from N-type materials. The oxide layer insulates the metal gate connection from the other layers. There are holes in the oxide layer left for the metallic contacts of the drain and source.

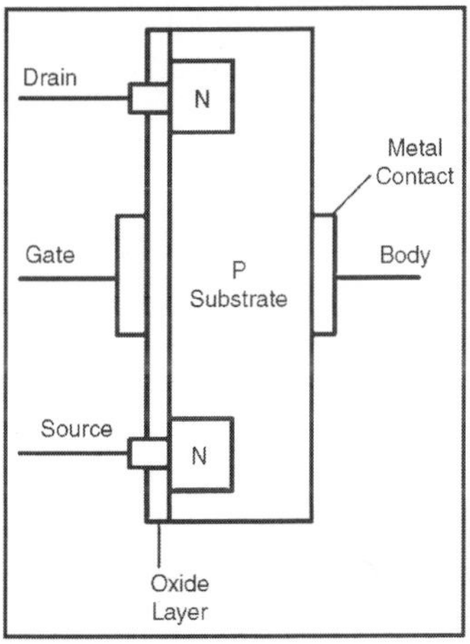

STRUCTURE

N-P junctions act as diodes, and because the layers of the mosfet have P-type and N-type materials in close contact, diodes are formed from the body to the drain or source, as shown in the drawing here of the 4-terminal mosfet.

If you were to leave the body connection floating, there could be no voltage flow through the two PN body diodes because they are reverse connected to each other.

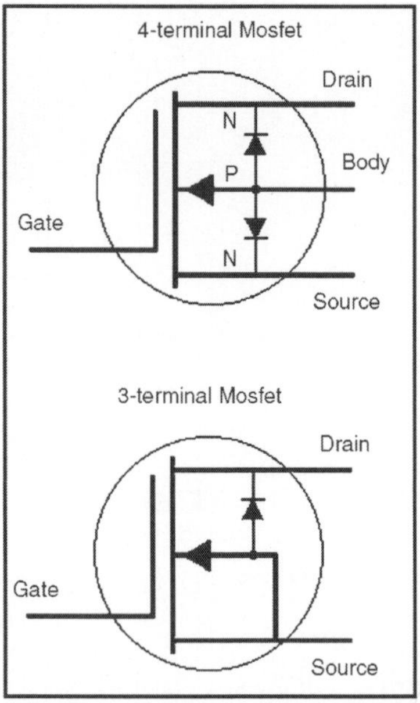

However, in the 3-terminal mosfets (like the BS170) that are commonly used in guitar effects pedals, the body and source are connected internally and this shorts out the PN junction of the lower diode and removes it from the circuit.

The top body diode can now conduct from drain to source because of the internal connection and therefore has to be taken into consideration during circuit design. This means that every 3-terminal mosfet has a body diode due the construction of the device. It is not added as a separate component inside the mosfet.

WORKING

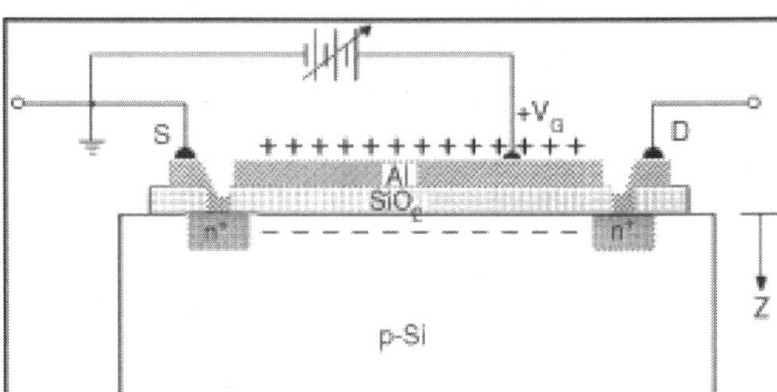

In MOSFET, inversion layers are formed at the interface between a semiconductor and an insulator or between two semiconductors, with one of them acting as an insulator.

The system in which the Quantum Hall Effect (QHE) was discovered has Si for the semiconductor, SiO_2 for the insulator. Figure above is a schematic side view of a silicon MOSFET showing the aluminum gate, the SiO_2 insulator and the p-type Si crystal substrate.

The principle of the inversion layer is quite simple. It is arranged that an electric field perpendicular to the interface attracts electrons from the semiconductor to it.

These electrons sit in a quantum well created by this field and the interface. The motion perpendicular to the interface is quantized and thus has a fundamental rigidity which freezes out motional degrees of freedom in this direction.

The result is a two-dimensional system of electrons. Further, the wavelengths of these electrons are long so that an effective mass approxiamtion with parabolic bands is quite good. The total self-consistent potential seen by the electrons is conveniently described by the picture of "band bending".

That is to say, the periodic lattice potential gives rise to energy bands, and the slowly varying electric potential then is regarded as bending these bands. Figure gives the schematic diagram for this process.(Charge density $\sim 10^{13}\,cm^{-2}$)

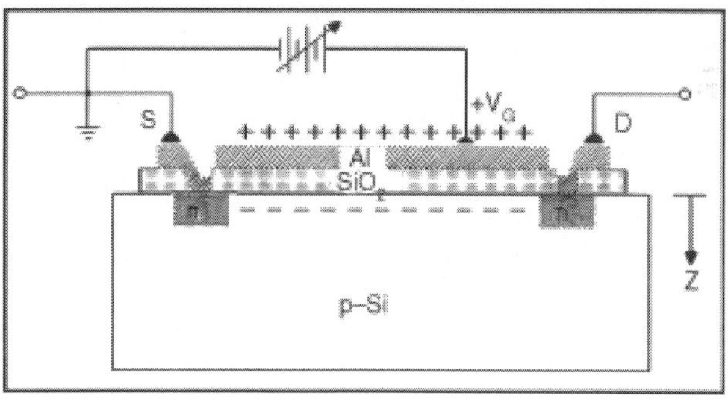

Fig. Electron Energy Level Diagram of a Silicon MOSFET with a Positive Voltage Applied to the Aluminum Gate

Another type of two-dimensional electron system is formed in the heterostructures of two semiconductors. Using molecular beam epitaxy(MBE) technique, people can grow two semiconductors alternately to form a one dimensional sandwich like structure.

Each layer has a width of about several nanometers. They are called superlattice. They can also be grown by metal-organic chemical vapour deposition(MOCVD). For example, in GaAs-$Ga_{1-x}Al_xAs$ superlattice, a certain controlled number of layers of GaAs is followed by an almost perfectly matched sequence of layers of GaAlAs.

The GaAlAs is deliberately doped n-type, which puts mobile electrons into its conduction band. These electrons will migrate to fill the few holes on the top of the GaAs valence band but most of them will end up in states near the bottom of the GaAs conduction band. However, there is a positive charge left on the donor impurities which attracts these electrons to the interface and bends the bands in the process.

This is the source of the electric field in this system. The transfer of electrons from GaAlAs to GaAs will continue until the dipole layer formed from the positive donors and the negative inversion layer is sufficiently strong.

This dipole layer gives rise to a potential discontinuity which finally makes the Fermi level of the GaAs equal to that of the GaAlAs. Figure shows the band structure. (Charge density $\sim 10^{11} cm^{-2}$)

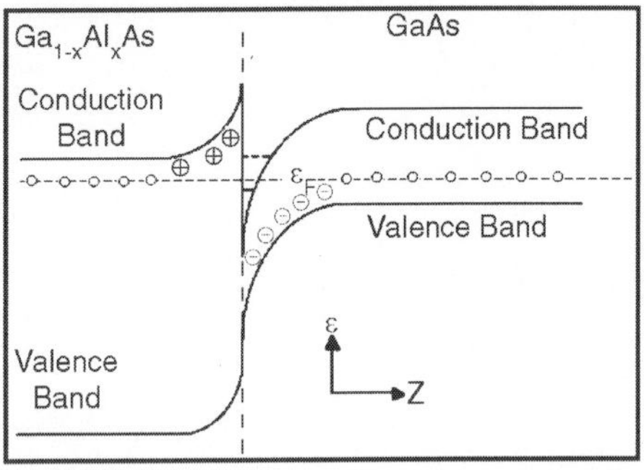

Fig. Electron Energy Level Diagram of a GaAs-AlGaAs Heterostructure Device

Two-dimensional electron system can also be formed in the surface of liquid Helium. There exists a potential barrier of about 1 eV in the surface of liquid Helium which prevents electrons from transmitting into the liquid.

On the other hand, the mirror potential attracts the electrons in the surface, resulting a 2D electron system.(Charge density $\sim 10^{9} cm^{-2}$)

WORKING

As can be seen on the figure the source and drain regions are identical. It is the applied voltages which determine which n-type region provides the electrons and becomes the source, while the other n-type region collects the electrons and becomes the drain.

The voltages applied to the drain and gate electrode as well as to the substrate by means of a back contact are refered to the source potential, as also indicated on the figure.

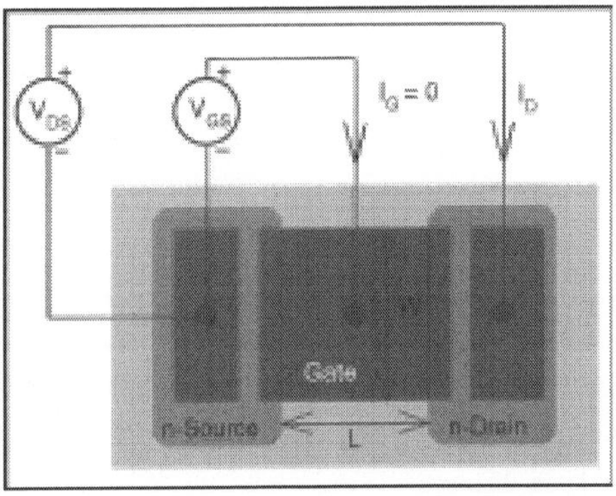

Fig. View of an n-type Metal-Oxide- Semiconductor-Field-Effect-Transistor

A top view of the same MOSFET is shown in Figure, where the gate length, *L*, and gate width, *W*, are identified. Note that the gate length does not equal the physical dimension of the gate, but rather the distance between the source and drain regions underneath the gate.

The overlap between the gate and the source and drain region is required to ensure that the inversion layer forms a continuous conducting path between the source and drain region.

Typically this overlap is made as small as possible in order to minimize its parasitic capacitance.

The flow of electrons from the source to the drain is controlled by the voltage applied to the gate. A positive voltage applied to the gate, attracts electrons to the interface between the gate dielectric and the semiconductor.

These electrons form a conducting channel between the source and the drain, called the inversion layer.

No gate current is required to maintain the inversion layer at the interface since the gate oxide blocks any carrier flow. The net result is that the current between drain and source is controlled by the voltage which is applied to the gate.

V-I CHARACTERISTICS

The typical current versus voltage (I-V) characteristics of a MOSFET are shown in the figure below. Implemented is the quadratic model for the MOSFET.

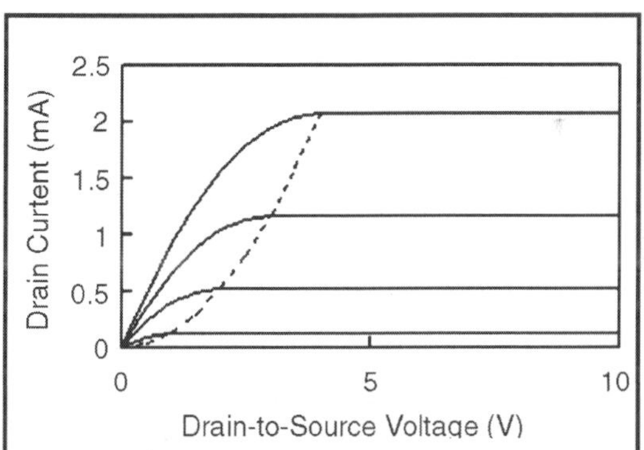

Fig. *I-V Characteristics of an n-type MOSFET*

We will primarily discuss the *n-type* or n-channel MOSFET. This type of MOSFET is fabricated on a p-type semiconductor substrate.

The complementary MOSFET is the *p-type* or p-channel MOSFET. It contains p-type source and drain regions in an n-type substrate. The inversion layer is formed when holes are attracted to the interface by a negative gate voltage. While the holes still flow from source to drain, they result in a negative drain current. CMOS circuits require both n-type and p-type devices.

MOSFET AS AN AMPLIFIER

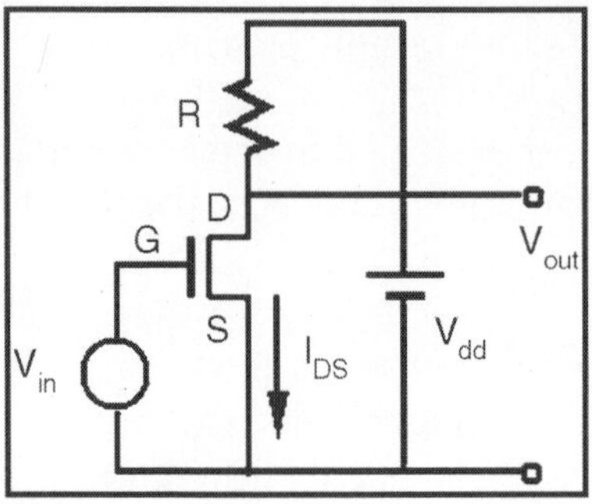

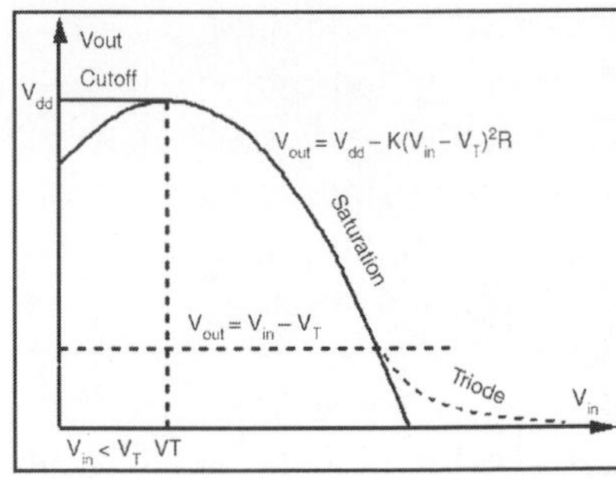

Assume in the circuit above $V_{in} = V_{GS} > V_T$ and the transistor is in saturation region, *i.e.*, $V_{DS} > V_{in} - V_T$, then we have,

$$\begin{cases} I_{DS} = K(V_{in} - V_T)^2 \\ V_{\infty\varepsilon} = V_{DS} = V_{sd} - I_{DS}R = V_{dd} - KR(V_{in} - V_T)^2 \end{cases}$$

The second equation relates the output V_{out} to the input V_{in}, as shown by the red segment of the curve in the plot above. As the transistor is in saturation region,

$$V_{out} = V_{dd} - KR(V_{in} - V_T)^2 \geq V_{in} - V_T$$

which can be solved for V_{in} to get:

$$V_{in} < \frac{-1 + \sqrt{1 + 4KRV_{dd}}}{2KR} + V_T$$

It can be seen that when the transistor is in saturation mode the slope of the curve indicates the ratio between input V_{in} and output V_{out}. And the voltage gain of the circuit is defined as:

$$\vartheta = \frac{dV_{out}}{dV_{in}} = \frac{d}{dV_{in}}\left(V_{dd} - K(V_{in} - V_T)^2 R\right) = -2KR(V_{in} - V_T)$$

Example: Assume V_{dd} = 10V, R = 10 kΩ, K = 0.5 mA/V^2, V_T = 1 V. For the transistor to be in saturation region, we need

$$V_{in} > V_T = 1V$$

$$V_{in} < \left|-1 + \sqrt{1 + 4KRV_{dd}}\right| / 2KR + / V_T = 2.32V$$

then we have

$$V_{out} = V_{dd} - K(V_{in} - V_T)^2 R = 10 - 5(V_{in} - 1)^2$$

and the voltage gain is a function of the input V_{in}:

$$g(V_{in}) = \frac{d}{dV_{in}} V_{out}(V_{in}) = -10(V_{in} - 1)$$

This nonlinear equation can be represented by the table below:

V_{in}	0	1	1.4	1.5	1.8	1.9	2.0	2.1	2.2	2.3	2.32	2.35	2.4
V_{out}	10	10	9.2	8.8	6.8	6.0	5.0	4.0	2.8	1.6	1.3	0.9	0.0

In particular, when the input V_{in} increases from 1.8 V to 1.9 V, the output V_{out} decreases from 6.8 V to 6 V, with a gaing(V_{in}). Also when the input V_{in} increases from 2.0 V to 2.1 V, the output V_{out} decreases from 5 V to 4 V, with a gain $g(V_{in})=g(2)=-10(2-1)=-10$.

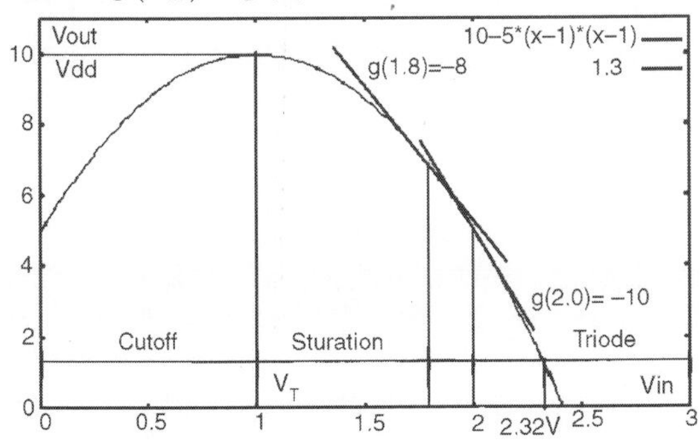

In summary, we see that

- When the transistor is in saturation mode, the circuit behaves as a voltage amplifier.
- The voltage gain is the slope of the tangent of the curve as a function of V_{in}.
- The value of the gain depends on the level of input V_{in}. When $V_T = 1V < V_{in} < 2.32V$, the gain $|g| > 1$ is greater than one.
- The output voltage is 180° out of phase with the input voltage (g < 0), as the slope of V_{out} V_{in} is negative.
- When $V_{in} < V_T = 1V$, the transistor is cutoff. On the other hand, when $V_{in} > 2.32V$ V_{out} is more than one V_T below V_{in}, for example, $V_{in} = 2.32$, $V_{out} = 1.3 < V_{in} - V_T = 2.32 - 1 = 1.32$, the transistor is in triode region. In either of the two cases, the transistor has no amplification capability.

Next we consider a MOSFED circuit with sinusoidal input. Assume the drain resistor is $R = 10k\Omega$, K = 1mA/V^2, V_T = 1V, V_{dd} = 5V and a sinusoidal input $U_{in}(t) = V_{\varepsilon i} + \sin(\omega t)$. If the bias voltage is V_{out} = 1.5V, the input voltage $u_{in}(t)$ varies between 1.4 V and 1.6 V. The output voltage can be found to be:

$$v_{out} = V_{dd} - RI_{DS} = V_{dd} - RK\left(V_{in} - V_T\right)^2 = 5 - 10\left(V_{in} - 1\right)^2$$

In particular, corresponding to $V_{in} = 1.4V, 1.5V, 1.6V$, the output voltage u_{out} and the current i_{DS} are, respectively, u_{out} = 3.4 V, 2.5, 1.4 V, and i_{DS} = 0.16 mA, 0.25 mA, 0.36 mA, as shown in the figure below:

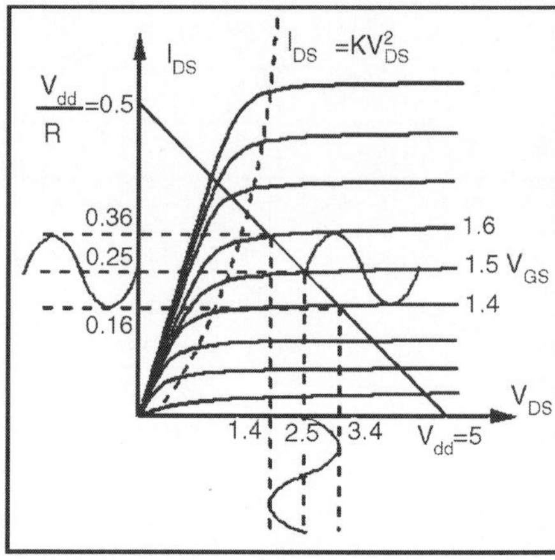

Biasing: In the example above, the DC offset of the input is at 1.5 V, so that the transistor is working in the saturation region when the magnitude of the AC input is limited. However, if this offset is either too high or too low, the gate voltage may go beyond the saturation region to enter either the triode or the cutoff region. In either case, the output voltage will be severely distorted, as shown below:

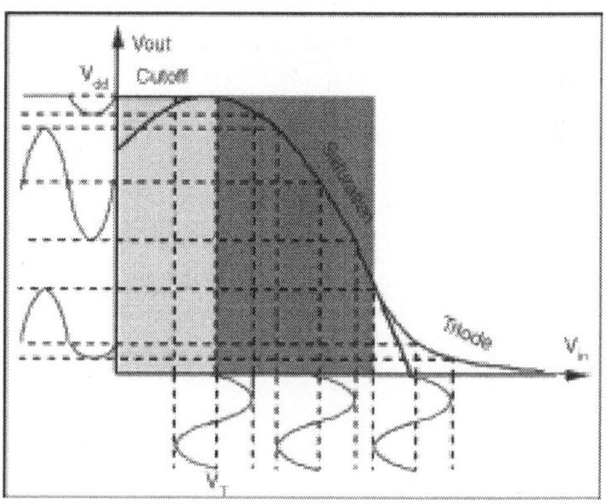

It is therefore clear that the DC offset or biasing gate voltage has to be properly setup to make sure the dynamic range of the input signal is within the saturation region.

Method 1: One way to provide the desired DC offset is to use two resistors R_1 and R_2 that form a voltage divider, as shown in the figure below (a). As the input resistance of a MOSFET transistor is very high, therefore the gate of the transistor does not draw any current, the DC offset voltage can simply obtained as:

$$V_{out} = V_{dd} \frac{R_1}{R_1 + R_2}$$

The input AC signal through the input capacitor is then superimposed on this DC offset.

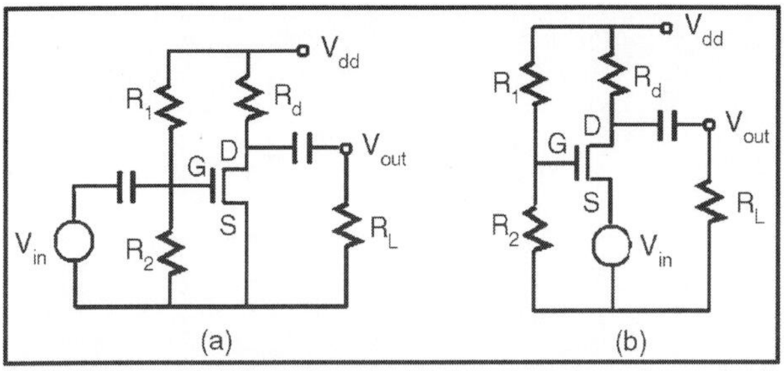

Method 2: Another way to set up the bias is the circuit shown in (b) above. Assume $R_1 = 84$ k, $R_2 = 16$ k, $V_{dd} = 10$ V, and $K = 0.5$ mA/V^2. The bias voltage can be found to be $V_B = V_{out} = 1.6$ V, and the voltage between gate and source is $V_{GS} = V_B - V_{in}$. The output voltage is:

$$V_{out} = V_{dd} - R_d K (V_{GS} - V_T)^2 = V_{dd} - R_d K (V_B - V_{in} - V_T)^2$$
$$= 10 - 10 \times (0.6 - V_{in})^2$$

When $V_{in} = 0$, $V_{out} = 6.4$ V.

To determine the dynamic range of the input V_{in}, recall the conditions for the transistor to be in saturation region:

- To avoid cutoff region: $V_{GS} - V_T \geq 0$. For this particular circuit,

$$V_{GS} - V_T - V_B - V_{in} - V_T = 0.6 - V_{in} \geq 0$$

Solving this we get $V_{in} \leq 0.6V$ with corresponding output $V_{out} \leq 10V$.

- To avoid triode region: $V_{DS} \geq V_{GS} - V_T$. For this particular circuit,

$$V_{DS} = V_{out} - V_{in} = (V_{dd} - R_d I_{DS}) - V_{in} = V_{dd} - R_d K$$
$$(V_B - V_{in} - V_T)^2 - V_{in} \geq V_B - V_{in} - V_T$$

that is,

$$V_{out} = V_{dd} - R_d K (V_B - V_{in} - V_T)^2 \geq V_B - V_T$$

i.e.,

$$V_{out} = 10 - 10 \times (0.6 - V_{in})^2 \geq 0.6$$

Solving this for V_{in} we get $V_{in} \geq -0.37V$, with corres-ponding output $V_{out} \geq 0.6V$

Therefore the overall dynamic range for the input is,

$$-0.37V \leq V_{in} \leq 0.6V$$

with the corresponding output range

$$0.6V \leq V_{out} \leq 10V$$

and the overall voltage gain is about g = 9.7. Note that the output voltage is in phase with the input voltage. *Source Follower*: If the output is taken from the source, instead of the drain of the transistor, the circuit is called a source follower.

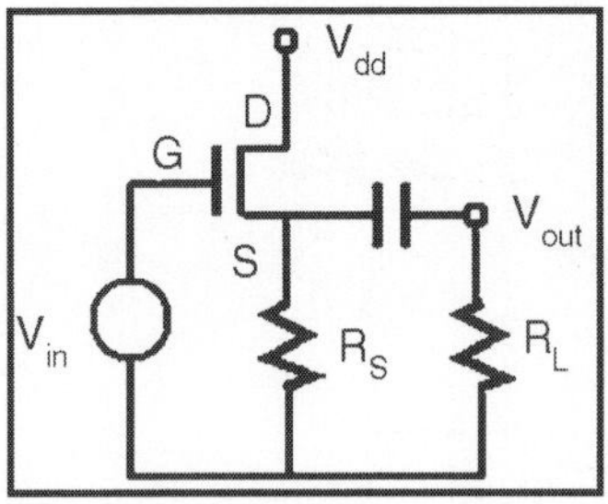

Assume $R_S = 10k\Omega$, $V_T = 1V$ and $K = 10mA/V^2$. To find the input and output voltages and the gain of the circuit, consider the current $I_{DS} = V_{out}/R_S$:

$$I_{DS} = K(V_{GS} - V_T)^2 = K(V_{in} - V_{out} - V_T)^2 = V_{out}/R_S$$

Plugging in the given values, we get,

$$(V_{in} - V_{out} - 1)^2 = V_{out}$$

If $V_{in} = 2V$, *this equation becomes*:

$$(V_{out} - 1)^2 = V_{out}$$

which can be solved to get $V_{out} = 2.6V$ or $V_{out} = 0.4V$. We take the smaller voltage in order for the transistor to be outside the cutoff region:

$$V_{GS} = V_{in} - V_{out} = 2 - V_{out} > V_T = 1$$

Similarly, if $V_{in} = 3V$, the equation becomes:

$$(V_{out} - 2)^2 = V_{out}$$

and we get $V_{out} = 1V$. The voltage gain of the source follower is

$$g = \frac{\Delta V_{out}}{\Delta V_{in}} = \frac{1 - 0.4}{3 - 2} = 0.6 < 1$$

MOSFET AS AN SWITCH

- There are several types of Metal Oxide Semi-conductor Field-Effect Transistors, also known as MOSFETs. MOSFETs are devices used as transistors in analog or digital circuits. They can either switch or amplify electronic signals. Power MOSFETs are specifically designed to manage large amounts of power and have high impedance levels.
- Power MOSFETs were first developed in the late 1970's, first being created by International Rectifier in 1978. While Power MOSFETs perform the same functions as Bi-polar transistors, Power MOSFETs are voltage controlled, while Bi-polar transistors are current controlled.
- Power MOSFETs are efficient at lower voltages, and are the most common low-voltage switches. They can be found in most power converters, power supplies and low-voltage controllers. Power

MOSFETs utilize a vertical structure, which allows the transistor to sustain high current with high-blocking voltage.

- MOSFETs operate under the application of power from a power source. When positive voltage comes from a power source, electrons are attracted to the positive voltage. If the voltage is above a certain level of voltage, enough electrons will be attracted and will accumulate to form a conducting channel across the body of the MOSFET. This channel allows electrons to flow in either direction.

COMPLEX SWITCHING CIRCUITS

All of the subsequent examples and problems will be typical of those which come up repeatedly in research. Many of the examples deal with learning experiments on animals because complex switching circuitry is even more extensively used in research on rats and pigeons in Skinner boxes than in the study of human behaviour. Probably this is a result of the fact that by verbal instruction it is possible to throw into operation a human subject's own built-in, logical machinery, thus replacing a lot of relays.

This chapter contains many problems, and it is important to solve them all. Solving a problem means drawing the solution on real paper. It often *seems* possible to read a problem and arrive at a solution without actually writing anything down. However, thinking up a way of solving a problem is easy. Solving it is the hard part.

Problem. John has a ball that weighs 3 pounds. It is hollow, and made of rubber weighing 63 pounds per cubic foot. The ball just barely floats in water. How thick are its walls?

Solution. Use division.

HOLDING RELAYS

There is one particular relay circuit that is used very extensively. It is called the holding, or lockup, circuit, and is used when a circuit element must remain activated after the activating circuit is turned off. In essence, a relay is made to close as soon as a switch is operated, hold itself closed even after the switch is opened, and release only when a second switch is operated. The basic circuit is illustrated in the following example.

Suppose you are given the problem of designing a circuit to measure the length of time it takes a rat to run from point X in a straight alley to another point Y farther down the alley. First, some device must be used to sense when the rat passes each of the two points. In solving the present problem, we will use the method mentioned in the previous chapter of hinging a section of the floor at X and another at Y, so that the rat's weight will momentarily operate a SPDT microswitch at each of the two points. The clock commonly used for this type of timing consists of an electric motor driving the hands through an electric clutch.

The motor runs constantly, but the hands turn only so long as the clutch is engaged. The clutch operates as long as a small current is being passed through it, and disengages as soon as the current is turned off.

To time a rat, then, a circuit is needed which will cause a current to begin to flow through the clutch when one SPDT switch (at *X*) is momentarily operated, and stop fl owing when another (at *Y*) is operated. Neither microswitch alone could operate the clutch in the desired way because, during the time the rat is between *X* and *Y*, both switches are open so that the clutch would not be engaged.

Suppose, instead, that the clutch is turned on and off through a relay. Now, if the relay could be made to close when switch *X* is closed, hold itself closed even after switch *X* is opened, and release when switch *Y* is operated, the problem would be almost solved. (The clutch could then be turned on and off by a pair of contacts on the relay.) The clutch connected through a relay, and the relay connected directly to one of the microswitches. This circuit is no improvement over connecting the clutch directly through the microswitch.

As soon as *X* is closed by the rat's running over that section of the alley, the relay closes, allowing current to flow in the clutch circuit. However, as soon as the rat leaves that section of the alley, the microswitch will open again, the relay will thus be released, and the clutch is turned off.

These contacts are connected in parallel with the normally open contacts of the microswitch.

The action of the circuit is as follows. At first, the microswitch and the relay are open so that no current flows anywhere (except, of course, through the clock motor). (Be sure that it is clear from the diagram that no current can flow. Try to find a closed path from one side of a power source to the other side of the same source. If there is no such path, then there can be no current flow, but as soon as there is such a path, there must be current flow.)

Now the rat steps on the hinged alley section, closing the normally open contacts of the microswitch. This furnishes a path for current to flow through the relay coil (dashed arrows), and the resulting magnetic field will cause the relay armature to pull down, closing the two pairs of normally open contacts. The top pair activates the clutch. The lower pair provides a second path for the current to flow through the relay coil (dotted arrows).

This time, when the rat leaves the section and the microswitch opens again, the relay will stay closed because there is still a path allowing current to flow through the coil. With the microswitch open, current will flow only along the dotted lines. Under these conditions, the relay is, in a sense, holding itself closed and once it has been closed, microswitch *X* no longer has any effect at all.

The clock will start running when the rat first crosses the hinged section of the alley and will keep running indefinitely, as the circuit is now drawn. Even if the rat were to retrace and go back and forth several times over this alley section, the clock would be unaffected.

The relay is your 115-volt a–c relay, and switch X is one of the microswitches. (Observe the same precautions as were discussed in the previous chapter: remember that 115 volts can be painful.) If the circuit is connected up correctly, the light should go on when you first press the microswitch and stay on after you let go. Now the bulb is lit (the clock is running). How do you turn it off? There are several ways. First, you can unscrew the bulb.

This is the equivalent of putting a switch in the clutch circuit and opening it. Suppose that this switch were actually the second microswitch *Y* in the alley, connected. That is, the normally closed pair is connected in series with the clutch. When the rat hits section *X*, the relay closes and the clock starts. When he hits section *Y*, the clock stops. If at that moment the clock could be read very quickly, the circuit problem would be solved. But as soon as the rat leaves section *Y*, the clock will start up again. Prove this by unscrewing the bulb and then putting it back again.

If the section *Y* switch could actually turn off the power to the *relay*, then the relay would stay open even after the section is returned to its normally closed condition, because the only things keeping the relay closed are its own contacts. In the circuit on your board, turn off the light by pulling out the wall plug. Now when the plug is connected again, the light will remain off and the whole circuit is restored to its initial condition.

On your board, connect up the normally closed contacts of the second microswitch. This switch will now do what you were doing when you momentarily disconnected the wall plug. If the circuit is working properly, the light will turn on when switch *X* is momentarily pressed, and stay on until switch *Y* is momentarily pressed.

This circuit, the holding relay circuit, is very fundamental to switching circuit design. It is used as it has been presented here or with only slight modifications in a very large proportion of the switching circuits found in behavioural research. For this reason, it is worthwhile for you to draw the circuit diagram over and over again until it becomes almost automatic. Be sure you understand its principles of operation. This will save you considerable time when you are solving many of the remaining problems in this book.

If the circuit were used in a straight alley way, it would permit the measurement of the total time between the first crossing of *X* and the first crossing of *Y* (*i.e.*, stepping on *Y* again would not restart the clock). Consider instead the problem of designing a circuit to measure the total time the animal spends in the portion of the alley between *X* and *Y*. In other words, if he runs out of this portion, past *Y*, and then turns around and runs back into it once more, the clock should start again.

Since this is probably the most difficult problem yet presented, it may be a good one with which to illustrate the sort of process used by people skilled at circuit design to solve relatively complicated problems.

Many of the processes involved in arriving at solutions to this kind of problem are difficult to verbalize in a coherent way. If prodded, people who are skilled at circuit design frequently will say that they knew by intuition where to start and what things to try. Intuition is probably an excellent word to describe what happens. Most people using such intuition will readily agree that it is developed through a combination of some intelligence and a lot of experience. Here, an attempt will be made to enumerate some of the submerged logic involved.

The first step in arriving at a solution is to state the circuit requirements as explicitly as possible. In this particular problem, the circuit must:

- Start the clock any time the animal (*a*) crosses *X* going towards *Y* or (*b*) crosses *Y* going towards *X*.
- Stop the clock any time he (*a*) crosses *X* going away from *Y* or (*b*) crosses *Y* going away from *X*.

(For any problem there are a number of solutions. For example, this problem may be easily solved, in theory at least, by hinging the entire section between *X* and *Y* and resting it on a microswitch. In practice, it is difficult to make a large section of alley operate that way. The solutions presented herein are chosen both because they are practical and because they illustrate various important aspects of circuit design. If you arrive at a solution different from one in the text-and this happens very often—that does not necessarily mean that yours is incorrect.)

Examination of the apparatus requirements listed above indicates that the circuit must sense the direction in which the rat is traveling as well as his presence when he is at point *X* or point *Y*. If there were two short, hinged floor sections at *X*, and if the circuit could determine the order in which these two sections were traversed, it would indicate which way the animal was going when he went by *X*, and either of the sections might also indicate his presence at *X*. In other words, it may be that if there were four hinged floor sections, each with a microswitch, two at *X* and two more at *Y*, it would be possible to solve the problem. A guess about what kinds of circuit elements might solve the problem can be considered the second step in the solution.

The third step might then be to see whether or not such a collection of elements really can be made to work. To do this, it is probably most efficient to begin by representing, on a large sheet of paper, each of the possible elements without any particular regard for how they will be connected up.

The clock will run so long as there is current flowing through its clutch coil, and it becomes clear and considering the original problem that the clutch must be operated through something that is not yet represented. This is true because, although the clutch must remain activated when the rat is, say, halfway between *X* and *Y*, there is nothing drawn yet that will act that way. As in the preceding problem, a holding relay circuit is strongly indicated. A relay is thus drawn in with at least enough contacts to hold itself, and connected so as to operate the clutch. Now the clutch no longer need be considered. The problem is to get the relay to be on all the time the rat is between *X* and *Y*. If this is done, the clock will necessarily do the same.

Figuring out additional connections requires consideration of the step-by-step operation of the circuit. Let the rat come out of the start box and begin to walk up the alley. The clutch is to engage, that is, the relay is to close, when the rat passes *X*. Therefore, either microswitch X_1, or X_2 must close the relay. Try X_1 first. With this circuit, when the rat walks past *X* and stands between *X* and *Y*, the clock keeps running (the relay is connected to hold itself on). Now, if the rat turns around and crosses *X* again, the clock are supposed to go off. Since X_2 is left over, try its normally closed contacts to turn off the clock.

This will stop the clock all right when the rat recrosses X_2 going towards the start box, but two other things are wrong. First, if he goes a little farther and crosses X_1 too, the clock will go back on again. Second, after he leaves the start box in the first place, he will start the clock at X_1, but stop it again as he goes by X_2. Therefore, try reversing the roles of X_1 and X_2 to see if that works any better. Now, when the rat crosses X_1 going towards Y, nothing happens. When he crosses X_2, the clock starts. If he then turns around and recrosses X_2 going the wrong way, nothing happens.

But as soon as he crosses X_1 going the wrong way, the clock stops. When he again turns around and recrosses X going towards Y, the clock starts once more. So far, then, this circuit is operating according to plan. All that remains is to make the switches at Y operate in a similar, but *opposite*, manner. In other words, it would seem plausible, as a first guess, to make Y1 start the clock and Y_2 stop it. (Crossing Y towards X should do the same thing as crossing X towards Y.) The circuit is connected this way and will solve the problem. In this circuit, either Y_1 or X_2 will start the clock, and either X_1 or Y_2 will stop it.

The clock will always be running when the rat is between X and Y because the rat must have most recently closed either X_2 or Y_1 to get there. And any time the rat leaves the section between X and Y, the clock must be stopped because the rat must have most recently opened either X_1 or Y_2. The next problem illustrates another modification of a holding relay circuit. A relay that is turned on and off repeatedly at a fairly fast rate will buzz. In the holding relay circuit, the relay holds itself on. If instead a relay is connected in such a way that it turns itself *off*, then it must buzz.

Before the switch is closed, the relay is open, so that the normally closed contacts of the relay are closed. As soon as the switch is closed, current will flow through the relay contacts and through the relay coil, causing the relay armature to pull in. But when the armature pulls in far enough to open the normally closed contacts, the circuit through the relay coil is broken, so the armature is released. This allows the contacts to close again, and the cycle automatically repeats itself.

Connect up this circuit on your board, using the sensitive relay (not the 115-volt one). Begin by using as low a voltage across the relay as is convenient (say 3 volts), and increase it, if necessary, until the relay buzzes. Be sure that you understand which terminals are connected to which parts of the relay (*i.e.*, which terminals on the body of the relay are internally connected to the relay coil) before you connect up the battery. If you cannot tell about some terminal by looking at the relay, check it with an ohmmeter to see to which part it is connected. Practically all buzzers commercially available operate on exactly the same principle as the one you have constructed. The only important difference between your buzzer and a doorbell buzzer is that the latter uses a much cheaper relay.

BIASING IN MOS AMPLIFIER CIRCUITS

Biasing is the static voltage and current settings for a given device for certain functions of operation. Depending on the needs of the circuit, there may be as many as six different biasing schemes. But for the most part there are four basic biasing schemes. They are Class A, Class B, Class C and Class AB.

Class A: This is where the device is biased at a point where it will conduct and amplify for the entire signal.

Class B: This is where the device is biased such that it will conduct for only half of the signal.

Class C: This is where the device will only conduct for a small portion of the peak of the signal. It can also be described as any biasing point where the device conducts for less than half of the signal. This biasing scheme is never used in audio and can be found in R.F. applications.

Class AB: This is where the device conducts for more than half of the signal but less than the full signal.

BIAS SETTING SOUND "BETTER" THAN ANOTHER

First, let us examine a simple single stage triode circuit with a standard bias and bypass capacitor:

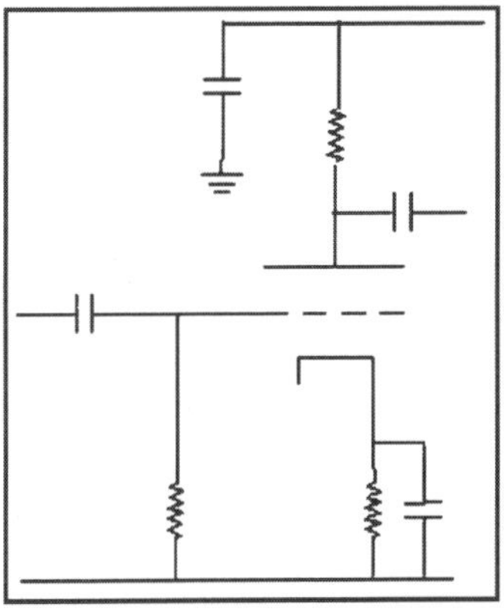

The above circuit is biased at the half points of the tube's maximum parameters. It has a nice sound to it and is loud but has some noise (hiss).

However, at strong loud bassy passages, it tends to seem limited but warmer. Sometimes it even seems to drop out. Why is this? Some would claim insufficient transient response from the power supply. It sounds slow, as many would say.

But let's take a closer look at this circuit. Let us rotate the circuit 90 degrees to the left. Now let's remove the grid circuit for a minute.

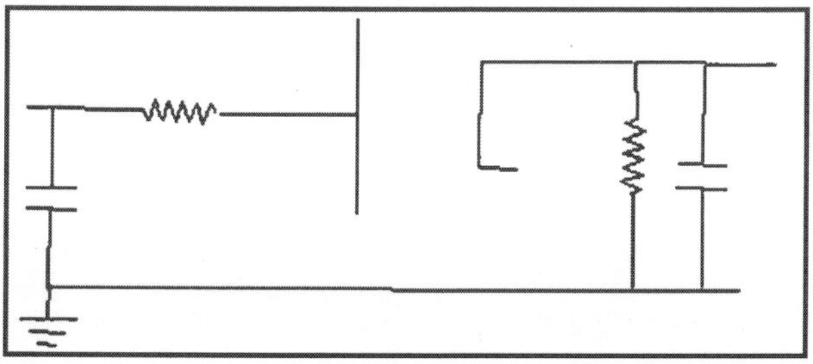

What does it look like now? That's right, a power supply rectifier! A half wave rectifier to be exact. So now what do the cathode resistor and bypass cap look like? A load and *filter*. The bypass cap sees a constant DC signal accompanied by an AC component. To a filter cap, or any capacitor for that matter, this appears as a fluctuating DC, or power supply ripple!

To get a bit deeper into the physics of the thing, if one were to calculate the time constant of the RC circuit at the cathode, one will find that for the above circuit the time constant is t = RC, or for the sake of argument, we will use the values I used for the actual circuit, 2200 x.000022 (22 microfarads).

The answer is 0.0484 seconds. What that means is that the capacitor resistor circuit requires about 0.0484 seconds to charge 63 per cent of its full capacity.

For the first 0.0484 seconds, if the capacitor were a 25 volt capacitor, and the bias were set to 2 volts, it would only take the first time constant to reach 2 volts. 0.0484 seconds factors out to a frequency of 20 hertz (1/t). If I took a 20 hertz signal of 0.7 volt RMS and inserted it, the positive peak will be 1 volt above the bias of 2 volts. Since the time constant is as small as it is, it wouldn't take much to increase the voltage of the capacitor.

This also simultaneously changes the bias of the circuit by about 0.4 volts (0.404 × Vpeak. The average of the positive portion of a sine wave). Looking at the curve chart below for a triode, this will change the characteristics of the tube which is then amplified through the tube. A larger capacitor will do this more and make things worse.

Looking at this chart, go to the 225 on the bottom line. This is the B+. Then go up to the 2.5 line from the left (the Y axis). This is the quiescent, or no input signal, current. For these conditions, it is showing a -1.4 volt grid bias. If the bias were anything different, then one would have to follow the line in the following chart to see the amount of voltage swing the input would accept before clipping.

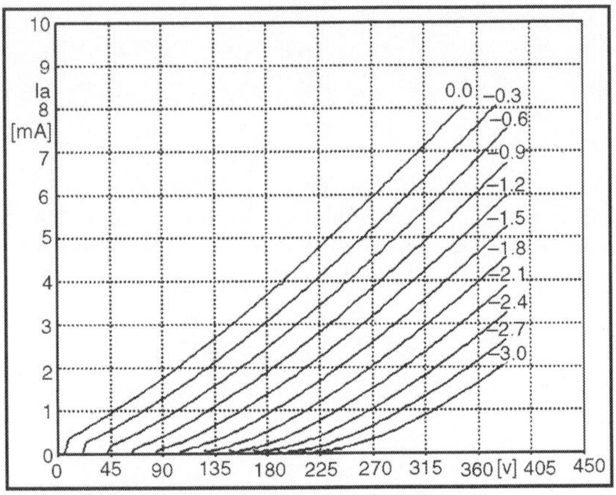

A method for biasing a tube is to draw a line from the B+ voltage to the desired current and choose a bias setting accordingly.

This line is drawn for the modified version of the above circuit I currently listen to. The B+ is 290 volts and the current is about 2 milliamps.

If you were to place a pointer at the bias point you designed the amp for, and follow the line, you will see the varying current and the voltage it will produce. Say I had biased the tube for 1.5 volts.

This would be a full peak to peak input swing. We can see that the voltage went to about 260 volts, where the current went down to about 0.3 mA. So the conclusion is that we use the above graph to set static voltages and predict to a relatively good degree the gain. So the output voltage swings about 210 volts peak to peak. This makes it about 74 volts RMS. So the gain can be calculated as Vout/Vin, or 74/1.06 = about 70. Theoretically, that is:

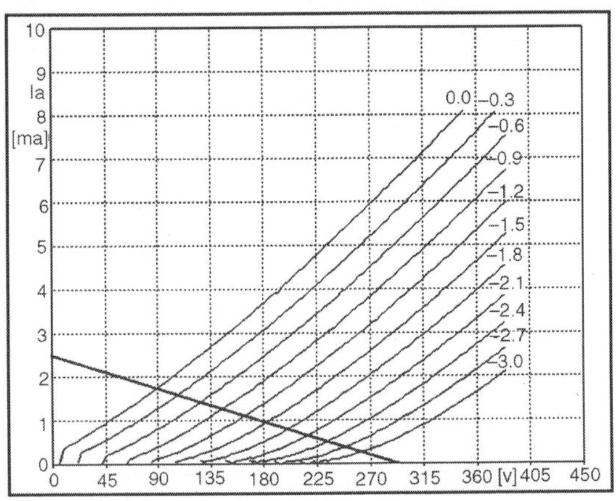

For the sound quality we want, we need to use a transfer chart in conjunction with this one. It is a chart which shows the relationship between grid voltage and plate current.

If we bias the tube along the most linear part of this line, and follow it with a nominal voltage swing across to the above chart, then one can find where clean signals end and clipping begins.

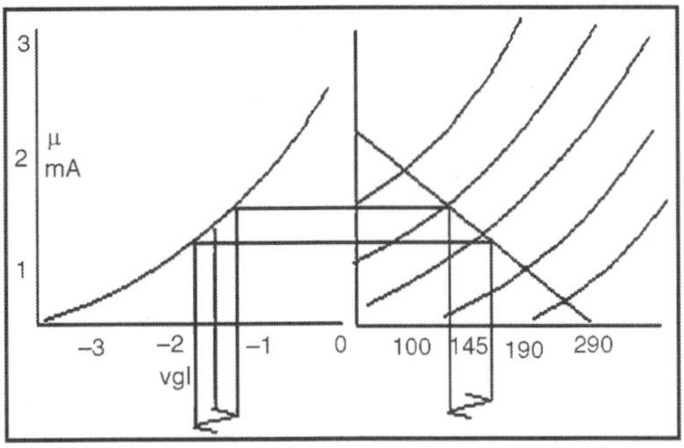

The diagram assumes accuracy, but it is not. The transfer curve is usually straighter in the area above the near cut-off region (depending on the tube type). However, if the circuit parameters are B+ 290 volts, Vp 145 volts, Vg -1.5 volts and Ip 1.5 milliamps, then a grid signal of about 0.75 volts peak to peak will output a plate voltage of about 45 volts peak to peak. That gives a voltage gain of about 60 according to the above chart. Actual gain is not like this, but circuit parameters might cause a higher or lower gain, depending on the design of the tube and the resistors chosen for biasing.

A fixed bias circuit could allow for the resistors to set gain by the simple gain equation G = Rp/Rk. If the plate resistance is 65 kilo ohms and the effective cathode resistor is 680 Ohms, the gain could be 100, and the grid bias can be exactly 1.5 volts. Certain tubes have a top gain (12AX7 and 6EU7 for example) of 100. This is known as the tube's amplification factor or mu.

The curve on the left, which is the transfer curve, is actually more curved as one approaches the more negative voltage level. Biasing down there will not yield a very linear response. So fixed biasing has its advantages, one being the possibility of not using a bypass resistor. Another is accurate biasing that doesn't change much with time or heat, providing better linearity. In audio, with one instrument or even the fundamental being at a high amplitude, the instant that the peak reaches this area, another signal of a higher frequency may suffer the effects of distortion. This is a form of modulation distortion.

Capacitors cause current to lag the voltage, so the bias will not follow the signal, hence altering the incoming signal's nature, and the overall sound quality. What this means is that a signal whose current component and voltage component are in phase will come out of a capacitor with current being 60 degrees behind the voltage signal.

This may not necessarily be a problem if there is no grid resistor (some circuits do this), but if there is a grid resistor, then the resistor will react to the varying current and add to the original voltage signal the resultant voltage signal across the resistor and cause a slight phase distortion.

This is because, not only do the biasing schemes mentioned at the outset of this discussion alter dramatically the sound quality, but also variations within the most popular, the class A, also alter the sound quality.

You will find that the schemes that use the lower voltages and currents are more popular because they are the least linear. This may seem like a contradiction to what an audiophile would want from a sound system, especially one that uses tube and expensive irons. It is.

In an amplifier configuration using a 6NS7 tube (a twin triode) where one half of the tube amplifies and the other half acts as a follower, a biasing scheme where the output is taken from the cathode and it doesn't amplify voltage but current, but the first half does the amplifying of voltage. This particular circuit had an open loop (no NFB) distortion rating of less than 0.2per cent! Astonishingly low.

PUSH PULL AMPLIFIER

One of the controversies about push pull biasing is that there exists a cross over distortion. What this means is that when the signal input to either tube causes the tube to cut off there is a less than linear transition for the other tube to "take up" when it begins to conduct for the other half of the wave.

In a pure class B biased circuit, this means that there is a gap between when one tube stops conducting and the other starts to conduct.

This makes for a wave form that kind of looks like this:

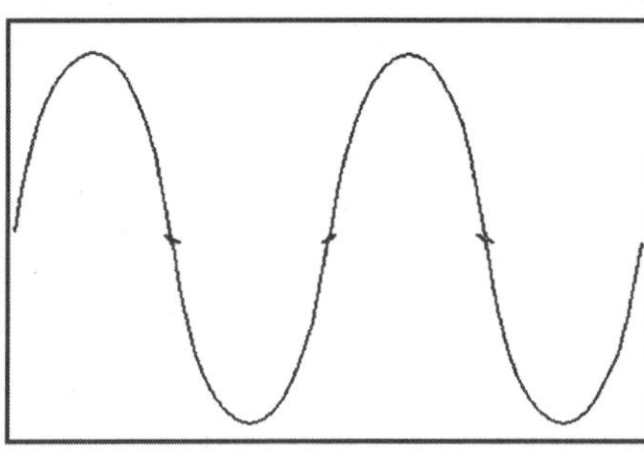

Notice the bends where the wave crosses the zero line. Class AB makes up for this by causing the tubes to conduct a little bit more of the cycle so as to overlap the cut off to turn on times. In other words, those bends are now in the signal past the zero crossing.

The result is a "purer" sine wave. The resulting spikes can then be filtered out with a small value capacitor that will not affect the audible high frequencies. Inverted feedback will also reduce this "crossover distortion". Also eliminated is the low volume level distorted sound of the crossover problem in pure class B operation. In essence, the tubes conduct as class A for low level signals, and as class B on loud level signals.

The real audible difference in a push-pull configuration, even in class A push-pull, is the fact of reduced even order harmonics. Or, increased odd order harmonics. Whichever textbook you wish to refer to, the fact is that odd order harmonics dominate in the push-pull stage. This is where the *real* sonic differences between push-pull and single ended amplifiers occur.

This is not to imply that the crossover distortion is not real, just that the distortion is virtually eliminated by biasing and filtering. Transformers, too, naturally filter out the spikes. Solid state amps don't have transformers to do this, so many persons complain about the buzz, grainy or raspy sound they hear.

It is actually the crossover spikes they are listening to. It could also be a higher amplitude high end which causes the high end to sound (subjectively speaking) like an icing on cake, and not the cake itself, which is what we want to hear.

Better amps usually use a small choke in line with the output to the speaker to reduce the spikes that NFB doesn't. OTL (output transformer less) tube amps usually work in class A anyway so there is no problem with cross-over distortion there either.

In many high end tube amplifiers there are two potentiometers per channel that are user accessible. They are labeled "bias".

These potentiometers are there to set the bias for the output tubes so that they match each other. This allows both tubes to amplify their respective half of the signal *equally*. This also causes the tubes to cancel the even order harmonics (or *whatever*).

Many audiophiles mis-align the biasing. They claim that the amp sounds better. What is really happening is that now even order harmonics are showing up because of not being as canceled out by the push pull configuration.

MOSFET INTERNAL CAPACITANCES

The first component of capacitive parasitics we will examine is the MOSFET capacitances. These parasitic components are mainly responsible for the intrinsic delay of logic gates, and they can be modeled with fairly high accuracy for gate delay estimation. The extraction of transistor parasitics from physical structure (mask layout) is also fairly straight forward.

The parasitic capacitances associated with a MOSFET are lumped elements between the device terminals.

Based on their physical origins, the parasitic device capacitances can be classified into two major groups:

- Oxide-related capacitances.
- Junction capacitances.

The gate-oxide-related capacitances are Cgd (gate-to-drain capacitance), Cgs (gate-to-source capacitance), and Cgb (gate-to-substrate capacitance).

Notice that in reality, the gate-to-channel capacitance is distributed and voltage dependent. Consequently, all of the oxide-related capacitances described here change with the bias conditions of the transistor.

Figure below shows qualitatively the oxide-related capacitances during cut-off, linear-mode operation and saturation of the MOSFET. The simplified variation of the three capacitances with gate-to-source bias voltage is shown

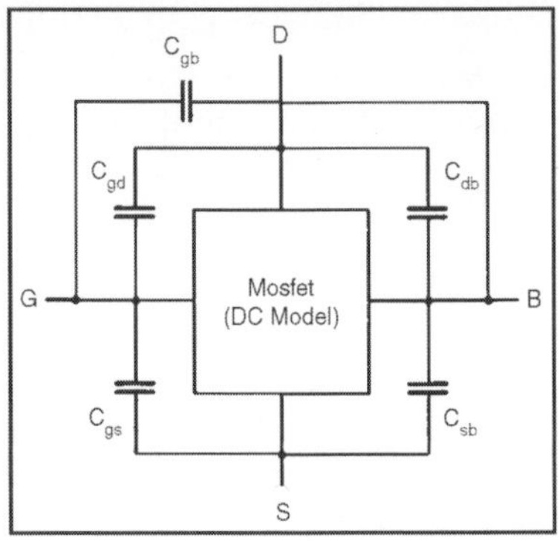

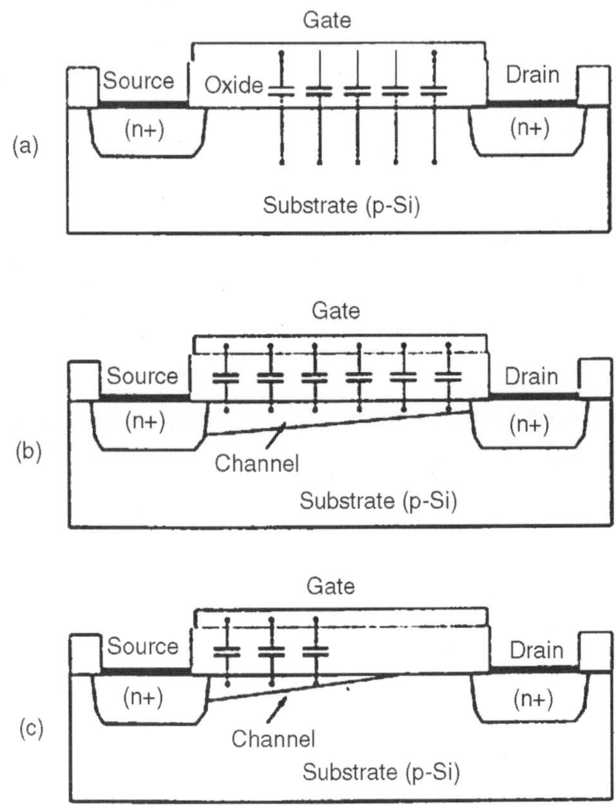

Fig: Schematic Representation of MOSFET Oxide Capacitances During (a) Cut-off, (b) Linear- mode Operation, and (c) Saturation.

We see that the total gate oxide capacitance is mainly determined by the parallel-plate capacitance between the polysilicon gate and the underlying structures.

Hence, the magnitude of the oxide-related capacitances is very closely related to:

- The gate oxide thickness,
- The area of the MOSFET gate.

Obviously, the total gate capacitance decreases with decreasing device dimensions (W and L), yet it increases with decreasing gate oxide thickness.

In sub-micron technologies, the horizontal dimensions (which dictate the gate area) are usually scaled down more easily than the horizontal dimensions, such as the gate oxide thickness.

Consequently, MOSFET transistors fabricated using sub-micron technologies have, in general, smaller gate capacitances.

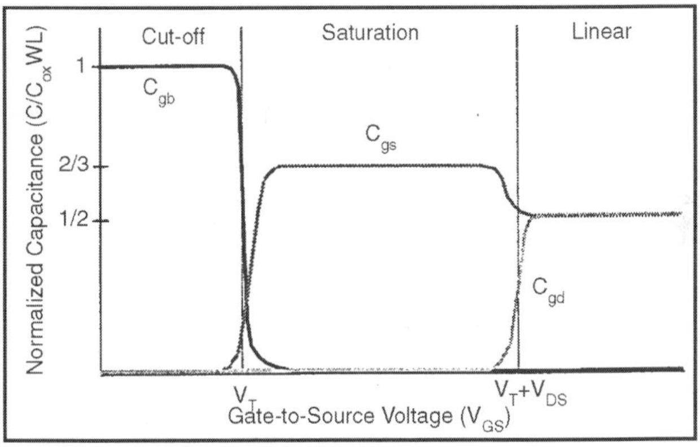

Fig: Variation of Oxide Capacitances as Functions of Gate-to-Source Voltage.

Now we consider the voltage-dependent source-to-substrate and drain-to-substrate capacitances, Csb and Cdb. Both of these capacitances are due to the depletion charge surrounding the respective source or drain regions of the transistor, which are embedded in the substrate.

Figure shows the simplified geometry of an n-type diffusion region within the p-type substrate. Here, the diffusion region has been approximated by a rectangular box, which consists of five planar pn-junctions.

The total junction capacitance is a function of the junction area (sum of all planar junction areas), the doping densities, and the applied terminal voltages.

Accurate methods for estimating the junction capacitances based on these data are readily available in the literature, therefore, a detailed discussion of capacitance calculations will not be presented here.

One important aspect of parasitic device junction capacitances is that the amount of capacitance is a linear function of the junction area. Consequently, the size of the drain or the source diffusion area dictates the amount of parasitic capacitance. In sub-micron technologies, where the overall dimensions of the individual devices are scaled down, the parasitic junction capacitances also decrease significantly.

It was already mentioned that the MOSFET parasitic capacitances are mainly responsible for the intrinsic delay of logic gates. We have seen that both the oxide-related parasitic capacitances and the junction capacitances tend to decrease with shrinking device dimensions, hence, the relative significance of intrinsic gate delay diminishes in sub-micron technologies.

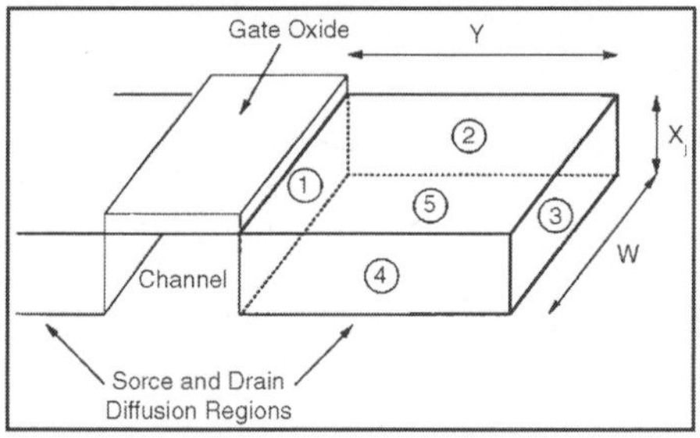

FREQUENCY RESPONSE OF CS AMPLIFIER

In a common-source amplifier, one of three basic single-stage field-effect transistor (FET) amplifier topologies, typically used as a voltage or trans conductance amplifier. The easiest way to tell if a FET is common source, common drain, or common gate is to examine where the signal enters, and leaves. The remaining terminal is what is known as "common". In this example, the signal enters the gate, and exits the drain. The only terminal remaining is the source. This is a common-source FET circuit.

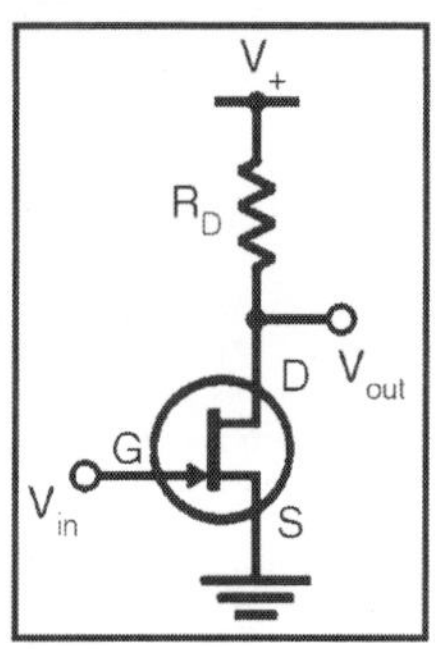

The analogous bipolar junction transistor circuit is the common-emitter amplifier. The common-source (CS) amplifier may be viewed as a trans conductance amplifier or as a voltage amplifier. .

As a trans conductance amplifier, the input voltage is seen as modulating the current going to the load. As a voltage amplifier, input voltage modulates the amount of current flowing through the FET, changing the voltage across the output resistance according to Ohm's law.

However, the FET device's output resistance typically is not high enough for a reasonable trans conductance amplifier (ideally infinite), nor low enough for a decent voltage amplifier (ideally zero). Another major drawback is the amplifier's limited high-frequency response.

Therefore, in practice the output often is routed through either a voltage follower (common-drain or CD stage), or a current follower (common-gate or CG stage), to obtain more favourable output and frequency characteristics. The CS–CG combination is called a cascode amplifier, *Consider a circuit*:

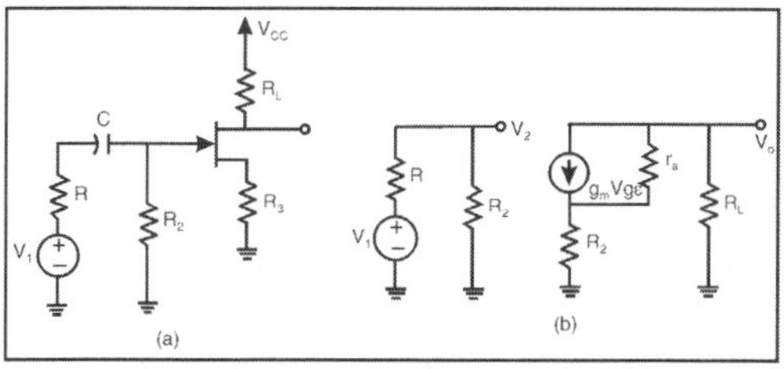

Fig.(a) A common-Source Amplilier. (b) An Equivalent Circuit for the Common Source Amplifier.

- The source voltage is determined from the following equations

$$\frac{V_D - V_2}{\gamma_d} + \frac{V_D}{R_1} + g_m V_{\varepsilon 2} = 0$$

$$and \frac{V_d}{R_2} + \frac{V_2 - V_0}{\gamma_d} = g_m V_{\varepsilon 2}$$

- Since the current leaving output node and source is zero
- The current through R_s The source voltage is

$$V_3 = \frac{V_0 - V_s}{R_L}$$

- The voltage gain is

$$\frac{V_0}{V_i} = \frac{-g_m R_L r_d}{R_L + r_d + R_5 (1 + g_m r_d)}$$

- If rd >> R_s and R_L we have the following relationship

$$A_V = \frac{-g_m R_L}{1 + g_m R_2}$$

- For gm R_s >> 1 we have

PAL ARCHITECTURE

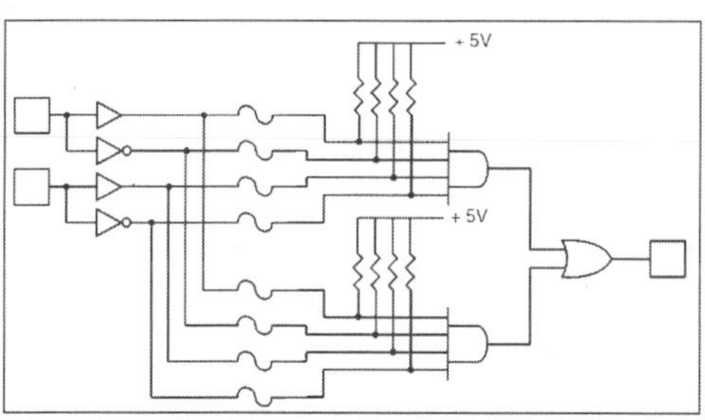

Fig. Simplified Programmable Logic Device

The PAL architecture consists of two main components: a logic plane and output logic macrocells.

Programmable Logic Plane

The programmable logic plane is a programmable (PROM) array that allows the signals present on the devices pins (or the logical complements of those signals) to be routed to an output logic macrocell.

PAL devices have arrays of transistor cells arranged in a"fixed-OR, programmable-AND" plane used to implement"sum-of-products" binary logic equations for each of the outputs in terms of the inputs and either synchronous or asynchronous feedback from the outputs.

Output Logic

The early 20-pin PALs had 10 inputs and 8 outputs. The outputs were active low and could be registered or combinational.

Members of the PAL family were available with various output structures called"output logic macrocells" or OLMCs. Prior to the introduction of the"V" (for"variable") series, the types of OLMCs available in each PAL were fixed at the time of manufacture.

The PAL16L8 had 8 combinational outputs and the PAL16R8 had 8 registered outputs. The PAL16R6 had 6 registered and 2 combinational while the PAL16R4 had 4 of each.

Each output could have up to 8 product terms (effectively AND gates), however the combinational outputs used one of the terms to control a bidirectional output buffer. There were other combinations that had fewer outputs with more product term per output and were available with active high outputs. The 16X8 family or registered devices had an XOR gate before the register. There were also similar 24-pin versions of these PALs.

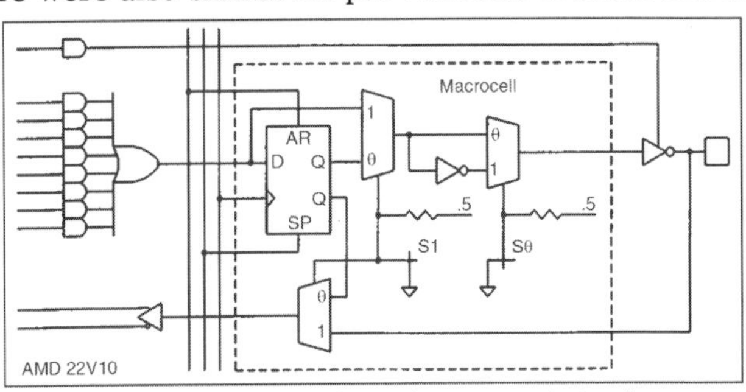

This fixed output structure often frustrated designers attempting to optimize the utility of PAL devices because output structures of different types were often required by their applications.

For example, one could not get 5 registered outputs with 3 active high combinational outputs. So, in June 1983 AMD introduced the 22V10, a 24 pin device with 10 output logic macrocells.

Each macrocell could be configured by the user to be combinational or registered, active high or active low. The number of product term allocated to an output varied from 8 to 16. This one device could replace all of the 24 pin fixed function PAL devices. Members of the PAL"V" ("variable") series included the PAL16V8, PAL20V8 and PAL22V10.

PROGRAMMING PALS

PALs were programmed electrically using binary patterns and a special electronic programming system available from either the manufacturer or a third-party, such as DATA/IO. In addition to single-unit device programmers, device feeders and gang programmers were often used when more than just a few PALs needed to be programmed.

For large volumes, electrical programming costs could be eliminated by having the manufacturer fabricate a custom metal mask used to programme the customers' patterns at the time of manufacture; MMI used the term"hard array logic" (HAL) to refer to devices programmed in this way.

Programming Languages

Though some engineers programmed PAL devices by manually editing files containing the binary fuse pattern data, most opted to design their logic using a hardware description language(HDL) such as Data I/O's ABEL, Logical Devices' CUPL, or MMI's PALASM. These were computer-assisted design (CAD) (now referred to as"design automation") programmes which translated (or"compiled") the designers' logic equations into binary fuse map files used to programme (and often test) each device.

PUSH-PULL AND GRID-GROUND CONNECTIONS

Most sensitive amplifiers, both direct- and capacitance-coupled, are designed for what is known as push-pull amplification.

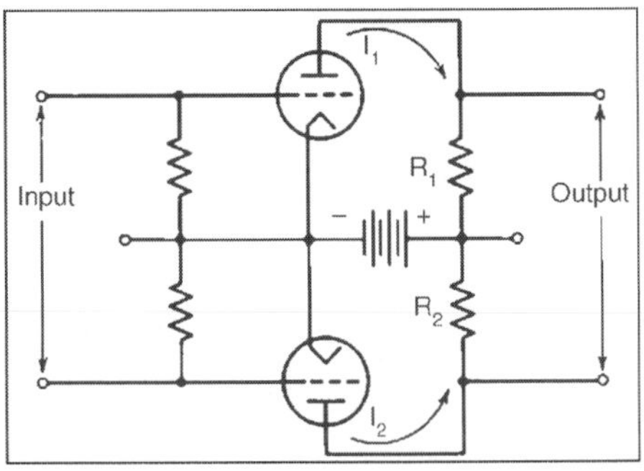

Fig. A One-stage Triode Amplifier with a Push-pull Input and Output.

The input voltage is impressed across a pair of grid resistors and two triodes connected in opposition. When the upper input terminal is positive with respect to the lower one, the upper triode passes more current through the output resistor R_1, and the lower triode less through R_2. Therefore, a difference in voltage appears across the output terminals that are proportional to the input signal. There are several fairly complicated reasons why push-pull circuits are advantageous, and two of them will be mentioned very briefly here. It is often convenient to amplify the algebraic difference between two voltages.

The push-pull circuit automatically does this when each of the two voltages is connected between one of the input terminals and the centre, ground terminal. Suppose, for example, that the occurrence of a stimulus is to be marked directly on a record of a psychogalvanic skin response (PGR). The PGR itself is fed between one of the input terminals and ground, and a pulse at the occurrence of the stimulus is fed between the other terminal and ground. The pulse appears on the record added to the PGR.

Because the push-pull circuit amplifies the algebraic difference between the two input voltages when they are connected, this kind of circuit may be used to reduce unwanted voltages relative to the desired ones. Any two equal voltages that appear with the same polarity at the same time on both sides of the input will cancel each other out, while differences are amplified. One all too common trouble in electronic apparatus is what is called 60cycle pickup. The power lines, lighting circuits, etc., that surround most laboratories actually broadcast fairly strong signals at 60 cps, and any sensitive amplifier will pick these signals out of the air and add them to the voltage that is supposed to be amplified.

For example, if an amplifier is connected to the brain to record brain waves, the record will often show a strong 60-cps wave. When a push-pull input circuit is used, and the central—ground—terminal is also connected to the animal, it is sometimes possible to find a location on the animal where the 60-cycle pickup is equally strong between that location and each of the other two electrodes. When this is the case, the 60-cps pickup is cancelled out, and the uncontaminated brain wave is all that appears at the output of the amplifier.

Most biological amplifiers, and many other amplifiers as well, have push-pull input stages. The input signal may then be connected either from one side to the other, or the signal may be connected between one of the grids and the ground, the other grid being connected directly to the ground. The first of these input connections is called a push-pull input and the second a grid-ground input.

INPUT IMPEDANCE

Consider the measuring of the voltage across a cell membrane. Say that the actual voltage generated is 50 microvolts. The circuit represents a suitable amplifier and a voltmeter. (Actually more than one stage of amplification would be needed, but the addition of other stages to this figure would just complicate things without changing anything basic.) Before the membrane is connected to the amplifier, there is a finite resistance between the two amplifier input terminals, consisting of the grid-to-cathode resistance of the tube in parallel with the grid resistor R_g.

The actual grid-to-cathode resistance of vacuum tubes is so very great that it may be considered infinite for these purposes (and for all but a very few special amplifier circuits). Therefore, the resistance that will be connected from one end of the cell membrane to the other will approximately equal the resistance of R_g. But the cell membrane and the electrodes themselves also have resistance, which may be diagrammed. The current that will flow when the amplifier is connected to the membrane equals $50 \times 10–6/R_g + R_m$. The voltage that the amplifier will "see" is the voltage between the grid and cathode of its first stage, and this is the voltage across R_g. Therefore, if R_g happens to be equal to R_m, for example, the amplifier will see only 25 microvolts.

In general, as R_g increases relative to the internal resistance of the source and electrodes, the proportion of source voltage that the amplifier will see increases. In addition, as the resistance of the entire input loop increases, the current drawn from the source decreases. If the source happens to be something like a cell, the less current drawn from it, the more normally it will behave. For these reasons, the resistance between the input terminals of an amplifier is important to know, and, at least for biological amplifiers, a large input resistance is desirable.

However, it cannot be made infinitely large because at very high values of Rg (*e.g.*, 40 megohms), the resistances of things such as the insulation between the pins in the base of the tube become comparable to Rg, and the voltage between the grid and cathode is reduced. The term "resistance" has been used in this discussion so far because it is a familiar concept. Actually the more general term "impedance" should be substituted for resistance. The impedance between the input terminals of an amplifier is called the input impedance. For any input signal whose frequency is not too high (say, up to a few thousand cycles per second), the input impedance of a direct-coupled amplifier essentially equals the resistance of the resistor connected between the grid and cathode of the first tube. For higher frequency signals and for capacitance-coupled amplifiers, the calculation or measurement of the input impedance is beyond the scope of this discussion.

OUTPUT IMPEDANCE

Any instrument that is driven by an amplifier (*e.g.*, a meter, relay, etc.) must draw a certain amount of current from the output of the amplifier, and that current must also pass through part of the output stage of the amplifier. The greater the impedance of that part of the amplifier, the more voltage will be lost across it, and the less will be available to drive the meter or relay. The impedance inside the amplifier through which the output current must pass is called the output impedance, and it is exactly analogous to the internal impedance of any power source, such as a battery or a living cell. As a general rule, the lower the output impedance, the better.

Most loudspeakers have fairly low impedances and draw relatively large amounts of current at low voltages, but most vacuum tubes have the opposite property, that is, they pass small currents at high voltages. Therefore vacuum tubes are not very suitable for directly driving loudspeakers. However, the characteristics of tubes and loudspeakers may be matched by placing a transformer between them.

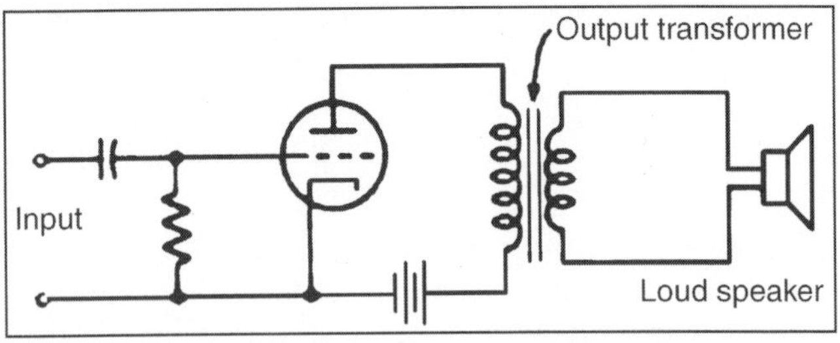

Fig. A One-stage Amplifier Designed to Drive a Loudspeaker. The Output is Transformer-coupled.

The (step-down) transformer is called an output transformer and is used in virtually all vacuum tube amplifiers that are designed to drive loudspeakers. As a rule the secondaries of such transformers have several taps, so that the amplifier may be fitted to speakers with different impedances.

Caution: When the output of an amplifier is transformer-coupled, the amplifier must never be turned on unless the loudspeaker or equivalent resistance is connected across the output terminals. If there is no load (*e.g.*, one of the loudspeaker wires is disconnected), very large voltages may be induced into the output stage of the amplifier, damaging the transformer or output tubes or both.

TRANSISTOR CIRCUITS

When a small signal must be amplified by a very large factor or when the output signal must be an undistorted replica of the input, it is best to buy a commercial amplifier. But there are many occasions when less exacting amplification is required, and for these situations an adequate and cheap transistor amplifier may be easily built.

The fundamental principles of operation of transistors will not be discussed here. They are too complicated and, in many ways, not well enough understood to be worth presenting in this book. What will be discussed is the design and construction of a one-transistor amplifier to be used where a small signal must reliably operate a relay.

Consider, as an example, a situation in which a rat must cross a charged grid to get to food, and the number of times he gets shocked during a 24-hour period is to be counted. Assume that the shock is a constant-current d-c shock of 1 milliampere. If a very sensitive relay, one that closes with a current of less than 1 milliampere, were connected in series with the rat, and the relay were to operate a counter, the counter would register the number of shocks. But relays that sensitive are expensive and hard to find.

If an amplifier were introduced, such as the vacuum-tube amplifier, the signal could be amplified sufficiently to operate a less expensive and more available relay. Although a vacuum-tube amplifier could be built to perform this function, the characteristics of transistors make them particularly suited to this application, and a transistor amplifier would be easier and cheaper to build.

All have three principle elements, an emitter, a base, and a collector. These elements may be considered as analogous to the three elements of a triode vacuum tube. The emitter corresponds to the cathode, the base to the grid, and the collector to the plate. The emitter emits charges which are collected by the collector after having been controlled in density by the base. When the voltage Ec is applied between the emitter and the collector, current will flow through that loop. This current is called the collector current. If the base is then made more positive with respect to the emitter, the collector current will increase, in a way analogous to the increase in plate current when the grid of a triode is made more positive with respect to the cathode. In a vacuum tube, the resistance of the path from the grid to the cathode is extremely high, and virtually no current flows through it. However, the resistance from the base to the emitter of a transistor is relatively low (the actual value depends upon the particular transistor), so the voltage between the base and emitter will cause an appreciable current to flow in that loop.

For a typical transistor, a change of 1 milliampere in the base current will produce a corresponding change of about 40 milliamperes in the collector current. (Common transistors range in amplification factors from about 20 to about 100.) If the transistor is connected into the shock circuit, the 1 milliampere that flows through the rat and the base-emitter circuit will change the current through the relay coil by 40 milliamperes. In this way, the transistor eliminates the need for a very sensitive and expensive relay. The two principle classifications of transistors are the p-n-p and the n-p-n. These letters refer to the physical structures of the transistors themselves. The two types perform in exactly the same way except that all the voltages impressed on a p-n-p must have the opposite polarity from those on an n-p-n. In the above examples, the polarities are correct for the n-p-n transistor. This one was chosen for discussion because the voltages have the same polarities as the corresponding vacuum-tube amplifier voltages. However, many of the transistors most useful for behavioural work are only made with the p-n-p construction.

When the rat is touching the grids, the shock box drives a 1-milliampere direct current through the base-to-emitter circuit of the transistor, making the base negative with respect to the emitter. This base-to-emitter current change produces a corresponding change of about 40 milliamperes in the emitter-collector circuit, which contains the relay coil. The 50,000-ohm variable resistance serves as a bias control. When the slider is all the way to the left, the base is virtually disconnected from the negative side of the battery (because, for this transistor, the base to emitter resistance is considerably lower than 50,000 ohms).

When the slider is all the way to the right, the base is almost the full 22½ volts negative with respect to the emitter. The 1000-ohm fixed resistor prevents the base from going so far negative that the collector current rating is exceeded. The variable resistor may thus be set so that, when there is no shock, the relay has some current flowing through it, but not enough to hold it closed. Then the actual change in collector current need only be big enough to change the relay from open to closed rather than all the way from zero current to closed. The particular transistor is a cheap and versatile one at the time of this writing. However, transistor technology is progressing at such a remarkable rate that it may be considered a collector's item by the time this chapter is printed. The transistor in this circuit is one of a class called power transistors, because they are capable of controlling a large amount of power without overheating. Other types, such as audio-amplifier transistors, are no cheaper and are much more easily damaged by circuit misconnections. For applications in which the output of the amplifier is to control a relay, power transistors are preferable.

Transistors are extremely reliable devices, and are not damaged by fairly severe mechanical shocks (*e.g.*, dropping on the floor). In reliability, size, basic circuit complexity, and efficiency, transistors are superior to vacuum tubes. At their present stage of development, they are inferior to vacuum tubes for two reasons that are revelant to typical behavioural applications. The first important disadvantage of transistors relative to vacuum tubes depends upon the fact that the interelement resistances of transistors are very much lower than the corresponding resistances for vacuum tubes.

Because of these relatively low resistances, the input and output elements and all the stages of a multistage amplifier interact with each other, making the quantitative design of multielement circuits very difficult. In almost all vacuum-tube amplifiers, for example, the current that flows in the plate circuit of one stage has virtually no effect on the functioning of the preceding stage (unless feedback is deliberately introduced), and each stage can be designed and understood more or less by itself. But changing the value of a resistor anywhere in a multistage transistor circuit changes the functioning of almost everything else in the entire circuit.

The second principle disadvantage of transistors is that they are very sensitive to heat. When soldering a circuit, great care must be taken that the transistors are not made hot. Sometimes a transistor can be firmly connected into a circuit without soldering anything directly to its leads. For example, some transistors fit into sockets. Solderless connections may be made to power transistors. The clips in this photograph are obtained by smashing a tube socket designed to fit a miniature vacuum tube. If wires must be soldered directly to the transistor leads, a pair of pliers or some other heat sink should be clamped between the body of the transistor and the place on the lead where it is to be soldered. Once the circuit is operating, a transistor's sensitivity to heat is manifested in the fact that its parameters (*e.g.*, amplification factor, base-emitter resistance, etc.) change with its temperature.

Therefore any transistor circuit that must be really stable in its characteristics must be carefully designed with special stabilizing circuitry. (Circuits to operate relays need only be designed so that the change in output current is normally well above the minimum necessary to close the relay.) There are many books and booklets on the design of transistor circuits for a variety of applications. Because of the very rapid development of transistor technology, it is not worthwhile to list special references here, but excellent booklets may be purchased at any electronics supply store.

Bibliography

Alvarez, A.R. Ed: *BICMOS Technology and Applications, second ed.*, Kluwer, 1993.

Arora, N: *MOSFET Models for VLSI Circuit Simulation*, Springer, 1993.

Bakoglu, H.B.: *Circuits, Interconnections and Packaging for VLSI*, Addison-Wesley, 1990.

Campbell, S.A.: *The Science and Engineering of Microelectronic Fabrication*, Oxford University Press, 1996.

Chandrakasan, A.P., Brodersen, R.W.: *Low Power Digital CMOS Design*, Kluwer Academic Publishers, 1995.

Cheng, Y., Chan, M., Hui, K., Jeng, M., Liu, ZH. Huang, J., Chen, K., Chen, J., TU, R., KO, P.K., and HU, CH.: *BSIM3v3 Manual*, University of California, Berkeley, 1996.

Chua, L.O., and LIN, P.M.: *Computer-Aided Analysis of Electronic Circuits: Algorithms & Computational Techniques*, Prentice-Hall, Englewood Cliffs, 1975.

Çilingirosglu, U: *Systematic Analysis of Bipolar and MOS Transistors*, Artech House, Boston, 1993.

Diaz, C.H., Kang, S.M. (STEVE), and Duvvury, C: *Modeling of Electrical Overstress in Integrated Circuits*, Kluwer, 1995.

El-Kareh, B: *Fundamentals of Semiconductor Processing Technologies*, Kluwer, Boston, 1995.

Ferry, D.K., Akers, L.A., Greeneich, E.W.: *Ultra Large Scale Integrated Microelectronics*, Prentice-Hall, 198A.

Franca, J.E., Tsividis, Y., Eds.: *Design of Analog-Digital VLSI Circuits for Telecommunications and Signal Processing*, 2nd ed, Prentice-Hall Inc., 1994.

Kosko, B.: *Neural Networks and Fuzzy Systems*, Prentice-Hall, 1992.

LIM, J.S., Oppenheim, and A.V.: *Advanced Topics in Signal Processing*, Prentice Hall, 198A.

LIU, D.: *Low Power Digital CMOS Design*, Dissertation, Linköping University, 1994.

Machado, G.A.S., Ed.: *Low-Power HF Microelectronics A Unified Approach*, IEE London, 1996.

Massobrio, G., and Antognetti, P.: *Semiconductor Device Modeling with Spice, second ed.*, McGraw-Hill, 1993.

Prince, B.: *Semiconductor Memories*, Wiley, 1991.

Rabaey, J.M., and Pedram, M., Eds.: *Low Power Design Methodologies*, Kluwer Academic Publishers, 1996.

Schwarz, A.F.: *Digital-Circuit Aspects and State of the Art, vol. II of Computer-Aided Design of Microelectronic Circuits and Systems*, Academic Press, 1987.

Schwarz, A.F.: *General Introduction and Analog Circuit Aspects, vol. I of Computer-Aided Design of Microelectronic Circuits and Systems*, Academic Press, 1987.

Stach, A.: *Simulation von MOSFET-Schaltungen, Diplomarbeit*, Technische Universität Wien, 1995.

SZE, S.M., Ed.: *VLSI Technology*, second ed, McGraw-Hill, 198A.

SZE, S.M.: *High-Speed Semiconductor Devices*, Wiley, New York, 1990.

Tsividis, Y.P.: *Operation and Modeling of the MOS Transistor*, McGraw-Hill, 1987.

Wang, N.: *Digital MOS Integrated Circuits*, Prentice-Hall, 198B.

Weste, N.H.E., and Eshraghian, K.: *Principles of CMOS VLSI Design, second ed.*, Addison Wesley, 1993.

Wolf, S.: *Silicon Processing for the VLSI Era*, Lattice Press, 1990.

Wolf, S.: *The Submicron MOSFET, vol. 3 of Silicon Processing for the VLSI Era*, Lattice Press, Sunset Beach, California, 1995.

Yourdon, E., and Constantine, L.L.: *Structured Design*, Prentice-Hall, 197B.

Index